Tolerance Graphs

CAMBRIDGE STUDIES IN ADVANCED MATHEMATICS 89

Editorial Board:
B. Bollobas, W. Fulton, A. Katok, F. Kirwan, P. Sarnak, B. Simon

TOLERANCE GRAPHS

MARTIN CHARLES GOLUMBIC
University of Haifa, Israel

ANN N. TRENK
Wellesley College, Massachusetts

CAMBRIDGE
UNIVERSITY PRESS

PUBLISHED BY THE PRESS SYNDICATE OF THE UNIVERSITY OF CAMBRIDGE
The Pitt Building, Trumpington Street, Cambridge, United Kingdom

CAMBRIDGE UNIVERSITY PRESS
The Edinburgh Building, Cambridge CB2 2RU, UK
40 West 20th Street, New York, NY 10011–4211, USA
477 Williamstown Road, Port Melbourne, VIC 3207, Australia
Ruiz de Alarcón 13, 28014 Madrid, Spain
Dock House, The Waterfront, Cape Town 8001, South Africa

http://www.cambridge.org

First published 2004

Printed in the United Kingdom at the University Press, Cambridge

Typeface Times 10/13 pt.   *System* LaTeX 2_ε [TB]

A catalog record for this book is available from the British Library

ISBN 0 521 82758 2 hardback

The publisher has used its best endeavors to ensure that the URLs for external websites
referred to in this book are correct and active at the time of going to press. However, the
publisher has no responsibility for the websites and can make no guarantee that a site will
remain live or that the content is or will remain appropriate.

Dedicated to
Lynn Pollak Golumbic and *Rick Cleary*

Contents

Preface   *page* xi

1   Introduction   1
1.1   Background and motivation   1
1.2   Intersection graphs and interval graphs   4
1.3   Tolerance graphs: definitions and examples   5
1.4   Chordal, comparability, interval graphs   7
1.5   Ordered sets   13
1.6   The hierarchy of permutation, parallelogram, trapezoid,
        function and AT-free graphs   15
1.7   Other families of graphs   20
1.8   Other reading and general references   24
1.9   Exercises   25
2   Early work on tolerance graphs   29
2.1   Notation and observations   29
2.2   Permutation graphs and interval graphs   31
2.3   Bounded tolerance graphs   33
2.4   Tolerance graphs are weakly chordal   36
2.5   Tolerance graphs are perfect   40
2.6   A first look at unit vs. proper   45
2.7   Classes of perfect graphs   48
2.8   Exercises   52
3   Trees, cotrees and bipartite graphs   53
3.1   Trees and cotrees   53
3.2   Bipartite tolerance graphs – the bounded case   60
3.3   Exercises   61
4   Interval probe graphs and sandwich problems   63
4.1   Physical mapping of DNA   63

4.2	Interval probe graphs	65
4.3	The hierarchy of interval, probe, and tolerance graphs	66
4.4	The trees that are interval probe graphs	71
4.5	Partitioned interval probe graphs	73
4.6	The enhancement of a partitioned probe graph is chordal	74
4.7	The Interval Graph Sandwich Problem	77
4.8	The NP-completeness of the Interval Probe Graph Sandwich Problem	80
4.9	Exercises	82
5	**Bitolerance and the ordered sets perspective**	84
5.1	The concept of a bounded tolerance order	84
5.2	Classes of bounded bitolerance orders	85
5.3	Geometric interpretations	91
5.4	Exercises	96
6	**Unit and 50% tolerance orders**	98
6.1	Unit tolerance orders with six or fewer elements	98
6.2	Unit vs. proper for bounded bitolerance orders	103
6.3	Width 2 bounded tolerance orders	107
6.4	Exercises	108
7	**Comparability invariance results**	109
7.1	Comparability invariance	109
7.2	Autonomous sets and Gallai's Theorem	111
7.3	Dimension is a comparability invariant	112
7.4	Bounded tolerance orders	113
7.5	Unit bitolerance and unit tolerance orders	115
7.6	Exercises	122
8	**Recognition of bounded bitolerance orders and trapezoid graphs**	124
8.1	Preliminaries	125
8.2	The order $\mathcal{B}(\mathcal{I})$ of extreme corners	127
8.3	The isomorphism between $B(P)$ and $\mathcal{B}(\mathcal{I}^*)$	130
8.4	The recognition algorithm and its complexity	132
8.5	Exercises	133
9	**Algorithms on tolerance graphs**	135
9.1	Tolerance and bounded tolerance representations	136
9.2	Coloring tolerance representations	137
9.3	Maximum weight stable set of a tolerance representation	140
9.4	Exercises	144
10	**The hierarchy of classes of bounded bitolerance orders**	146
10.1	Introduction	146

10.2	Equivalent classes	148
10.3	Bipartite orders	152
10.4	Separating examples	158
10.5	Exercises	163
11	**Tolerance models of paths and subtrees of a tree**	164
11.1	Introduction	164
11.2	Intersection models	164
11.3	Discrete models	165
11.4	Neighborhood subtrees	169
11.5	Neighborhood subtree tolerance (NeST) graphs	173
11.6	Subclasses of NeST graphs	176
11.7	The hierarchy of NeST graphs	183
11.8	A connection with threshold and threshold tolerance graphs	187
11.9	Exercises	191
12	**ϕ-tolerance graphs**	193
12.1	Introduction	193
12.2	ϕ-tolerance chain graphs	195
12.3	Archimedean ϕ-tolerance graphs	201
12.4	Polynomial functions ϕ	209
12.5	Every graph can be represented by an Archimedean polynomial	210
12.6	Construction of a universal Archimedean tolerance function	213
12.7	Unit and proper representations	215
12.8	Exercises	217
13	**Directed tolerance graphs**	219
13.1	Ferrers dimension 2	220
13.2	Bounded bitolerance digraphs	222
13.3	Recognition of bounded bitolerance digraphs	224
13.4	Characterizations of bounded bitolerance digraphs	225
13.5	The digraph hierarchy	228
13.6	Cycles	234
13.7	Trees	237
13.8	Unit vs. proper	243
13.9	Exercises	248
14	**Open questions and further directions of research**	249
	References	253
	Index of symbols	260
	Index	262

Chapter dependencies

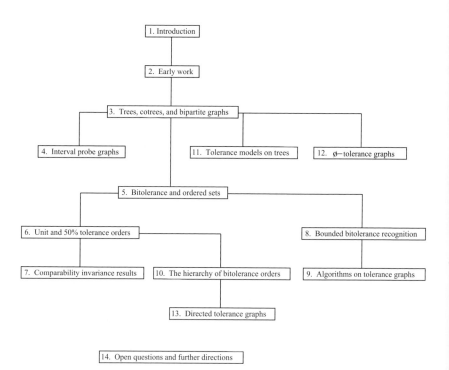

1. Introduction

2. Early work

3. Trees, cotrees, and bipartite graphs

4. Interval probe graphs

11. Tolerance models on trees

12. ∅−tolerance graphs

5. Bitolerance and ordered sets

6. Unit and 50% tolerance orders

8. Bounded bitolerance recognition

7. Comparability invariance results

10. The hierarchy of bitolerance orders

9. Algorithms on tolerance graphs

13. Directed tolerance graphs

14. Open questions and further directions

Preface

At the 13th Southeastern Conference on Combinatorics, Graph Theory and Computing (Boca Raton, 1982), a mathematical model of tolerance, called *tolerance graphs*, was introduced by Golumbic and Monma in order to generalize some of the well known applications associated with interval graphs. Their motivation was the need to solve scheduling problems in which resources such as rooms, vehicles, support personnel, etc. may be required on an exclusive basis, but where a measure of flexibility or tolerance would be allowed for sharing or relinquishing the resource when total exclusivity prevented a solution. An example of such an application opens our Chapter 1.

During the ensuing years, properties of tolerance graphs have been studied, and quite a number of variations have appeared in the literature, including bitolerance graphs, bounded tolerance orders, NeST graphs, ϕ-tolerance graphs, tolerance digraphs and others. This continues to be an interesting and active area of investigation. At the 30th Southeastern Conference on Combinatorics, Graph Theory and Computing (Boca Raton, 1999), Ann delivered an invited survey talk on the subject, and together we organized a special session on tolerance graphs and related topics. The following year, Marty gave a largely complementary survey talk at the Fields Institute Workshop on Structured Families of Graphs (Toronto, 2000). In July 2001, DIMACS sponsored a workshop on Intersection Graphs and Tolerance Graphs.

It seems to us that the time is ripe to collect and survey the major results on tolerance graphs, presenting them in one volume. Many mathematical scientists around the world have carried out the research which has enabled us to do this, and we salute them. Yet there are still various basic unanswered questions concerning tolerance graphs. Tolerance graphs have not yet been characterized, nor are there recognition algorithms. Other open questions appear in Chapter 14. We hope that this book helps to inspire others to pursue these topics further.

What started as a survey paper has grown into a three year project and a 300-page manuscript. Even so, we have had neither time nor space to include all the topics we would have liked. In particular, we have not covered interval digraphs or the literature on tolerance competition graphs, or very recent results which have not had the opportunity to appear in a journal.

This book is intended primarily for researchers and graduate students, although some of it should be accessible to advanced undergraduates. We have included exercises to facilitate the use of this book in a seminar course. Algorithms and applications are presented in addition to the theory of tolerance graphs. In general, we have tried to include proofs whenever possible, omitting them only when they already appear in other books or sometimes when they are quite long. In several chapters we include hierarchies of structured families of graphs. Naturally, we have tried to catch all errors. We hope our readers will be tolerant of those that inevitably remain, and will report these errors to us.

We would like to thank our families for their support in our project, especially our spouses Lynn and Rick to whom we dedicate this book. Ann's parents were very helpful in finding us places to work in New York City several times, and Marty's daughters demonstrated unbounded tolerance for his traveling too much. We would like to acknowledge several colleagues who have worked with us on tolerance problems over the years and during the writing of this monograph: Ken Bogart, Garth Isaak, Robert Jamison, Haim Kaplan, Marina Lipshteyn, Clyde Monma, Kathryn Nyman, Uri Peled, Ron Shamir, Randy Shull, Assaf Siani, and Tom Trotter. We are grateful to Mike Fisher for a thorough reading of a draft of our manuscript and to Peter Fishburn, Ross McConnell and Jerry Spinrad for useful comments, pointers, references and encouragement. Phil Hirschhorn, Greta Pangborn, and Michael Wagner provided invaluable computer assistance, and Wellesley College students Charlotte Henderson and Jue Wang were a tremendous help in the editing stage. We also thank the institutions which have provided some of the support for our joint work: The American Association of University Women, Bar-Ilan, Cornell, Haifa, Rutgers, and Wellesley.

Chapter 1

Introduction

1.1 Background and motivation

Our mathematical adventure begins with a collection of intervals on the real line. The intervals may have come from an application, for example, they could represent the durations of a set of events on a time line, or fragments of DNA on the genome, or sectors of consecutive elements of a linearly ordered set. Some of the intervals may intersect one another, and others may be disjoint. No matter what they may represent, intervals are familiar to us as mathematical entities. There are many relationships between these intervals that we could study. In this book, we deal mostly with intersection.

When two intervals intersect, we might interpret this positively as their having something important in common, like an opportunity to share information. For example, if each interval represented the time period during which a group of school children would be visiting a science museum, then two groups whose intervals intersect could participate in a joint activity. We might then ask, how many times would we need to flash the new Artificial Bolt of Lightning so that each group would get to see it? Or we might interpret intersection negatively as having a major conflict, like competing for a resource that cannot be shared. For example, in a one-television household, when a parent wants to watch the News and at the same time a teenager wants to watch an old movie on a different channel, we have a temporal conflict.

In graph theory, the family of *interval graphs* was introduced to study such problems of intersecting intervals on the line. In this model, each vertex v in a graph $G = (V, E)$ is associated with an interval I_v, and two vertices are connected by an edge in G if their associated intervals have nonempty intersection. Formally, $uv \in E(G) \iff I_u \cap I_v \neq \emptyset$, for all $u, v \in V(G)$. The graph G is called an interval graph.

In our museum example, there is a well-defined minimum number α of how many times that the lightning must be flashed, and it is easy to calculate the number α and an optimal schedule for the flashes. Well, at least it is "easy" for the authors since we have been teaching students about interval graphs for a long time. But it is also "easy" in a computational sense since there are well-known linear time algorithms to do this.

But what do you do if the electricity requirements allow only $\alpha - 2$ flashings? Either some of the groups will be disappointed, or they will have to reschedule the time of their visit. Similarly, in our television example, when one spouse wants to watch a game show and the other spouse wants a basketball game, it is fair game to assume that a compromise is needed.

In this book, we study the class of tolerance graphs, which are a generalization of interval graphs. Tolerance graphs are constructed from intersecting intervals in a manner similar to interval graphs, but putting an edge between two vertices depends on measuring the size of the intersection of their two intervals before declaring that an edge exists. Informally, if both intervals are willing to "tolerate" or ignore the intersection, then no edge is added between their vertices in the graph.

Tolerance graphs were introduced by Golumbic and Monma (1982) to generalize some of the well-known applications associated with interval graphs. Their original motivation was the need to solve scheduling problems in which resources such as rooms, vehicles, support personnel, etc. may be needed on an exclusive basis, but where a measure of flexibility or tolerance would allow for sharing or relinquishing the resource if a solution is not otherwise possible. Let's look at simple example.

A motivating example

On a typical morning, six parliamentary or corporate meetings are to convene according to a fixed schedule, where meeting m_i is scheduled for the time interval $I_i = [a_i, b_i]$. Each meeting must be assigned a meeting room. Let us consider the example:

$$I_1 = [8:00\text{–}9:45], \quad I_2 = [9:00\text{–}11:30], \quad I_3 = [8:30\text{–}11:15],$$
$$I_4 = [10:00\text{–}11:00], \quad I_5 = [10:15\text{–}12:00], \quad I_6 = [10:45\text{–}12:30].$$

In our example, meeting m_1 could use the same room as either m_4 or m_5 or m_6 since its time interval I_1 does not intersect with the time intervals I_4, I_5 or I_6. Being very strict with these intervals, we see that at 10:50 five rooms are needed simultaneously (see Figure 1.1). But suppose there are only four meeting rooms! Should we cancel one of the meetings? Probably not. Rather,

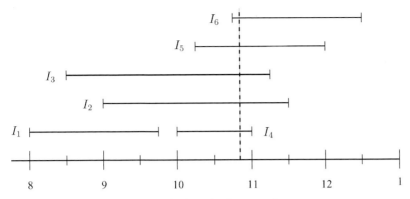

Figure 1.1. A motivating example.

we should try to identify some flexibility in these time constraints which may allow us to find an acceptable assignment of rooms.

The tolerance graph model, which we will formally define below, provides a mechanism for associating a numerical tolerance to each meeting to indicate the degree of its flexibility in allowing some intersection with other intervals. In this way, it may be possible to give an assignment of rooms to all the meetings by sharing the room for a short period or by moving the start or finish time. In our example, if both I_4 and I_6 were willing to tolerate an overlap of more than 15 minutes, then there would be a four room solution.

Resource assignment problems of this nature arise in many contexts: motorcycles for delivering express mail (or pizza), nurses for operating rooms, waterfront space for picnics, ovens for warming a caterer's dishes, etc. In a real world situation, some meetings or deliveries may indeed have strict deadlines which must be met, while others may be more flexible. By taking these tolerances into account, solutions can often be found which would otherwise not exist under the strict constraints. There would be a great benefit to having algorithmic methods for automatically resolving such conflicts.

This example, and the discussion on intersecting intervals, briefly motivates the topic of our book. The volume and scope of research in this area has expanded significantly both from the mathematical and algorithmic points of view. Many special families of graphs and ordered sets will be encountered along the way. Each will depend on the specific tolerance model being discussed.

In this chapter, we provide the formal definition of a tolerance graph and give some elementary properties. We also give a brief review of many of the important families of graphs which are related in some way to tolerance graphs.

1.2 Intersection graphs and interval graphs

Let $\mathcal{F}$ be a collection of sets. The *intersection graph* of $\mathcal{F}$ is the graph obtained by assigning a distinct vertex to each set in $\mathcal{F}$ and joining two vertices by an edge precisely when their corresponding sets have a nonempty intersection. When the types of sets allowed in $\mathcal{F}$ is limited, interesting classes of graphs result.

Most important to us will be the *interval graphs* which arise when the sets in $\mathcal{F}$ are intervals in the real line, that is, a graph $G = (V, E)$ is an *interval graph* if each vertex $v \in V$ can be assigned a real interval I_v so that $xy \in E \iff I_x \cap I_y \neq \emptyset$. The set of intervals $\{I_v \mid v \in V\}$ is an *interval graph representation* of G.

Interval graphs are important for their applications to scheduling problems, microbiology, and VLSI circuit design. In our previous motivating example (Figure 1.1), the intervals represented fixed time slots for a set of meetings which needed to be assigned rooms. The interval graph for this example is shown in Figure 1.2. Finding a consistent assignment of rooms can be viewed as a *coloring problem* on the interval graph, where the meeting rooms are the colors and adjacent vertices must be assigned different colors. There are efficient algorithms for coloring the vertices of an interval graph using a minimum number of colors (Golumbic, 1980). In our example, there cannot be a solution with four rooms since the interval graph has a *clique* (or *complete subgraph*) of size 5. Indeed, the only subsets that could be colored by the same color in this example are $\{1, 4\}$ or $\{1, 5\}$ or $\{1, 6\}$. A *stable set* (or *independent set*) is a subset of vertices no two of which are connected by an edge. Here there is no stable set larger than size 2.

In this book, we also consider other families of intersection graphs, such as *trapezoid graphs* and *parallelogram graphs* which are intersection graphs of trapezoids (resp. parallelograms) having two of their sides on two fixed

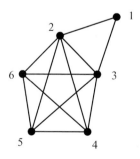

Figure 1.2. The interval graph for our motivating example.

parallel lines. Later in this chapter, we discuss permutation graphs which can be interpreted as intersection graphs of line segments in a matching diagram. Also, in Chapter 11, we present a variety of intersection graphs involving subtrees and paths in trees.

All of these families of intersection graphs satisfy the *hereditary property*, namely, if a graph $G = (V, E)$ is the intersection graph of a certain type (e.g., intervals, trapezoids, etc.), then every *induced subgraph* G_X of G is also an intersection graph of that same type, where $V(G_X) = X \subseteq V(G)$ and $E(G_X) = \{uv \in E(G) \mid u, v \in X\}$.

1.3 Tolerance graphs: definitions and examples

A graph $G = (V, E)$ is a *tolerance graph* if each vertex $v \in V$ can be assigned a closed interval I_v and a tolerance $t_v \in \mathbf{R}^+$ so that $xy \in E$ if and only if $|I_x \cap I_y| \geq \min\{t_x, t_y\}$. Such a collection $\langle \mathcal{I}, t \rangle$ of intervals and tolerances is called a *tolerance representation* where $\mathcal{I} = \{I_x \mid x \in V\}$ and $t = \{t_x \mid x \in V\}$. If graph G has a tolerance representation with $t_v \leq |I_v|$ for all $v \in V$, then G is called a *bounded tolerance graph* and the representation is called a *bounded tolerance representation*.

Consider once again our motivating example. If each of the tolerances were to be 5 minutes, then the tolerance graph would be the same as the interval graph since all of the nonempty intersections are longer than 5 minutes. However, if the tolerances of I_4 and I_6 were 20 minutes (or anything greater than 15 minutes) and each of the others 5 minutes, then the tolerance graph would have no edge between v_4 and v_6, as shown in Figure 1.3. In this case, the vertices of the tolerance graph can be colored using 4 colors, which provides a consistent assignment of meeting rooms.

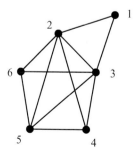

Figure 1.3. The tolerance graph for our motivating example, where I_4 and I_6 have a tolerance of 20 minutes and each of the others 5 minutes.

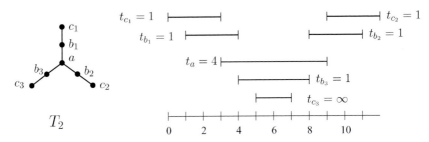

Figure 1.4. The graph T_2 and a tolerance representation of it.

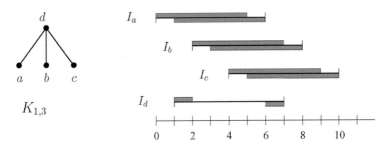

Figure 1.5. The graph $K_{1,3}$ and a bounded tolerance representation of it.

We next look at some additional examples of tolerance graphs. For tolerance representations, we draw the interval assigned to each vertex and list its tolerance next to it, as in the representation of the tree T_2 in Figure 1.4. Notice that the vertex c_3 has infinite tolerance. In fact, any tolerance greater than $|I_{c_3}|$ would work equally well. In Chapter 3, we will see that every tolerance representation of T_2 must have some vertex whose tolerance is greater than its interval length.

For bounded tolerance representations, the tolerance assigned to vertex v is at most the length of the interval $I_v = [L(v), R(v)]$ assigned to v. In this case, we sometimes find it clearer to show the tolerances visually using shading. We shade in the interval from $L(v)$ to $L(v) + t_v$ above I_v and shade in the interval from $R(v) - t_v$ to $R(v)$ below I_v. Figure 1.5 shows a bounded tolerance representation of the graph $K_{1,3}$ in which tolerances are indicated by shading.

The exercises at the end of this chapter will help the reader to become familiar with the concepts presented. Our formal study of tolerance graphs begins in Chapter 2. The remainder of this chapter is devoted to definitions, background and classical results.

1.4 Chordal graphs, comparability graphs, and properties of interval graphs

1.4.1 Chordal graphs and split graphs

A graph G is a *chordal graph* if every cycle of length greater than or equal to 4 has a *chord*, that is, an edge connecting two vertices that are not consecutive on the cycle. For example, the graph in Figure 1.3 is chordal, and the edge (3,5) is a chord of the cycle [3,4,5,6,3]. The chordal graphs are a well known classical family of graphs, and they appear in many interesting applications including relational databases, matrix theory, statistics, and biology. In the literature, chordal graphs are also called *triangulated graphs* (Berge, 1973; Golumbic, 1980) or *rigid circuit graphs* (Roberts, 1976). The family of chordal graphs includes all interval graphs but does not include all tolerance graphs.

There are several interesting characterizations of chordal graphs which we will now review. We present their equivalence below in Theorem 1.1.

A vertex v is called *simplicial* if its *neighborhood* $\mathcal{N}(v) = \{w \in V(G) \mid \forall w \in E(G)\}$ is a clique, that is, every pair of neighbors of v is connected by an edge of the graph. Let $\sigma = [v_1, v_2, \ldots, v_n]$ be an ordering of the vertices $V(G)$, and let $G_i = G_{\{v_i,\ldots,v_n\}}$ denote the subgraph remaining after deleting $\{v_1, \ldots, v_{i-1}\}$ from G. We define σ to be a *perfect elimination ordering (peo)* if v_i is a simplicial vertex in the graph G_i, for all i. For example, two possible perfect elimination orderings for the graph in Figure 1.3 are [4,6,5,1,3,2] and [1,4,3,5,2,6], but [3,4,5,6,1,2] is *not* a perfect elimination ordering for this graph.

A *maximum cardinality search (MCS)* of a graph G is done as follows: Initially all vertices are unnumbered and have counters set to zero. Choose an unnumbered vertex with largest counter, give it the next number, and add 1 to the counters of each of its neighbors. Continue doing this until all the vertices have been numbered. Suppose that the vertices were numbered in this way: $[x_1, x_2, \ldots, x_n]$, then we will call it a *maximum cardinality search ordering*. Such an MCS ordering for the graph in Figure 1.3 is [1,2,3,4,5,6].

Theorem 1.1. *The following conditions are equivalent:*

(i) *G is a chordal graph.*
(ii) *G has a perfect elimination ordering.*
(iii) *The reversal $[x_n, \ldots, x_2, x_1]$ of any MCS ordering of G is a perfect elimination ordering.*
(iv) *G is the intersection graph of a family of subtrees of a tree.*

The equivalence (i)⇔(ii) is due to Dirac; (i)⇔(iii) to Tarjan; (i)⇔(iv) independently to Buneman, Gavril and Walters; see Brandstädt, Le, and Spinrad

(1999), Golumbic (1980), Golumbic (1984), McKee and McMorris (1999) for a proof of this theorem and for additional references.

Both conditions (ii) and (iii) suggest algorithms for recognizing chordal graphs. Using (ii), one would repeatedly look for and eliminate a simplicial vertex, breaking ties arbitrarily, until either all vertices are eliminated (success) or no simplicial vertex can be found (failure). This greedy method is correct since once a vertex becomes simplicial, it remains simplicial in any induced subgraph. Using (iii), one would carry out a maximum cardinality search while testing its reversal to verify that it is a perfect elimination ordering (success) or is not a peo (failure). The latter method gives a more efficient algorithm, having complexity $O(n + e)$ for a graph with n vertices and e edges, see Berry, Blair, and Heggernes (2002), Golumbic (1980), Golumbic (1984) and Tarjan and Yannakakis (1984).

There are also efficient, polynomial time algorithms for finding a minimum coloring, maximum clique, maximum stable set, or a minimum clique cover of a chordal graph. In general, these graph problems are NP-complete, which means that chordal graphs are indeed a very special family of graphs.

We conclude this section by defining and characterizing the class of split graphs. A graph $G = (V, E)$ is called a *split graph* if its vertex set can be partitioned $V = X \cup Y$ into a stable set X and a clique Y. The graph in Figure 1.3 is a split graph with partition $X = \{1, 4\}$ and $Y = \{2, 3, 5, 6\}$.

The *complement* $\overline{G}$ of G is the graph where $V(\overline{G}) = V(G)$ and $E(\overline{G}) = \{xy \mid xy \notin E(G), x \neq y\}$. Since a stable set in G is a clique in the complement $\overline{G}$, and vice versa, G is a split graph if and only if $\overline{G}$ is a split graph. Földes and Hammer (1977) have given the following characterization of split graphs.

Theorem 1.2. *The following conditions are equivalent:*

(i) *G is a split graph.*
(ii) *G and $\overline{G}$ are chordal graphs.*
(iii) *G contains none of the graphs $2K_2$, C_4, C_5 as an induced subgraph, (see Figure 1.6).*

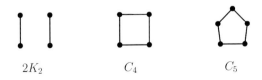

$$2K_2 \qquad C_4 \qquad C_5$$

Figure 1.6. The forbidden subgraphs characterizing split graphs.

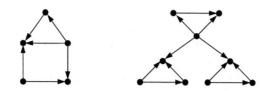

Figure 1.7. Some transitive orientations.

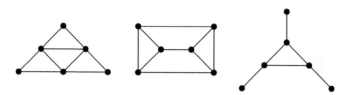

Figure 1.8. Some graphs which are *not* transitively orientable.

For a proof of this theorem and for further reading on chordal graphs and split graphs, see Brandstädt, Le, and Spinrad (1999), Golumbic (1980) and McKee and McMorris (1999). We will see split graphs again in Chapter 11.

1.4.2 Comparability graphs and transitive orientations

A *transitive orientation* F of graph $G = (V, E)$ is an assignment of a direction, or orientation, to each edge in E such that if $xy \in F$ and $yz \in F$ then $xz \in F$. A graph is called a *comparability graph* if it has a transitive orientation. For example, the even length chordless cycles $C_{2k}(k \geq 2)$ are comparability graphs, but the odd length chordless cycles C_5, C_7, etc. are not comparability graphs. Comparability graphs are also known as *transitively orientable (TRO)* graphs. Additional examples of comparability graphs and their transitive orientations can be found in Figure 1.7. Figure 1.8 shows several graphs which have no transitive orientation. Gallai (1967) gave a list of forbidden subgraphs that characterizes the class of comparability graphs, (see also Duchet, 1984). The name "comparability" graph comes from the observation that relation F is a strict partial ordering of V whose comparability relation is precisely E. We will discuss more about ordered sets in Section 1.5.

Comparability graphs can be recognized, and a transitive orientation can be produced, using the following well known greedy method. (a) Choose an orientation of an arbitrarily chosen edge. (b) Propagate all other orientations forced by this and all subsequently oriented edges (usually called the *implication class*). If at some point an edge is forced in both opposite directions, exit with failure. (c) When no other orientations are forced, add the oriented edges to F

and remove them from E. If the graph still has some edges, repeat this sequence of steps. When this algorithm finishes, F will be a transitive orientation. The reader unfamiliar with this topic is referred to Golumbic (1980, 1984). This method can be implemented to run in $O(n \cdot e)$ time for a graph with n vertices and e edges, or by a more careful counting $O(\Sigma_{v \in V} d_v^2)$, where d_v is the degree of v. (The *degree* of a vertex v is the number of edges that have v as an endpoint, that is, $d_v = |\mathcal{N}(v)|$.)

Asymptotically faster algorithms for recognizing comparability graphs, which use a technique called modular decomposition, have been given in McConnell and Spinrad (1999). In that paper, the authors show how to find an orientation F of an arbitrary graph G such that F is a TRO of G if and only if G is a comparability graph. This is very good if there is other information guaranteeing that G is a comparability graph. However, this alone does not recognize comparability graphs, since the algorithm simply produces an orientation which is *not* transitive when G is *not* a comparability graph. Hence, to complete it to a recognition algorithm, one must test F to determine whether it is transitive. The complexity of their method uses $O(n + e)$ time to produce F and $O(n^\alpha)$ to test whether F is transitive, where $O(n^\alpha)$ is the complexity to perform transitive closure or $n \times n$ matrix multiplication (currently $n^{2.376}$).

The complements of comparability graphs, called *cocomparability graphs*, are of particular interest in this book since, as we will see in the next chapter, all bounded tolerance graphs are cocomparability graphs. Cocomparability graphs also have a characterization as the intersection graphs of function diagrams Golumbic, Rotem, and Urrutia (1983), which we present in Section 1.6.

1.4.3 Interval graphs

We defined interval graphs in Section 1.2 as being the intersection graphs of intervals on a line. Interval graphs have several important characterizations which we review here. One of these is the equivalence of interval graphs and the graphs that are both chordal and cocomparability. A second relates to the notion of an asteroidal triple of vertices which we now define.

Three vertices $v_1, v_2, v_3 \in V(G)$ form an *asteroidal triple* (*AT*) of G if, for all permutations i, j, k of $\{1,2,3\}$, there is a path from v_i to v_j which avoids using any vertex in the *closed neighborhood* $\mathcal{N}[v_k] = \{v_k \cup \mathcal{N}(v_k)\}$. An easy way to verify this for v_k is to delete $\mathcal{N}[v_k]$ and test whether v_i and v_j remain in the same connected component of $G - \mathcal{N}[v_k]$. It also follows from the definition that the three vertices of an asteroidal triple are pairwise nonadjacent. For example, $\{c_1, c_2, c_3\}$ is an asteroidal triple in the tree T_2 in Figure 1.4.

A graph is called *asteroidal triple free* (*AT-free*) if it contains no asteroidal triple. Golumbic, Monma, and Trotter (1984) showed that every cocomparability graph is AT-free, which we prove in Theorem 1.13. More recently, Corneil, Olariu, and Stewart (1997) have given other mathematical and algorithmic properties characterizing AT-free graphs. The connection with interval graphs is given in the following theorem. Additional characterizations of interval graphs can be found in Golumbic (1980) and Brandstädt, Le, and Spinrad (1999).

Theorem 1.3. *The following conditions are equivalent:*

(i) *G is an interval graph.*
(ii) *G is chordal and a cocomparability graph.*
(iii) *G is chordal and AT-free.*

The equivalence (i)⇔(ii) is due to Gilmore and Hoffman (1964), and (i)⇔(iii) is due to Lekkerkerker and Boland (1962) who also gave a list of forbidden subgraphs which characterize interval graphs.

Efficient algorithms which run in $O(n + e)$ time are known for recognizing an interval graph with n vertices and e edges, as well as for solving the coloring, clique and stable set problems on interval graphs (Booth and Lueker, 1976). As an illustration, let's consider the case where we are given an interval representation $\mathcal{I} = \{I_v \mid v \in V\}$ for $G = (V, E)$, and we want to color the intervals using a minimum number of colors so that intersecting intervals are assigned different colors. This is equivalent to coloring the vertices of G so that adjacent vertices get different colors. The following procedure handles the coloring.

Algorithm for coloring a set of intervals Sort the intervals according to their left endpoints. Sweep across the representation from left to right, assigning colors in a first fit manner, that is, when a new interval is encountered, always assign the lowest numbered available color, and when an interval is finished, its color becomes available again.

It is an easy exercise, or a good exam question, to show that this "greedy" coloring algorithm is optimal. In particular, during the left to right sweep, just at the point where the highest numbered color k is used, we will find a clique of size k.

We will see this greedy coloring algorithm again in Chapter 4 being applied to representations of probe graphs. We will also show how to color a tolerance representation in Chapter 9.

1.4.4 Unit interval graphs and proper interval graphs

An interval graph G that has a representation in which each interval has the same (unit) length is called a *unit interval graph*. Similarly, if G has a representation in which no interval properly contains another interval, G is called a *proper interval graph*. Clearly, a unit representation is also proper. It is easy to verify that the bipartite graph $K_{1,3}$ does not have a proper interval representation. The following classical result of Roberts (1969) tells us that the unit interval graphs are equivalent to the proper interval graphs, and they are further equivalent to the $K_{1,3}$-free interval graphs.

Theorem 1.4. *The following conditions are equivalent:*

(i) *G is a unit interval graph.*
(ii) *G is a proper interval graph.*
(iii) *G is an interval graph and is $K_{1,3}$-free.*

We conclude this section with a very useful lemma, which will allow us to assume certain canonical properties of an interval representation, for example, distinct endpoints. We write $I_x \ll I_y$ to mean that the interval I_x is completely to the left of interval I_y.

Lemma 1.5. *A set of intervals $\mathcal{I} = \{I_v \mid v \in V\}$ can be transformed into another set $\mathcal{I}' = \{I_v' \mid v \in V\}$ in which all interval endpoints are distinct, and this transformation preserves the following relationships:*

(i) $I_x \ll I_y \iff I_x' \ll I_y'$
(ii) $I_x \subset I_y \iff I_x' \subset I_y'$
(iii) $|I_x| = |I_x'|$

In particular, (ii) shows that the transformation preserves the "proper" property and (iii) implies that it preserves the "unit" property.

Proof. Let $\mathcal{I} = \{I_v \mid v \in V\}$ be a set of intervals where $I_v = [L(v), R(v)]$ for all $v \in V$. If there is a repeated endpoint, let $S = \{L(v), R(v) \mid v \in V\}$ be the set of endpoints in the representation, let ϵ be the smallest positive difference between elements of S, and let s be the smallest repeated endpoint in S. If there exist $x \in V$ with $R(x) = s$, pick the one whose interval I_x is the longest and replace I_x by $I_x' = [L(x) + \epsilon/2, R(x) + \epsilon/2]$. Otherwise, pick $x \in V$ with $L(x) = s$ and $|I_x|$ as large as possible and replace I_x by $I_x' = [L(x) - \epsilon/2, R(x) - \epsilon/2]$. It is not hard to see that this new collection satisfies (i), (ii) and (iii) and there is one fewer pair of elements sharing an endpoint. If necessary, recompute ϵ and repeat until all endpoints are distinct. $\qquad\square$

1.5 Ordered sets

An *ordered set* $P = (X, \prec)$ consists of a ground set X and a binary relation $\prec$ on X which is irreflexive, transitive and therefore asymmetric. Two elements $x, y \in X$ are *comparable* in P if $x \prec y$ or $y \prec x$; otherwise x and y are *incomparable*, which we denote $x \parallel y$. We say that y *covers* x if $x \prec y$ and there is no z with $x \prec z \prec y$. Ordered sets (also known as *orders* or *posets*) are often depicted by their *Hasse diagrams* in which edges implied by transitivity are not drawn. For example, Figure 1.9 shows the Hasse diagram of the order P whose only comparabilities are $a \prec b, a \prec c, b \prec d, c \prec d$ and $a \prec d$.

A *linear order* (or *chain*) is one with no incomparabilities and an *antichain* is an order with no comparabilities. The *dual* of the ordered set $P = (X, \prec)$ is the order $P^d = (X, \prec^d)$ with $x \prec y \iff y \prec^d x$.

Two graphs are naturally associated with the order $P = (V, \prec)$. The *comparability graph* $G = (V, E)$ of P has edge set $E = \{xy \mid x \prec y \text{ or } y \prec x\}$ and the *incomparability graph* $\overline{G} = (V, \overline{E})$ has edge set $\overline{E} = \{xy \mid x \parallel y\}$. Figure 1.9 shows an order P and its comparability graph G and its incomparability graph $\overline{G}$. Note that the incomparability graph of any order is always a cocomparability graph and conversely, any cocomparability graph is the incomparability graph of an order.

1.5.1 Interval orders

An ordered set $P = (V, \prec)$ is an *interval order* if each element $v \in V$ can be assigned a real interval I_v so that $x \prec y \iff I_x$ is completely to the left of I_y. The set of intervals $\{I_v \mid v \in V\}$ is an *interval order representation* of P. The same set of intervals also provides an interval graph representation of the incomparability graph $\overline{G}$ of P since $I_x \cap I_y \neq \emptyset \iff x \parallel y$ in P, as illustrated in Figure 1.9. Note, however, that different interval orders may give rise to the same incomparability graph. For example, the set of intervals in Figure 1.10 gives an interval representation of the order P' and the incomparability graph $\overline{G}$.

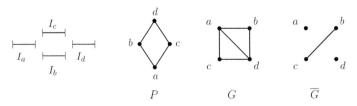

$$P \qquad\qquad G \qquad\qquad \overline{G}$$

Figure 1.9. An interval representation of ordered set P, its comparability graph G, and its incomparability graph $\overline{G}$.

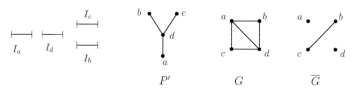

Figure 1.10. A different interval order P' with the same comparability graph and incomparability graph as P in Figure 1.9.

Figure 1.11. The order $\mathbf{2} + \mathbf{2}$.

The name "interval order" first appears in Fishburn (1970) and Fishburn (1985) gives a modern treatment of the subject. However, its origins go back to Norbert Weiner (1914) whose definition of a interval order (which he called a *relation of complete sequence*) was not then known to Fishburn, see Fishburn and Monjardet (1992).

Interval orders have a well-known forbidden suborder characterization which we give below. The order $\mathbf{2} + \mathbf{2}$ consists of four elements a, b, c, d whose only comparabilities are $a \prec b$ and $c \prec d$ (see Figure 1.11). More generally, the order $\mathbf{r} + \mathbf{s}$ consists of two chains: one with r elements, the other with s elements, and everything in the first chain is incomparable to everything in the second chain.

Theorem 1.6. *(Fishburn, 1970) An ordered set is an interval order if and only if it has no suborder isomorphic to $\mathbf{2} + \mathbf{2}$.*

1.5.2 Dimension and interval dimension

The intersection of orders $P_1 = (X, \prec_1)$, $P_2 = (X, \prec_2)$, ..., $P_k = (X, \prec_k)$ with the same ground set is the order $P = (X, \prec)$ where $x \prec y \iff x \prec_i y$ for $i = 1, 2, \ldots, k$.

A *linear extension* of $P = (X, \prec)$ is a linear order $L = (X, \prec_L)$ so that $x \prec_L y$ whenever $x \prec y$. Thus a linear extension of P has all the comparabilities of P plus additional comparabilities to make L linear. One can show that any order is the intersection of all its linear extensions (Exercise 1.12). This makes the following notion of dimension well-defined.

A *linear realizer* of an order P is a set of linear orders whose intersection is P. The *dimension* of P (denoted dim(P)) is the size of a smallest linear realizer

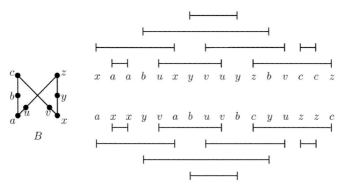

Figure 1.12. The order B with idim$(B) = 2$ and an interval realizer of it.

of P. It is easy to see that dim$(P) = 2$ and dim$(P') = 2$ for the examples in Figures 1.9 and 1.10. Every transitive orientation of a comparability graph is an order, so it has a well-defined dimension. An important result, which we will prove in Section 7.3, is that every transitive orientation of a comparability graph G has the same dimension (i.e., dimension is a comparability invariant.) Thus, we can denote this common value by dim(G). For a comprehensive treatment of dimension theory of ordered sets, see Trotter (1992).

Similarly, an *interval realizer* of an order P is a set of interval orders whose intersection is P, and the *interval dimension* of P (denoted idim(P)) is the size of a smallest interval realizer of P. For example, the order B in Figure 1.12 has idim$(B) = 2$ and an interval realizer of it is shown. The order B can not have interval dimension 1 since it contains suborders isomorphic to $2 + 2$.

The interval dimension is also known to be a comparability invariant (Habib, Kelly, and Möhring, 1991). Since linear orders are interval orders, interval dimension is well-defined and idim$(P) \leq$ dim(P) for all P. In Chapters 5 and 10, we will be interested in the class of orders P with idim$(P) \leq 2$.

1.6 The hierarchy of permutation, parallelogram, trapezoid, function, and AT-free graphs

In this section, we survey a hierarchy of well-known graph classes arising from intersection diagrams.

A graph $G = (V, E)$ is a *permutation graph* if there is a permutation π of $V = \{1, 2, 3, \ldots, n\}$ so that for vertices i, j we have $ij \in E$ if and only if the order of i and j are reversed in π. For example, the path P_4 with edge

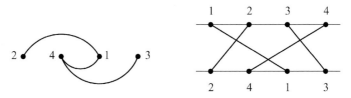

Figure 1.13. The path P_4 as a permutation graph.

set $\{(1, 2), (1, 4), (3, 4)\}$ is a permutation graph using $\pi = [2, 4, 1, 3]$ (see Figure 1.13). If graph G is a permutation graph using π, then its complement $\overline{G}$ is also a permutation graph using the reversal of π. We record this fact as a remark.

Remark 1.7. A graph G is a permutation graph if and only if its complement $\overline{G}$ is a permutation graph.

Alternatively, a permutation graph can be viewed as the intersection graph of line segments in a matching diagram as follows. Write the elements $\{1, 2, 3, \ldots, n\}$ in order on a horizontal line L_1, and underneath write them in the order of π on another horizontal line L_2. For $k = 1, 2, 3, \ldots, n$, connect the two occurrences of k with a straight line segment S_k. Then $ij \in E(G) \iff S_i \cap S_j \neq \emptyset$ (see Figure 1.13). We call such a representation a *permutation diagram*.

Permutation graphs are characterized by the following theorem. The equivalence (i) $\Leftrightarrow$ (ii) is due to Pnueli, Lempel, and Even (1971) and (ii) $\Leftrightarrow$ (iii) is due to Dushnik and Miller (1941). For a proof of this and a more comprehensive treatment of permutation graphs, see Golumbic (1980).

Theorem 1.8. *The following are equivalent.*

 (i) *G is a permutation graph.*
(ii) *G is both a comparability graph and a cocomparability graph.*
(iii) $\dim(G) = 2$.

We now successively generalize permutation diagrams and permutation graphs to other geometric forms. Figure 1.14 shows the hierarchy of these classes together with a sample diagram for each.

Let L_1 and L_2 be two horizontal lines with L_1 above L_2. A *parallelogram diagram* consists of L_1, L_2 and a set of n parallelograms $\{P_i \mid i = 1, \ldots, n\}$ where each P_i has parallel sides along L_1 and L_2. A *trapezoid diagram* consists of L_1, L_2 and a set of n trapezoids $\{T_i \mid i = 1, \ldots, n\}$ where the parallel sides of each T_i lie on L_1 and L_2. We allow degenerate trapezoids (and parallelograms),

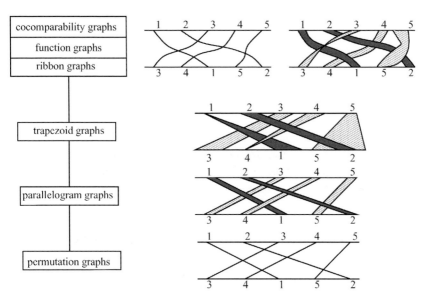

Figure 1.14. A hierarchy of graph classes and their associated intersection diagrams.

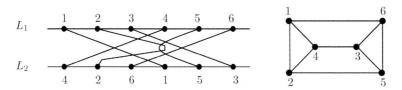

Figure 1.15. A function diagram and its intersection graph (which is isomorphic to $\overline{C_6}$).

that is, the sides along L_1 and/or L_2 may be points, in which case the resulting trapezoid may be a triangle or a straight line segment. Thus, a permutation diagram is also a parallelogram diagram, which in turn is also a trapezoid diagram.

A continuous curve f connecting a point on L_1 with a point on L_2 is called a *function line* if, whenever two points (x, y) and (x', y') on f have the same horizontal value $y = y'$, the points must be equal, i.e., $x = x'$. A *function diagram* consists of L_1, L_2 and a set of n function lines connecting points on L_1 and L_2. The function diagram in Figure 1.15 has six function lines. Finally, we define a *ribbon* to be the area bounded by two function lines, and a *ribbon diagram* to consist of L_1, L_2 and a set of n ribbons. We note that a trapezoid is a ribbon whose bounding function lines are straight.

Definition 1.9. If R_i and R_j are ribbons (trapezoids, parallelograms), we write $R_i \ll R_j$ if R_i and R_j do not intersect and R_i is completely to the left of R_j; or formally, for every horizontal line L, cutting through the diagram, all points on the interval $R_i \cap L$ are to the left of all points on the interval $R_j \cap L$.

We now define the classes of *parallelogram* graphs, *trapezoid* graphs, *function* graphs and *ribbon* graphs to be the family of intersection graphs of their respectively named diagrams.

Remark 1.10. Clearly, these graph families satisfy the containments: permutation $\subseteq$ parallelogram $\subseteq$ trapezoid $\subseteq$ ribbon.

In Figure 2.8 we will see these classes again as part of a larger hierarchy in which separating examples are given. The next result justifies the placement of "cocomparability", "function" and "ribbon" graphs in the same box of Figure 1.14 by proving these classes are equivalent.

Consider the following special type of function diagram in which the curves are piecewise linear. Let $L_1, L_2, \ldots, L_{k+1}$ be horizontal lines each labeled from left to right by a permutation of the numbers $1, 2, \ldots, n$. For each i $(1 \le i \le n)$ the curve f_i consists of the union of the k straight line segments which join i on L_t with i on L_{t+1} $(1 \le t \le k)$. When $k = 1$, this is just a permutation diagram; when $k \ge 2$, it is called the *concatenation* of k permutation diagrams (see Figure 1.16).

In the following theorem, the equivalences (i) $\Leftrightarrow$ (iii) $\Leftrightarrow$ (iv) are due to Golumbic, Rotem, and Uruttia (1983) and their equivalence with (ii) was observed in Golumbic and Lewenstein (2000).

Theorem 1.11. *The following are equivalent.*

(i) *G is a function graph.*
(ii) *G is a ribbon graph.*
(iii) *G is a cocomparability graph.*
(iv) *G is the intersection graph of a concatenation of permutation diagrams.*

Proof. (iv) $\Longrightarrow$ (i) $\Longrightarrow$ (ii): This is immediate since a concatenation of permutation diagrams is a function diagram, and a function diagram is a ribbon diagram where each pair of bounding curves is equal.

(ii) $\Longrightarrow$ (iii): Let G be the intersection graph of the ribbon diagram whose set of ribbons is $R_1, R_2, \ldots, R_n$. Since i and j are adjacent in the complement $\overline{G}$ if and only if R_i and R_j do *not* intersect, we may define an orientation F of

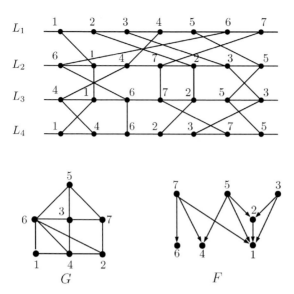

Figure 1.16. A concatenation of three permutation diagrams, its intersection graph G and a transitive orientation F of the complement $\overline{G}$.

$\overline{G}$ as follows

$$ij \in F \Longleftrightarrow R_i \ll R_j$$

where $R_i \ll R_j$ is defined in Definition 1.9. The orientation F is obviously transitive, so $\overline{G}$ is a comparability graph.

(iii) $\Longrightarrow$ (iv): Let $\overline{G}$ be the comparability graph of an order $P = (X, \prec)$, and let $\mathcal{L} = \{L_1, L_2, \ldots, L_{k+1}\}$ be a realizer of P. We may assume, without loss of generality, that $X = \{1, 2, \ldots, n\}$. We will build a concatenation of permutation diagrams whose intersection graph will be G.

For each linear order L_i $(1 \le i \le k+1)$, draw a horizontal line and label n points on the line with the elements of X from left to right according to the order L_i. We also use L_i to denote this line and its labeled points. We stack these $k+1$ horizontal lines as shown in the example in Figure 1.16. Let the curve f_i consist of the union of the k straight line segments which join i on L_t with i on L_{t+1} $(1 \le t \le k)$. We will show that this concatenation of permutation diagrams represents G. If $ij \in E(G)$, then i and j are not comparable in P, so there are linear orders $L_r, L_s \in \mathcal{L}$ such that $i \prec_r j$ and $j \prec_s i$. Therefore, f_i and f_j intersect somewhere within the area between the horizontal lines L_r and L_s. Otherwise, if $ij \in E(\overline{G})$, then either $i \prec_t j$ for all $L_t \in \mathcal{L}$ and f_i lies completely to the left of f_j, or $j \prec_t i$ for all $L_t \in \mathcal{L}$ and f_i lies completely to the right of f_j. In either case, f_i and f_j do not intersect, which completes the proof of the theorem.   □

By choosing $\mathcal{L}$ to be a minimum realizer above, Golumbic, Rotem, and Uruttia (1983) proved the following result which we state as a remark.

Remark 1.12. If ℓ is the minimum value for which G is the intersection graph of a concatenation of ℓ permutation diagrams, then $\dim(\overline{G}) = \ell + 1$.

We conclude this section by adding asteroidal triple free (AT-free) graphs to our hierarchy of Figure 1.14, showing that

$$\text{cocomparability} \subseteq \text{AT-free.}$$

Indeed, this inclusion is proper because the chordless cycle C_5 is AT-free but is not a cocomparability graph.

Theorem 1.13. *(Golumbic, Rotem, and Urrutia, 1983) All cocomparability graphs are AT-free.*

Proof. If $G = (V, E)$ is a cocomparability graph, then G is the intersection graph of a function diagram D, by Theorem 1.11. Suppose, for a contradiction, that G has an asteroidal triple $\{a, b, c\}$, and consider their associated function lines f_a, f_b, f_c in the diagram D. Since a, b, c are pairwise nonadjacent, the curves f_a, f_b, f_c do not intersect one another. Therefore, one of them, say f_b, lies totally between the other two.

Now consider what happens if we remove f_b and all curves which intersect it. We will obtain a function diagram for $G_{V-\mathcal{N}[b]}$ in which a and c are separated into distinct connected components. This contradicts the assumption that $\{a, b, c\}$ is an asteriodal triple, and proves the theorem.   □

1.7 Other families of graphs

1.7.1 Weakly chordal graphs

Weakly chordal graphs, as the name suggests, are a generalization of chordal graphs. They have gained interest in the recent literature, and will play an important role in our study of tolerance graphs in the next chapter.

Hayward (1985) introduced the class of *weakly chordal* graphs (also called *weakly triangulated*) as those with no induced subgraph isomorphic to C_n or to $\overline{C_n}$ for $n \geq 5$. Since $\overline{C_5} = C_5$, and $\overline{C_n}$ contains induced copies of C_4 for $n \geq 6$, the class of weakly chordal graphs contains the class of chordal graphs.

We will call vertices x and y a *two-pair* if every chordless path between x and y has exactly two edges. The weakly chordal graphs have been characterized using two-pairs as follows.

Theorem 1.14. *The following are equivalent.*

(i) *G is a weakly chordal graph.*
(ii) *Every induced subgraph of G is either a clique or has a two-pair.*
(iii) *If edges are repeatedly added between two-pairs in G, the result is eventually a clique.*

The implication (ii) $\Longrightarrow$ (i) follows from the observation that nonadjacent vertices in C_k or $\overline{C_k}$ (for $k \geq 5$) are not a two-pair (a good exercise). The implication (i) $\Longrightarrow$ (ii) is due to Hayward, Hoàng, and Maffray (1990), and (i) $\Longleftrightarrow$ (iii) is due to Spinrad and Sritharan (1995). The latter equivalence also leads to an $O(n^4)$ recognition algorithm for weakly chordal graphs.

1.7.2 Strongly chordal graphs

The strongly chordal graphs have been studied only recently and specialize chordal graphs in several ways. We will encounter these graphs in Chapters 11 and 12. We next define strongly chordal graphs and give additional definitions which will be used in Theorem 1.16 to characterize strongly chordal graphs using chords of a cycle, forbidden subgraphs and elimination orderings.

Let $C = [u_1, u_2, \ldots, u_{2k}, u_1]$ be a cycle of even length $2k \geq 6$. A chord $u_i u_j \in E(G)$ is called an *odd chord* if one of i and j is even and the other is odd, that is, it divides C into two even length cycles.

A graph G is defined to be *strongly chordal* if it is chordal and every cycle of even length greater than or equal to 6 has an odd chord. The graph in Figure 1.3 is strongly chordal, however, the graph S_3 in Figure 1.17 is not strongly chordal since the even cycle $[a, d, b, e, c, f, a]$ has no odd chord.

A vertex x is called a *simple* vertex if the following condition holds for closed neighborhoods: for every pair of neighbors y and z of x, either $\mathcal{N}[y] \subseteq \mathcal{N}[z]$ or $\mathcal{N}[z] \subseteq \mathcal{N}[y]$. An ordering of the vertices $[v_1, v_2, \ldots, v_n]$ is called a *simple elimination ordering* for G if v_i is a simple vertex in the graph G_i, for all i, where, as before, $G_i = G_{\{v_i, \ldots, v_n\}}$ denotes the subgraph of G remaining after deleting $\{v_1, \ldots, v_{i-1}\}$. Note that the graph S_3 in Figure 1.17 has no simple vertex, so it does not have a simple elimination ordering.

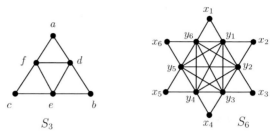

Figure 1.17. The suns S_3 and S_6.

A *strong elimination ordering* is defined to be an ordering of the vertices $[v_1, v_2, \ldots, v_n]$ where, for all $i < j < k < \ell$, if $v_i v_k, v_i v_\ell, v_j v_k \in E(G)$ then $v_j v_\ell \in E(G)$.

Remark 1.15. It is an easy exercise to verify that simple elimination orderings and strong elimination orderings are special cases of perfect elimination orderings (Exercise 1.10).

The graph S_3 is one of a family of forbidden subgraphs characterizing strongly chordal graphs. They are known in the literature both as suns and as trampolines. The *k-sun* S_k $(k \geq 3)$ consists of $2k$ vertices, a stable set $X = \{x_1, x_2, \ldots, x_k\}$ and a clique $Y = \{y_1, y_2, \ldots, y_k\}$, and edges $E_1 \cup E_2$ where $E_1 = \{x_1 y_1, y_1 x_2, x_2 y_2, y_2 x_3, \ldots, x_k y_k, y_k x_1\}$ forms the *outer cycle* and $E_2 = \{y_i y_j \mid i \neq j\}$ forms the inner clique. Figure 1.17 shows the graphs S_3 and S_6 and motivates the name sun. The suns are split graphs, so they are chordal by Theorem 1.2, but they are not strongly chordal since the outer cycle has no odd chord.

The next theorem, due to Farber (1983), summarizes the characterizations of strongly chordal graphs.

Theorem 1.16. *The following are equivalent.*

(i) *G is a strongly chordal graph.*
(ii) *G has a simple elimination ordering.*
(iii) *G is chordal and sun-free.*
(iv) *G has a strong elimination ordering.*

For further reading on strongly chordal graphs, and additional characterizations, see McKee and McMorris (1999) and Brandstädt, Le, and Spinrad (1999).

1.7.3 Threshold graphs

A graph $G = (V, E)$ is called a *threshold graph* if there exist positive weights a_i ($i \in V$) and a threshold $t > 0$ such that

$$S \subseteq V \text{ is a stable set} \iff \sum_{s \in S} a_s \leq t.$$

We will see threshold graphs again in Chapters 4, 11 and 12. The class of threshold graphs was introduced by Chvátal and Hammer (1977) who proved the next characterization theorem. A vertex which is adjacent to every other vertex is called *universal*; a vertex which is adjacent to no other vertex is called *isolated*.

Let $0 < \delta_1 < \delta_2 < \cdots < \delta_m < |V|$ be the vertex degrees of the nonisolated vertices of G, where the δ_i are distinct and there may be many vertices of degree δ_i; further, let $\delta_0 = 0$, even if there are no isolated vertices. The *degree partition* of V is given by $V = D_0 \cup D_1 \cup \cdots \cup D_m$, where D_i is the set of all vertices of degree δ_i. Only D_0 is possibly empty.

Theorem 1.17. *The following are equivalent.*

   (i) *G is a threshold graph.*
  (ii) *$\overline{G}$ is a threshold graph.*
 (iii) *There exist positive weights w_i ($i \in V$) and a threshold $\theta > 0$ such that $xy \in E \iff w_x + w_y > \theta$.*
 (iv) *Repeatedly removing either a universal or an isolated vertex from G results eventually in the empty set.*
  (v) *G does not contain any of P_4, C_4 or $2K_2$ as an induced subgraph.*
 (vi) *For all distinct vertices $x \in D_i$ and $y \in D_j$, we have $xy \in E \iff i + j > m$.*

Additional equivalent conditions and proofs can be found in Mahadev and Peled (1995). Theorem 1.17 immediately implies the following.

Theorem 1.18. *Threshold graphs are chordal, co-chordal, comparability and cocomparability graphs; hence, they are also interval, split and permutation graphs.*

Proof. Let G be a threshold graph. The chordality of G follows from (v) and co-chordality then follows from the equivalence of (i) and (ii). To show G is a comparability graph, we fix an ordering $\prec$ on $V(G)$ using (iii) where $x \prec y$ whenever $w_x < w_y$. Now orient $E(G)$ according to $\prec$. This orientation will be transitive using (iii). Thus G is a comparability graph, and it is also a cocomparability graph using the equivalence of (i) and (ii). The remaining conclusions follow from Theorems 1.3, 1.2, and 1.8.   □

1.8 Other reading and general references

In this book, it would be impossible to present all of the topics in graph theory that would be of interest to a researcher studying tolerance graphs. For further reading and reference we offer a modest list of important works that should be consulted.

- M. C. Golumbic, *Algorithmic Graph Theory and Perfect Graphs,* Academic Press (1980) provides an introduction to classes of perfect graphs such as comparability graphs, chordal graphs and interval graphs. In addition to the mathematical foundations, there is an emphasis on applications as well as algorithms and complexity.

Four books have appeared recently which cover advanced research in this area. They are the following, and are a must for any graph theory library.

- A. Brandstädt, V. B. Le, and J. P. Spinrad, *Graph Classes: A Survey*, SIAM, Philadelphia (1999) is an extensive and invaluable compendium of the current status of complexity and mathematical results on hundreds of families of graphs. It is comprehensive with respect to definitions and theorems, and citing over 1100 references.
- T. A. McKee and F. R. McMorris, *Topics in Intersection Graph Theory*, SIAM, Philadelphia (1999) is a focused monograph on structural properties, presenting definitions, major theorems with proofs and many applications.
- N. V. R. Mahadev and U. N. Peled, *Threshold Graphs and Related Topics*, North-Holland (1995) is a thorough and extensive treatment of all research done in the past years on threshold graphs, threshold dimension, and orders, and a dozen new concepts which have emerged.
- W. T. Trotter, *Combinatorics and Partially Ordered Sets*, Johns Hopkins University Press, Baltimore (1992) is a valuable book which covers new directions of investigation and research on ordered sets with an emphasis on dimension theory.

Other important classical books are Roberts (1976) and Fishburn (1985). All these references illustrate the many uses of the intersection graph model, which has become a necessary and important tool for solving real-world problems, and the rich mathematical structures motivated by them.

Temporal Reasoning. One of the "traditional" applications of interval graphs is reasoning about time intervals, which started with the original questions of Hajós in 1957 and Benzer in 1959 (see Golumbic (1980) page 171). Temporal reasoning is an essential part of many applications in artificial intelligence (AI). Given a set of explicit relationships between certain events, we would like to be able to infer additional relationships which are implicit in those given. For example, the transitivity of "before" and "contains" may allow us to derive information regarding the sequence of events. Seriation problems ask for a mapping of temporal events onto the time line such that all the given relations are satisfied, that is, a consistent scenario. Similarly, there are problems of scheduling, planning, and story understanding in which one is interested in constructing a time line where each particular event or phenomenon or task corresponds to an interval representing its duration.

Allen (1983) introduced a model for temporal reasoning using the thirteen primitive interval relations obtained by considering all possible orderings of their four endpoints. Several authors working in AI have studied and adapted Allen's model further, and have incorporated such models into reasoning systems. The paper by Golumbic and Shamir (1993) has provided a bridge linking some of these temporal reasoning notions from the AI community with those of the combinatorics community and extending results in both disciplines. We also refer the reader to Golumbic (1998) which is a survey paper[1] in the same spirit as this book. It describes a number of directions of current work on reasoning about time, many of which employ graph algorithms.

1.9 Exercises

Exercise 1.1. Let $\mathcal{I} = \{I_i\}$ for $i = 0, \dots, 6$ where $I_i = [i, 8 + 6i - i^2]$.

(a) What is the interval graph represented by $\mathcal{I}$?
(b) If $t_i = 2i + 1$, what is the tolerance graph represented by $\langle \mathcal{I}, \{t_i\} \rangle$?
(c) If $t'_i = 7 - i$, what is the tolerance graph represented by $\langle \mathcal{I}, \{t'_i\} \rangle$?
(d) What is the size of the largest clique in each of these graphs?
(e) What is the size of the largest stable set in each of these graphs?

Exercise 1.2. Find a tolerance representation for the chordless 4-cycle C_4 in Figure 1.6.

---

[1] This survey paper also includes some of that author's newest illustrative stories, "Will Allan get to Judy's in time?" and "Goldie and the Four Bears".

Exercise 1.3. Find a maximum cardinality search (MCS) ordering for each of the graphs in Figure 1.8. Check whether the reversal of these MCS orderings are perfect elimination orderings. Explain your findings in terms of Theorem 1.1.

Exercise 1.4. Prove Theorem 1.2.

Exercise 1.5. (a) Give a transitive orientation (TRO) for the graph in Figure 1.3.

(b) Give an argument for why each of the graphs in Figure 1.8 does not have a transitive orientation.

Exercise 1.6. Let $G = (V, E)$ be a graph, and let $\overline{G} = (V, \overline{E})$ be its complement. Prove the following:

If F_1 is a TRO of G and F_2 is a TRO of $\overline{G}$, then $F_1 \cup F_2$ is transitive, i.e., a TRO of the complete graph.

Exercise 1.7. At the Center for Disease Research each new researcher (i.e., doctoral student) visits the Germ Exposure Room once during the first day of the semester, and is exposed to all the bacteria of everyone who is there at the time. How can we assign the researchers to a minimum number of offices in such a way that no one will be exposed to a new person? Give a graph theoretic solution.

Exercise 1.8. Let $G_{20} = (V, E)$ be a graph with vertices $\{v_1, v_2, \ldots, v_{20}\}$ and edges $(v_i, v_j) \in E \iff i + j \geq 18$.
(a) What is the size of the largest clique of G_{20}?
(b) Prove that G_{20} is an interval graph.
(c) Find a perfect elimination ordering for the vertices of G_{20}.

Exercise 1.9. What graph is represented by the intersection diagrams in Figure 1.14? Show that this graph is not a threshold graph.

Exercise 1.10. Show that all simple elimination orderings and all strong elimination orderings are perfect elimination orderings.

Exercise 1.11. LALE Airline has published the following schedule and has exactly four B737 and two B757 aircraft available.

Flight	Departs TelAviv	Arrives TelAviv	Aircraft
TelAviv–Athens–TelAviv #1	7:00	12:30	B757
TelAviv–Athens–TelAviv #2	11:30	17:00	B737
TelAviv–Athens–TelAviv #3	13:00	18:30	any
TelAviv–Athens–TelAviv #4	16:00	21:30	any
TelAviv–Rome–TelAviv #5	9:00	19:30	B757
TelAviv–Cairo–TelAviv #6	10:30	15:00	B737
TelAviv–Istanbul–TelAviv #7	19:00	23:50	any
TelAviv–Amman–TelAviv #8	16:30	19:30	B737
TelAviv–Milan–TelAviv #9	15:00	23:50	B757

(a) Assume that minimum "ground time" between flights is 75 minutes. Can LALE meet its schedule above? Explain why.

(b) What is the minimum number of B757 aircraft required if LALE adds the three additional flights below? Explain your answer in terms of interval graphs.

Additional Flights	Departs TelAviv	Arrives TelAviv	Aircraft
TelAviv–Bucharest–TelAviv #1	6:30	13:30	B757
TelAviv–Athens–TelAviv #2	14:30	20:00	B757
TelAviv–Eilat–TelAviv #3	21:00	23:30	B757

Exercise 1.12. Show that any order is the intersection of all its linear extensions.

Exercise 1.13. Give a transitive orientation F for the chordless 6-cycle C_6, and draw the associated Hasse diagram for this order. Prove that this order has dimension 3 and interval dimension 3.

Exercise 1.14. The graph G in Figure 1.16 is a cocomparability graph since its complement $\overline{G}$ has a transitive orientation. Does G have a transitive orientation? Is G a permutation graph? Why?

Exercise 1.15. Let G be a chordal graph and $n = |V(G)|$. Show that the number of maximal cliques in G is at most n. (Hint: Let $[v_1, v_2, \ldots, v_n]$ be a perfect elimination ordering, and consider the sets $\{v_i\} \cup [\mathcal{N}(v_i) \cap V(G_i)]$).

Exercise 1.16. Algorithm for maximum stable set of intervals

Consider the following algorithm, applied to a set of intervals with distinct endpoints:

1. Initialize empty sets *Stable* and *Active* and let $i = 0$.
2. Sweep across the representation from left to right.
3.   When a left endpoint (new interval) is encountered,

4.   add its interval to the *Active* set.
5.   When a right endpoint is encountered,
6.   add its interval to the *Stable* set, increase i by one,
7.   define the set $K_i = Active$, and reset *Active* to empty.
8. After completing the sweep,
9.   set S and k to be the final values of X and i, respectively.

   (a) Prove that S is a maximum stable set, and that $K_1 \cup \cdots \cup K_k$ is a minimum clique cover.

   (b) What is the complexity of the algorithm?

Exercise 1.17. (a) Prove the following: if G is a comparability graph but not a permutation graph, then G is not a trapezoid graph.

   (b) Show that the graph H in Figure 2.5 is not a trapezoid graph.

Chapter 2

Early work on tolerance graphs

In this chapter we discuss the definitions, results and open questions that appear in Golumbic and Monma (1982) and Golumbic, Monma, and Trotter (1984), the papers which introduced the topic of tolerance graphs. We also present consequences of these results and related topics from more recent literature.

2.1 Notation and observations

Recall from Section 1.3 that a graph G is a *tolerance graph* if each vertex $v \in V(G)$ can be assigned a closed interval I_v and a tolerance $t_v \in \mathbf{R}^+$ so that $xy \in E(G)$ if and only if $|I_x \cap I_y| \geq \min\{t_x, t_y\}$. If graph G has a tolerance representation with $t_v \leq |I_v|$ for all $v \in V(G)$, then G is called a *bounded tolerance graph*.

Many important graph properties are inherited by all induced subgraphs and thus called *hereditary* properties. Given a (bounded) tolerance representation $\langle \mathcal{I}, t \rangle$ of a graph G, for any subset of vertices $W \subseteq V(G)$ the intervals $\{I_w \mid w \in W\}$ and tolerances $\{t_w \mid w \in W\}$ give a representation of G_W. Thus, induced subgraphs of tolerance graphs are also tolerance graphs and induced subgraphs of bounded tolerance graphs are also bounded tolerance graphs. We record this as a remark.

Remark 2.1. The property of being a tolerance graph (resp. bounded tolerance graph) is hereditary.

In a tolerance representation of a graph G, we may have intervals of the form $I_x = [a_x, a_x]$. In this case, for any other interval I_y in the representation, $|I_x \cap I_y| = 0 < \min\{t_x, t_y\}$ since tolerances are positive. Thus $xy \notin E(G)$. However, in a tolerance representation that is bounded, it is impossible to have any interval I_v be a point because $|I_v| \geq t_v > 0$. We summarize this as follows:

Remark 2.2. (i) If $\langle \mathcal{I}, t \rangle$ is a tolerance representation of G and I_v is a point, then the corresponding vertex v is an isolated vertex of G.
(ii) If $\langle \mathcal{I}, t \rangle$ is a bounded tolerance representation of G, then no interval I_v is a point.

At times we will need to refer to the endpoints of intervals in a tolerance representation. We denote the left endpoint of interval I_v by $L(v)$ and the right endpoint by $R(v)$. It is often convenient to have a tolerance representation that satisfies one or more of the following additional properties.

(1) Any tolerance larger than the length of its corresponding interval is set to infinity.
(2) All tolerances are distinct (except for those set to infinity).
(3) No two different intervals share an endpoint.

A tolerance representation satisfying all three of these properties is called a *regular representation* and in Lemma 2.3 we prove that every tolerance graph has such a representation. (Our notation differs slightly from that in Golumbic, Monma, and Trotter (1984) where they include a fourth condition in the definition of regular.) Lemma 5.18 contains a more general result.

Lemma 2.3. *Every tolerance graph has a regular representation.*

Proof. Let $G = (V, E)$ be a tolerance graph and fix a representation $\langle \mathcal{I}, t \rangle$ of G. Let $I_v = [L(v), R(v)]$ for each $v \in V$.
Proof of (1). If $t_x > |I_x|$ for some vertex $x \in V$, then $xy \in E \iff |I_x \cap I_y| \geq t_y$. The same is true if $t_x = \infty$.
Proof of (2) and (3). Let ϵ be the smallest positive number appearing in the union of the sets (i) – (vi) below where x and y are taken over all vertices in G.

(i) $\{|L(x) - L(y)|\}$.
(ii) $\{|R(x) - R(y)|\}$.
(iii) $\{|L(x) - R(y)|\}$.
(iv) $\{t_x\}$.
(v) $\{|t_x - t_y|\}$.
(vi) $\{t_x - |I_x \cap I_y|\}$.

If x and y are distinct vertices with $t_x = t_y$, then choose one of them, say x, replace t_x by $t'_x = t_x - \epsilon/2$, and leave t_y unchanged. We show that this gives a representation of G with one fewer repeated tolerance. If $xz \in E$, then $|I_x \cap I_z| \geq \min\{t_x, t_z\} \geq \min\{t'_x, t_z\}$. If $xz \notin E$, then $|I_x \cap I_z| < \min\{t_x, t_z\}$ and by our choice of ϵ we know $\min\{t_x, t_z\} - |I_x \cap I_z| \geq \epsilon$. Thus $|I_x \cap I_z| \leq$

$\min\{t_x, t_z\} - \epsilon < \min\{t'_x, t_z\}$ as desired. If necessary, recompute ϵ and repeat the process until all tolerances are distinct.

Now suppose two different intervals share an endpoint. Let $S = \{L(v), R(v) \mid v \in V\}$ be the set of endpoints in the representation, and let s be the smallest repeated endpoint in S. Let x and y be distinct elements of V for which s is an endpoint of I_x and I_y. If there exist $x \in V$ with $R(x) = s$, pick the one whose interval I_x is the longest and replace I_x by $I'_x = [L(x) + \epsilon/2, R(x) + \epsilon/2]$. Otherwise, pick $x \in V$ with $L(x) = s$ and $|I_x|$ as large as possible and replace I_x by $I'_x = [L(x) + \epsilon/2, R(x) + \epsilon/2]$. It is not hard to see that this gives a representation of G with one fewer pair of elements sharing an endpoint. All tolerances are still distinct. If necessary, recompute ϵ and repeat this process until all endpoints are distinct. $\qquad\square$

Remark 2.4. The transformation in the proof of Lemma 2.3 maintains the lengths of the intervals.

2.2 Permutation graphs and interval graphs

In Golumbic and Monma (1982), the authors show that the class of bounded tolerance graphs is a simultaneous generalization of interval graphs and permutation graphs. The results appear in our next two theorems.

Theorem 2.5. *The following are equivalent statements about a graph G.*

(i) *G is an interval graph.*
(ii) *G is a tolerance graph with constant tolerances.*
(iii) *G is a bounded tolerance graph with constant tolerances.*

Proof. (i) $\implies$ (ii): Let G be an interval graph with a representation in which interval I_v is assigned to vertex v. Let c be any positive number less than $\min\{|I_x \cap I_y| : |I_x \cap I_y| > 0, x \neq y, \text{ and } x, y \in V(G)\}$. Then the intervals $\{I_v \mid v \in V(G)\}$ together with tolerances $t_v = c$ for all $v \in V(G)$ give a tolerance representation of G with constant tolerances.

(ii) $\implies$ (iii): Let $\langle \mathcal{I}, t \rangle$ be a tolerance representation of G with $t_v = c$ for all $v \in V(G)$. For those vertices $v \in V(G)$ with $|I_v| \geq c$, define $I'_v = I_v$. If $|I_x| < c$ then x is an isolated vertex of G and we define I'_x to be an interval of length c on the real line that does not intersect any other I'_y. This gives a bounded tolerance representation of G with constant tolerances.

(iii) $\implies$ (i): Let $\langle \mathcal{I}, t \rangle$ be a bounded tolerance representation of G with $t_v = c$ for all $v \in V(G)$. Denote I_v by $[L(v), R(v)]$. For those vertices $v \in V(G)$ with $R(v) - L(v) \geq c$, define $I'_v = [L(v) + c/2, R(v) - c/2]$. Otherwise, if

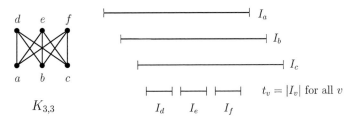

Figure 2.1. A bounded tolerance representation of $K_{3,3}$ in which the tolerance of each vertex is equal to the length of its interval.

$R(v) - L(v) < c$, then v is an isolated vertex of G and we define I'_v to be a point on the real line that does not intersect any other I'_z. The intervals $\{I'_v \mid v \in V\}$ give an interval representation of G. □

The next theorem shows that permutation graphs (defined in Section 1.6) are another special class of tolerance graphs. Permutation graphs are equivalent to interval containment graphs, which we next define.

An *interval containment graph* is one that can be represented by a set of real intervals $\{I_i \mid i \in V(G)\}$ so that $ij \in E(G)$ precisely when one of I_i, I_j contains the other. Such a representation is called an *interval containment representation*. Figure 2.1 gives such a representation for the complete bipartite graph $K_{3,3}$. The equivalence of (ii) and (iii) in the following theorem first appeared in Dushnik and Miller (1941).

Theorem 2.6. *The following are equivalent statements about a graph G.*

(i) *G has a tolerance representation with $t_i = |I_i|$ for all $i \in V(G)$.*
(ii) *G is an interval containment graph.*
(iii) *G is the comparability graph of an order of dimension at most 2.*
(vi) *G is a permutation graph.*

Proof. (i) $\Longleftrightarrow$ (ii): A graph G has a tolerance representation $\langle \mathcal{I}, t \rangle$ with $t_i = |I_i|$ for all $i \in V(G)$ if and only if it has an interval containment representation, using the same intervals.

(ii) $\Longrightarrow$ (iii): Let $\mathcal{I}$ be an interval containment representation of G. Without loss of generality, we may assume all endpoints of intervals are distinct (Lemma 1.5). Let $L_1 = (V, \prec_1)$ be the linear order of left endpoints of intervals in $\mathcal{I}$ and $L_2 = (V, \prec_2)$ be the *reverse* of the linear order of right endpoints of intervals in $\mathcal{I}$. Let $P = L_1 \cap L_2$ and G' be the comparability graph of P. Then it is easy to check that $ij \in E(G') \Longleftrightarrow I_i \subseteq I_j$ or $I_j \subseteq I_i$.

(iii) $\Longleftrightarrow$ (iv): Given a permutation graph G represented by a permutation π of $V(G) = \{1, 2, 3, \ldots, n\}$, let L_1 be the linear order $1 \prec_1 2 \prec_1 3 \prec_1 \cdots \prec_1 n$, and L_2 be the reverse of the order π. It is easy to check that G is the comparability graph of $L_1 \cap L_2$. Conversely, suppose G is the comparability graph of $L_1 \cap L_2$ where L_1 and L_2 are linear orders. Label the set of vertices $V(G) = \{1, 2, 3, \ldots, n\}$, so that $1 \prec_1 2 \prec_1 3 \prec_1 \cdots \prec_1 n$. Then G is a permutation graph with π as the ordering defined by the reverse of L_2.

(iv) $\Longrightarrow$ (ii): Let $V(G) = \{1, 2, 3, \ldots, n\}$ and let π be a permutation of $V(G)$ so that $ij \in E(G)$ if and only if the order of i and j is reversed in π. On a horizontal line, write the numbers $1, 2, 3, \ldots, n$ in increasing order, followed by the same set written in the order of π. For $i \in V(G)$, let the interval I_i have its endpoints at the two occurrences of i on the line. This gives an interval containment representation of G in which all endpoints of intervals are distinct. □

2.3 Bounded tolerance graphs

Recall from Chapter 1 that a graph G is a *comparability graph* if each edge $ij \in E(G)$ can be assigned a direction so that the resulting oriented graph (V, F) satisfies the following condition: $ij \in F$ and $jk \in F$ imply $ik \in F$ for all $i, j, k \in V(G)$. Such an orientation is called a *transitive orientation* of G and when such an orientation exists we say G is *transitively orientable*. Such a transitive orientation of graph G gives rise to an ordered set $P = (V, \prec)$ where $x \prec y$ if and only if $xy \in F$, and G is the comparability graph of P. Thus a graph is a comparability graph if and only if it is the comparability graph of an ordered set.

A *cocomparability graph* is a graph whose complement is a comparability graph, and thus it is the incomparability graph of an ordered set. There exist graphs which are comparability graphs but not cocomparability graphs such as the graph T_2 in the next example. Its complement $\overline{T_2}$ is therefore a cocomparability graph but not a comparability graph.

Example 2.7. The graph T_2 in Figure 2.2 is a comparability graph but not a cocomparability graph.

Proof. Orienting all edges of T_2 away from b_1, b_2 and b_3 gives a transitive orientation, hence T_2 is a comparability graph.

We show that T_2 is not a cocomparability graph. Refer to the labeling in Figure 2.2, and suppose $\overline{T_2}$ had a transitive orientation F. Reversing the orientation of each edge would give another transitive orientation, hence we may assume, without loss of generality, that at least two arcs are oriented away from vertex a in F.

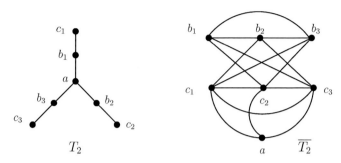

Figure 2.2. The graph T_2 and its complement $\overline{T_2}$.

By symmetry we may assume $ac_2 \in F$ and $ac_3 \in F$. The first of these implies $b_3 c_2 \in F$ which in turn implies $c_3 c_2 \in F$. The second implies $b_2 c_3 \in F$, which in turn implies $c_2 c_3 \in F$, a contradiction.   □

It is easy to check that the graphs in Figure 3.1 are neither comparability graphs nor cocomparability graphs. The following theorem enables us to find examples of graphs that are not bounded tolerance graphs by using graphs that are not cocomparability graphs, such as T_2 and the graphs in Figure 3.1.

Theorem 2.8. *Bounded tolerance graphs are cocomparability graphs.*

We choose to follow the proof of this theorem given in Bogart, Fishburn, Isaak, and Langley (1995). This proof uses parallelogram graphs which provide another way to think about bounded tolerance graphs. Fix two horizontal lines L_1 and L_2 with L_1 above L_2. Recall from Section 1.6 that a graph G is a *parallelogram graph* if each vertex $i \in V(G)$ can be assigned a parallelogram P_i with parallel sides along L_1 and L_2 so that G is the intersection graph of $\{P_i \mid i \in V(G)\}$. By shifting the line L_1 to the left if necessary, we may ensure that the non-horizontal sides of each parallelogram have negative slope. We allow degenerate parallelograms, that is, the sides along L_1 and L_2 may be points, in which case the resulting parallelogram is a line.

Figure 2.3 shows a bounded tolerance representation and a parallelogram representation of the graph C_4. The correspondence between these types of representations was first observed in Langley (1993), and is given in the next theorem.

Theorem 2.9. *A graph is a bounded tolerance graph if and only if it is a parallelogram graph.*

Proof of Theorem 2.9. Given a bounded tolerance representation $\langle \mathcal{I}, t \rangle$ of graph $G = (V, E)$ where $I_v = [L(v), R(v)]$, we associate a parallelogram with

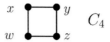

Tolerance representation of C_4

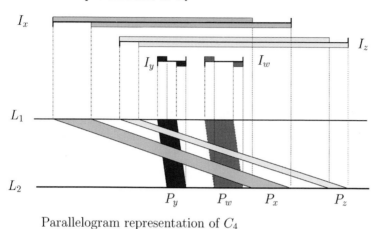

Parallelogram representation of C_4

Figure 2.3. A bounded tolerance representation and a parallelogram representation of C_4. The shading above and below each interval indicates the size of the tolerance.

each vertex $v \in V$ as follows. The parallelogram P_v has top edge $[L(v), R(v) - t_v]$ on L_1 and bottom edge $[L(v) + t_v, R(v)]$ on L_2 (see Figure 2.3). It is easy to check that P_v is indeed a parallelogram for each $v \in V$ and that $xy \in E(G) \iff P_x \cap P_y \neq \emptyset$. Thus, G is the intersection graph of $\{P_v \mid v \in V(G)\}$.

Conversely, suppose G is a parallelogram graph with representation $\{P_v \mid v \in V(G)\}$ for which the non-horizontal sides have negative slope. Let $[a_v, c_v]$ denote the edge of P_v along L_1 and let $[b_v, d_v]$ denote the edge of P_v along L_2. Finally, let $t_v = b_v - a_v$. Since P_v is a parallelogram, $t_v = b_v - a_v = d_v - c_v$ and $t_v > 0$ since the slope of the non-horizontal sides is negative. The collection of intervals $I_v = [a_v, d_v]$ and tolerances t_v for $v \in V(G)$ give a bounded tolerance representation of G. $\square$

Proof of Theorem 2.8. Let $\{P_i \mid i \in V(G)\}$ be a representation of G as a parallelogram graph. In order to show G is a cocomparability graph, we give a transitive orientation of its complement $\overline{G}$. If $ij \in \overline{G}$ then $ij \notin E(G)$ and thus $P_i \cap P_j = \emptyset$. In this case, one of the parallelograms P_i, P_j lies to the left of the other. Orient the edge $ij \in E(\overline{G})$ from i to j if P_i is to the left of P_j, and from j to i if P_j is to the left of P_i. This orientation is transitive, proving the result.   □

Algebraic proofs of Theorem 2.8 are given in Golumbic and Monma (1982) and Bogart, Fishburn, Isaak, and Langley (1995). In these alternative proofs, the authors fix a bounded tolerance representation of graph G and orient an edge xy in $\overline{G}$ according to the order of the right endpoints (resp. midpoints) of the intervals I_x, I_y, in the representation of G. In Exercise 2.3, the reader is asked to reconstruct these proofs.

2.4 Tolerance graphs are weakly chordal

Recall that a graph is *chordal* if it has no induced subgraph isomorphic to C_n for $n \geq 4$. Interval graphs are chordal, but tolerance graphs are not necessarily chordal. Indeed, the 4-cycle C_4 is often given as an example of a tolerance graph that is not an interval graph (see Figure 2.3). However, tolerance graphs are "almost" chordal, which we shall make more precise. In Section 1.7.1 we introduced the class of *weakly chordal* graphs (also called *weakly triangulated*) as those with no induced subgraph isomorphic to C_n or to $\overline{C_n}$ for $n \geq 5$. The rest of the section is devoted to the proof from Golumbic, Monma, and Trotter (1984) that tolerance graphs are weakly chordal.

We begin with a lemma about orienting 4-cycles in the complement of a cycle.

Lemma 2.10. *If G has an orientation which is transitive on each induced 4-cycle, then G contains no induced $\overline{C_n}$ for $n \geq 6$.*

Proof. Let F be an orientation of G which is transitive on each induced 4-cycle. Suppose, for a contradiction, that G has an induced subgraph which is isomorphic to $\overline{C_n}$ for some $n \geq 6$. Let $V(\overline{C_n}) = \{v_1, v_2, \ldots, v_n\}$ where $v_i v_j \in E(\overline{C_n})$ if and only if i and j differ by more than 1 modulo n. Without loss of generality, we may assume $v_1 v_3 \in F$. Applying the transitivity of F on the 4-cycle induced by v_n, v_1, v_3, v_4 we conclude $v_1 v_4$. Similarly, for $i = 4, 5, \ldots, n-2$ we apply the transitivity of F on the 4-cycle induced by v_1, v_2, v_i, v_{i+1} to conclude $v_1 v_{i+1} \in F$. Thus vertex v_1 is a source, that is, all edges of G incident to v_1 are oriented away from v_1 by F.

The same argument shows that whenever $v_i v_j \in F$, then v_j is a sink, that is, all edges of G incident to v_j are oriented towards v_j by F. Therefore, v_3 and v_{n-1} are both sinks, but this is a contradiction because $v_3 v_{n-1} \in E(G)$ for $n \geq 6$.   □

Corollary 2.11. *The graph $\overline{C_n}$ is not a comparability graph for $n \geq 5$.*

Proof. For $n = 5$, $\overline{C_5} = C_5$ is not a comparability graph. For $n \geq 6$, if $\overline{C_n}$ were a comparability graph, it would have a transitive orientation F which naturally will be transitive on each induced 4-cycle. This contradicts Lemma 2.10. □

Lemma 2.12. *Let $\langle \mathcal{I}, t \rangle$ be a regular representation of tolerance graph $G = (V, E)$. If $I_x \subseteq I_y$ and $xy \notin E$, then $\mathcal{N}(x) \subseteq \mathcal{N}(y)$.*

Proof. Since $xy \notin E$, we must have $t_x = \infty$. For any vertex $z \in \mathcal{N}(x)$ we have $t_z = \min\{t_x, t_z\} \leq |I_x \cap I_z| \leq |I_y \cap I_z|$, so $yz \in E$. □

Lemma 2.13. *Let $\langle \mathcal{I}, t \rangle$ be a regular representation of tolerance graph $G = (V, E)$. If replacing t_x by $\min\{t_x, |I_x|\}$ does not give a tolerance representation of G, then there exists a $y \in V(G)$, distinct from x with $\mathcal{N}(x) \subseteq \mathcal{N}(y)$.*

Proof. Since the representation is regular, it must be the case that $t_x = \infty$, and replacing it by $|I_x|$ can only change the graph represented by adding edges of the form xy where $I_x \subseteq I_y$. The conclusion follows from Lemma 2.12. □

Lemma 2.14. *The cycle C_n is not a tolerance graph for $n \geq 5$.*

Proof. By Corollary 2.11, $\overline{C_n}$ is not a comparability graph for $n \geq 5$, thus C_n is not a cocomparability graph for $n \geq 5$. Applying Theorem 2.8 we know that C_n is not a bounded tolerance graph for $n \geq 5$. Suppose, for a contradiction, that $C_n = (V, E)$ is a tolerance graph for some $n \geq 5$, and fix a regular tolerance representation $\langle \mathcal{I}, t \rangle$ of C_n that has the smallest possible number of vertices having infinite tolerance. Since C_n is not a bounded tolerance graph, there must be a vertex $x \in V$ for which $t_x = \infty$. If t_x were changed to $|I_x|$, the representation would have one fewer vertex with infinite tolerance, and hence it would no longer be a representation of C_n. Now applying Lemma 2.13, there exists a vertex $y \in V(G)$ with $x \neq y$ and $\mathcal{N}(x) \subseteq \mathcal{N}(y)$. This is a contradiction since there is no pair of vertices in C_n for which the neighborhood of one vertex is contained in the neighborhood of the other. □

In proving the analogous result for complements of cycles, we require a lemma about orienting edges of a tolerance graph according to tolerances. In Golumbic, Monma, and Trotter (1984), the authors define the *tolerance orientation* F associated with a regular representation $\langle \mathcal{I}, t \rangle$ of $G = (V, E)$ as follows: $xy \in F$ if and only if $xy \in E$ and $t_x < t_y$. This is well-defined since $xy \in E$ implies that at least one of t_x, t_y is finite.

Tolerance orientations are always acyclic, however, they are not always transitive. The next lemma shows that they induce transitive orientations on 4-cycles.

Lemma 2.15. *The tolerance orientation F of any regular representation of a tolerance graph G is transitive on each induced 4-cycle.*

Proof. Let C be a 4-cycle where $V(C) = \{a, b, c, d\}$ and $E(C) = \{ab, ac, bd, cd\}$, and fix a regular representation $\langle \mathcal{I}, t \rangle$ of G. We will show that F is transitive on C. By renaming the vertices if necessary, we may assume $t_a < t_b < t_c$ and $t_a < t_d$. It suffices to show $t_d < t_b$. For a contradiction, assume $t_b < t_d$.

First we show that neither of I_a, I_d contains the other. If $I_a \subset I_d$ then $|I_c \cap I_a| \leq |I_a| = |I_a \cap I_d| < t_a = \min\{t_a, t_c\}$, a contradiction since $ac \in E$. Similarly, if $I_d \subset I_a$ then $|I_c \cap I_d| \leq |I_d| = |I_a \cap I_d| < t_a < \min\{t_c, t_d\}$, a contradiction since $cd \in E$. So without loss of generality we may assume $L(a) < L(d)$ and $R(a) < R(d)$.

Next we show that each of I_b, I_c intersects each of $(I_a - I_d)$ and $(I_d - I_a)$. If $I_b \cap (I_a - I_d) = \emptyset$, then $t_a \leq |I_b \cap I_a| \leq |I_d \cap I_a|$, a contradiction since $ad \notin E$. We arrive at the same contradiction for I_c. If $I_b \cap (I_d - I_a) = \emptyset$, then $t_a < t_b \leq |I_b \cap I_d| \leq |I_a \cap I_d|$, a contradiction since $ad \notin E$. Similarly, for I_c.

Thus we may conclude that $L(b), L(c) < L(d)$ and $R(b), R(c) > R(a)$. Finally, $t_b = \min\{t_b, t_d\} \leq |I_b \cap I_d|$ so $[L(d), L(d) + t_b] \subseteq I_b$, and $t_b < \min\{t_c, t_d\} \leq |I_c \cap I_d|$ so $[L(d), L(d) + t_b] \subseteq I_c$.

But then $|I_b \cap I_c| \geq |[L(d), L(d) + t_b]| = t_b$, a contradiction since $bc \notin E$. $\square$

Corollary 2.16. *The graph $\overline{C_n}$ is not a tolerance graph for $n \geq 5$.*

Proof. The case $n = 5$ is covered by Lemma 2.14. For $n \geq 6$, we combine the results of Lemmas 2.15 and 2.10. $\square$

Now combining the results of Lemma 2.14 and Corollary 2.16 together with Remark 2.1 yields the following result.

Theorem 2.17. *(Golumbic, Monma, and Trotter, 1984) Tolerance graphs are weakly chordal.*

The term weakly chordal does not appear in Golumbic, Monma, and Trotter (1984) since it was not defined until Hayward (1985). An alternative proof of Theorem 2.17 can be obtained by first showing that every tolerance graph is a domination graph (Rusu and Spinrad, 2001) and then showing that every domination graph is weakly chordal (Dahlhaus, Hammer, Maffray, and Olariu, 1994).

A graph G is called *alternately orientable* if there is an orientation F of G which is transitive on every chordless cycle of length greater than or equal to 4 i.e., the directions of the oriented edges must alternate. (These graphs

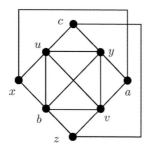

Figure 2.4. The Berlin graph $\overline{B}$.

are also known as *alternatingly orientable*.) Clearly, an alternately orientable graph may have only even length chordless cycles. Alternately orientable graphs generalize comparability graphs since the latter must have an orientation which is transitive on the entire graph. The following is an easy consequence of our previous results.

Theorem 2.18. *Tolerance graphs are alternately orientable.*

Proof. By Lemma 2.14, the only chordless cycles permitted in a tolerance graph are of length 4. Applying Lemma 2.15, the tolerance orientation of any regular representation is transitive on every chordless 4-cycle, so G is alternately orientable. □

We can use Theorem 2.18 to find examples of graphs which are not tolerance graphs, such as the following from Felsner (1998).

Example 2.19. The Berlin graph $\overline{B}$ in Figure 2.4 is not alternately orientable. Hence, it is not a tolerance graph.

Proof. Suppose that graph $\overline{B}$ in Figure 2.4 has an orientation F which alternates on the chordless 4-cycles:

$$[a, x, b, y], \ [b, y, c, z], \ [c, z, v, u], \ [v, u, x, a]$$

Then we would have the following implications:

$$ax \in F \Longleftrightarrow by \in F \Longleftrightarrow cz \in F \Longleftrightarrow vu \in F \Longleftrightarrow xa \in F$$

which produce a contradiction. □

The complement of the Berlin graph appears again in Chapters 10 and 13. Felsner also shows the following result which we state without proof.

Theorem 2.20. *(Felsner, 1998) If G is both alternately orientable and a co-comparability graph, then G is a trapezoid graph.*

We conclude this section with a result which first appeared in Corneil and Kamula (1987).

Theorem 2.21. *Trapezoid graphs are weakly chordal.*

Proof. Suppose G is not weakly chordal. Then by definition, G contains C_n or $\overline{C_n}$ as an induced subgraph for some $n \geq 5$. We show that the cycle C_n and its complement $\overline{C_n}$ are not trapezoid graphs for $n \geq 5$.

If C_n were a trapezoid graph then it would be a ribbon graph (Figure 1.14) and therefore a cocomparability graph (Theorem 1.11). Thus $\overline{C_n}$ would be a comparability graph, which contradicts Corollary 2.11.

Similarly, if $\overline{C_{2n+1}}$ were a trapezoid graph for $n \geq 2$, it would be a co-comparability graph and thus C_{2n+1} would be a comparability graph. This is a contradiction since odd cycles are not transitively orientable. Likewise, we will see that $\overline{C_{2n}}$ is not a trapezoid graph by combining Theorem 5.25 and Example 5.15.   □

2.5 Tolerance graphs are perfect

A graph G is *perfect* if for all induced subgraphs H of G, the chromatic number of H (denoted $\chi(H)$) equals the number of vertices in the largest clique in H (denoted $\omega(H)$). Perfect graphs are important for their applications and because certain decision problems that are NP-complete in general have polynomial-time algorithms when the graphs under consideration are perfect (Berge and Chvátal, 1984; Brandstädt, Le, and Spinrad, 1999; Golumbic, 1980; Grötschel, Lovász, and Schrijver, 1981). Lovász (1972) proved the following major result originally conjectured by Berge.

Theorem 2.22. *(The Weak Perfect Graph Theorem). A graph is perfect if and only if its complement is perfect.*

As this book goes to press, Chudnovsky, Robertson, Seymour and Thomas have announced their proof of the following longstanding (50 year) conjecture of Berge (see Mackenzie (2002)).

Theorem 2.23. *(The Strong Perfect Graph Conjecture/Theorem). A graph is perfect if and only if it contains no induced chordless odd cycle C_{2k+1} or its complement $\overline{C_{2k+1}}$, for $k \geq 2$.*

Cocomparability graphs are perfect and thus Theorem 2.8 implies that bounded tolerance graphs are perfect. In Golumbic, Monma, and Trotter (1984), the authors prove the stronger result that all tolerance graphs are perfect (our Theorem 2.28).

There are several ways to prove this theorem, as discussed in Brandstädt, Le, and Spinrad (1999). Golumbic, Monma, and Trotter (1984) rely on a result from Berge and Chvátal (1984) that a class of graphs called perfectly orderable graphs are perfect. They show that the complement of a tolerance graph is perfectly orderable, and appeal to Lovász's Weak Perfect Graph Theorem to conclude that tolerance graphs are perfect. This is the route we will follow since it is most related to tolerance graphs.

First we outline four alternatives. The first three involve showing that weakly chordal graphs are perfect and then applying Theorem 2.17. In the fourth alternative, one proves that alternately orientable graphs are perfect and then applies Theorem 2.18. While these methods are somewhat longer, they are interesting in their use of properties of weakly chordal graphs.

The first uses the characterization of weakly chordal graphs by two-pairs. Given a graph G, two vertices are called a *two-pair* if every chordless path between them contains exactly two edges. More generally, two vertices are called an *even pair* in G if every chordless path between them contains an even number of edges. Hayward, Hoàng, and Maffray (1990) prove that every weakly chordal graph is either a clique or contains a two-pair. Meyniel (1987) proves that minimally imperfect graphs do not contain any even pairs. Combining these results with the fact that the property of being weakly chordal is hereditary yields the desired result that weakly chordal graphs are perfect.

The second approach uses star-cutsets. A *cutset* of a graph $G = (V, E)$ is a set $N \subset V$ so that $G - N$ has more components than G. A *star-cutset* is a cutset N in which there is a vertex $v \in N$ that is adjacent to every other vertex in N. The Star-Cutset Lemma (Chvátal, 1985) says that a minimally imperfect graph G can not have a star-cutset and neither can its complement $\overline{G}$. However, if G is weakly chordal with at least three vertices, then either G or $\overline{G}$ will have a star-cutset (Hayward, 1985). Combining these results, we again conclude that weakly chordal graphs are perfect.

The third alternative follows from the observation that the result *weakly chordal graphs are perfect* reduces to a corollary of the Strong Perfect Graph Conjecture/Theorem as soon as the proof of that Theorem is confirmed and published. Finally, the fourth alternative again relies on Chvátal's Star-Cutset Lemma and a theorem of Hoàng (1987) which shows that alternately orientable graphs are perfect.

Next we present our proof that tolerance graphs are perfect. We begin with the background needed to show perfectly orderable graphs are perfect. Following Chvátal (1984), we define an *ordered graph* to be a graph $G = (V, E)$ together with a linear ordering $\prec$ on V. Label the vertex set $V = \{v_1, v_2, \dots, v_n\}$ so that $v_i \prec v_j \iff i < j$. Consider the vertices in the order $v_1, v_2, \dots, v_n$ and assign a value $f(v_j)$ to v_j as follows: $f(v_j)$ is the smallest positive integer not already assigned to any of v_j's lower indexed neighbors. The *Grundy number* of an ordered graph is the maximum integer in the set $\{f(v_1), f(v_2), \dots, f(v_n)\}$. We will denote the Grundy number of an ordered graph $(G, \prec)$ by $gr(G, \prec)$.

Any linear ordering $\prec$ of $V(G)$ produces a function f which is a proper coloring of $V(G)$ using $gr(G, \prec)$ colors. Algorithmically, the Grundy number is the number of colors used by a greedy, first fit algorithm according to the given ordering $\prec$. Thus $gr(G, \prec) \geq \chi(G)$. The inequality may be strict, although there will always be some ordering which gives equality. For example, when the path $P_4 = (V, E)$ with $V = \{a, b, c, d\}$ and $E = \{ab, bc, cd\}$ is given an ordering $\prec$ in which $a \prec b$ and $d \prec c$, then $gr(P_4, \prec) = 3$ while $\chi(P_4) = 2$. The six resulting orderings of P_4 are called *obstructions* in Chvátal (1984). Furthermore, Chvátal defines a linear order $\prec$ on a graph to be *admissible* if it has no induced obstructions and *perfect* if for every induced subgraph H we have $gr(H, \prec) = \chi(H)$ under the same ordering $\prec$. If a graph has a perfect order it is called *perfectly orderable*. The class of *co-perfectly orderable* graphs consists of graphs that are the complements of perfectly orderable graphs.

Clearly an ordering $\prec$ which is perfect is also admissible. The converse is also true, as we shall see in Theorem 2.26, but first we provide an example.

Example 2.24. The graph $\overline{C_6}$, shown in Figure 2.5, has no admissible ordering.

Proof. Suppose $\prec$ were an admissible ordering of $V(\overline{C_6})$, that is, it has no induced obstruction. Without loss of generality, we may assume $a \prec b$, since the graph is symmetric. This, together with the induced P_4 *abce* forces $c \prec e$, which in turn forces $f \prec a$ (using *cefa*), and then $d \prec c$ (using *fadc*) and

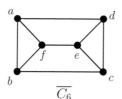
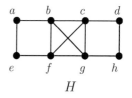

Figure 2.5. The graphs $\overline{C_6}$ and H, neither of which has an admissible ordering.

finally, $b \prec f$ (using $dcbf$). But now the three comparabilities $a \prec b$, $b \prec f$ and $f \prec a$ contradict the assumption that $\prec$ is a linear ordering. □

Similarly, one can show that the graph H shown in Figure 2.5 has no admissible ordering (Exercise 2.4). The graph H can be found in Hougardy (1998) as part of an extensive chart of 96 classes of perfect graphs given with inclusions and separating examples.

Lemma 2.25. *(Chvátal, 1984) Let $G = (V, E)$ be a graph and C a clique in G so that each $w \in C$ has a neighbor $p(w) \notin C$. Furthermore, suppose this set of neighbors $\{p(w) \mid w \in C\}$ is an independent set in G. If there is an admissible order $\prec$ so that $p(w) \prec w$ for all $w \in C$, then some $p(w)$ is adjacent to every vertex in C.*

Proof. We use induction on $|C|$. For each $w \in C$ we can apply the induction hypothesis to $G - w$ to conclude that there exists a $w^* \in C - w$ so that $p(w^*)$ is adjacent to each vertex in $C - w$. If $w\, p(w^*) \in E$ for some $w \in C$, we are done, so for a contradiction, assume $w\, p(w^*) \notin E$ for each $w \in C$. If $w^* = u^*$ for distinct vertices $u, w \in C$, then $w\, p(w^*) = w\, p(u^*) \in E$, a contradiction. Thus the function defined by $w \longmapsto w^*$ is a one-to-one function mapping C to itself, and hence is a bijection.

Let v be the vertex of C that appears first in the admissible order $\prec$. Since the function mapping w to w^* is a bijection, there exist $b, c \in C$ with $b^* = v$ and $c^* = b$. Note that $b \neq c$ since $b = c^* \in C - c$ by definition of c^*. We next show that the vertices $a = p(b), b, c$, and $d = p(v)$ induce a P_4 in G with edge set $\{ab, bc, cd\}$.

First we focus on the edges. By hypothesis, $p(b)b \in E$, so $ab = p(b)b \in E$. There is an edge between b and c since they are vertices of the clique C in G. Finally, $cd \in E$ because $d = p(v) = p(b^*)$ and, as noted before, $p(b^*)$ is adjacent to all elements in $C - b$, which includes c. Thus $\{ab, bc, cd\} \subseteq E$ and it remains to show that ac, bd, ad are not edges of G. We know $ad \notin E$ since $ad = p(b)p(v)$ and by hypothesis the set $\{p(w) \mid w \in C\}$ is an independent set. We have assumed $w\, p(w^*) \notin E$ for all $w \in C$, thus $bd = bp(v) = bp(b^*) \notin E$ and $ac = p(b)c = p(c^*)c \notin E$ as desired.

The P_4 induced by the vertices in $\{a, b, c, d\}$ is oriented so that $a = p(b) \prec b$ and $d = p(v) \prec v \prec c$ by our assumption about the ordering $\prec$ and our choice of v. This produces an obstruction in G, a contradiction. □

Theorem 2.26. *(Chvátal, 1984) Let $G = (V, E)$ be a graph. A linear ordering $\prec$ on V is perfect if and only if it is admissible. Moreover, if G is perfectly orderable, then G is a perfect graph.*

Proof. We have already noted that perfect orderings are admissible. For the converse, we proceed by induction on $|V(G)|$. Assume the result is true for graphs with fewer than n vertices and let $G = (V, E)$ be a graph with $|V| = n$ and let $\prec$ be an admissible ordering of V. By induction, we know that $gr(H, \prec) = \chi(H)$ for each proper induced subgraph H of G. It remains to show $gr(G, \prec) = \chi(G)$. Since $gr(G, \prec) \geq \chi(G) \geq \omega(G)$, it suffices to show $gr(G, \prec) \leq \omega(G)$, that is, we show that there exists a clique with $gr(G, \prec)$ vertices in G.

Let f be the function defined earlier and, for ease of notation, let $k = gr(G, \prec)$. This means there will be a vertex assigned to the integer k by f, that is, there exists a $w_k \in V$ with $f(w_k) = k$. Consider the smallest i so that there is a clique C with $V(C) = \{w_{i+1}, w_{i+2}, \ldots, w_k\}$ and $f(w_j) = j$ for each j. If $i = 0$, we are done, so assume $i \geq 1$. By the definition of f, each $w_j \in C$ has a neighbor $p(w_j) \notin C$ with $p(w_j) \prec w_j$ and $f(p(w_j)) = i$. The set $\{p(w_j) \mid w_j \in C\}$ is an independent set since each element has the same f value. Thus the conditions of Lemma 2.25 are satisfied and there exists a vertex $p(w) \notin C$ with $f(p(w)) = i$. This vertex could be added to C, contradicting the minimality of i.

Now to prove the last sentence of the theorem, let G be perfectly orderable and H an induced subgraph of G. We have just shown that for any perfect order $\prec$ on $V(G)$ we can apply the argument in the preceding paragraph to H to obtain $gr(H, \prec) \leq \omega(H)$. Thus $\chi(H) \leq gr(H, \prec) \leq \omega(H) \leq \chi(H)$ so equality holds throughout and $\chi(H) = \omega(H)$. This proves that G is perfect. □

In the proofs of Lemmas 2.15 and 2.16 we oriented the edges of a tolerance graph according to the tolerances in a fixed representation. In the proof of Theorem 2.27 we again orient edges according to a feature of a fixed tolerance representation. This time we orient edges of the complement of a tolerance graph and we orient according to right endpoints of intervals in the representation.

Theorem 2.27. (*Golumbic, Monma, and Trotter, 1984*) *If G is a tolerance graph, then $\overline{G}$ is perfectly orderable.*

Proof. Fix a regular representation $\langle \mathcal{I}, t \rangle$ of tolerance graph $G = (V, E)$ and let $\prec$ be the linear order of V determined by right endpoints: $x \prec y \iff R(x) < R(y)$. This gives an orientation F to the edges of $\overline{G}$ as follows:

$$xy \in F \iff xy \notin E \quad \text{and} \quad x \prec y.$$

The oriented edge $xy \in F$ is called Type 1 if $I_x \nsubseteq I_y$ and Type 2 if $I_x \subseteq I_y$. We show that $(\overline{G}, \prec)$ is perfectly ordered.

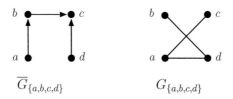

Figure 2.6. The oriented graph $\overline{G}_{a,b,c,d}$ and the graph $G_{a,b,c,d}$.

Suppose for a contradiction that the vertices a, b, c, d induce an obstruction in $(\overline{G}, \prec)$, with $a \prec b$, $d \prec c$ and without loss of generality, $b \prec c$ (see Figure 2.6). If ab is of Type 2 then $t_a = \infty$ because $ab \notin E(G)$. Lemma 2.12 implies that $N(a) \subseteq N(b)$ in G, a contradiction. Thus ab must be of Type 1.

If bc is also of Type 1, then right and left endpoints are ordered $R(a) < R(b) < R(c)$ and $L(a) < L(b) < L(c)$, so

$$|I_a \cap I_c| \leq \min\{|I_a \cap I_b|, |I_b \cap I_c|\} < \min\{t_a, t_c\}$$

a contradiction since $ac \in E(G)$. Thus bc must be Type 2. Now Lemma 2.12 implies $N(b) \subseteq N(c)$ in G, a contradiction.   □

Theorems 2.26 and 2.27 together imply that the complements of tolerance graphs are perfect. Now applying the Weak Perfect Graph Theorem (Theorem 2.22) yields our main result of the section.

Theorem 2.28. *(Golumbic, Monma, and Trotter, 1984) Tolerance graphs are perfect.*

2.6 A first look at unit vs. proper

One of the questions posed in Golumbic, Monma, and Trotter (1984) concerns an analog of unit and proper interval graphs. If an interval graph G has a representation in which no interval is properly contained in another, it is called a *proper interval graph*. Moreover, if G has a representation in which each interval has the same length, then G is called a *unit interval graph*.

The terms "unit" and "proper" also apply to tolerance representations in an analogous manner. A *unit tolerance graph* is one that has a tolerance representation in which all intervals have the same length and a *proper tolerance graph* is one that has a tolerance representation in which no interval is properly contained in another. Clearly the class of unit tolerance graphs is a subset of the class of proper tolerance graphs. Figure 2.7 shows the graph M_2 (which will be discussed further in Chapter 4) and a unit tolerance representation for it.

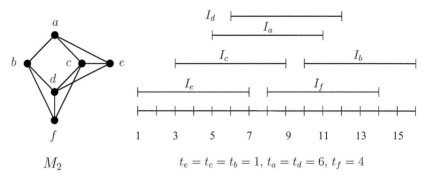

Figure 2.7. A graph and a unit tolerance representation for it.

Proposition 2.29. *Without loss of generality, any unit or proper tolerance representation may be assumed to have bounded tolerances.*

Proof. Fix a unit or proper tolerance representation $\langle \mathcal{I}, t \rangle$. By Remark 2.4, we may assume all endpoints in this representation are distinct. Replace t_x by $|I_x|$ for each $x \in V$ with $t_x > |I_x|$. Since there are no containments of intervals, this will not change any adjacencies. $\square$

Proposition 2.29 shows that the class of proper tolerance graphs is contained in the class of bounded tolerance graphs. The following is a separating example.

Example 2.30. The graph $K_{3,3}$ is a bounded tolerance graph, but not a proper tolerance graph.

Proof. Figure 2.1 gives a bounded tolerance representation of $K_{3,3}$, thus we need only show that $K_{3,3}$ is not a proper tolerance graph. Suppose there were a proper tolerance representation for $K_{3,3}$. We may assume this representation is regular, for if not, the arguments used in the proof of Lemma 2.3 can be applied to construct a regular representation $\langle \mathcal{I}, t \rangle$ of $K_{3,3}$ which is still proper.

By symmetry in $K_{3,3}$ we may assume the left endpoints of intervals in this representation satisfy $L(a) < L(b) < L(c)$, $L(d) < L(e) < L(f)$, and $L(a) < L(f)$. We may also assume $t_a < t_f$, for if not, we could reflect the entire representation about the y-axis and interchange the roles of a and f, of b and e and of c and d to achieve this inequality.

Since $af \in E(K_{3,3})$ and $ab \notin E(K_{3,3})$, we must have $L(f) < L(b)$ since otherwise we would have $|I_a \cap I_b| > |I_a \cap I_f| \geq \min\{t_a, t_f\} = t_a$, a contradiction. Thus the left endpoints satisfy $L(d) < L(e) < L(f) < L(b) < L(c)$. Now

$|I_d \cap I_c| < |I_b \cap I_c| < t_c$ and $|I_c \cap I_d| < |I_e \cap I_d| < t_d$, which is contradiction since $cd \in E(K_{3,3})$. $\square$

In Theorem 1.4 we saw that the classes of unit and proper interval graphs are equal and characterized as those interval graphs which are $K_{1,3}$-free. In Golumbic, Monma, and Trotter (1984), the authors pose the question of whether the class of unit tolerance graphs is equal to the class of proper tolerance graphs. The question is answered in Bogart, Fishburn, Isaak, and Langley (1995) where it is shown that the class of unit tolerance graphs is strictly contained in the class of proper tolerance graphs. The authors construct an infinite family of separating examples. The difficult part of this proof (which we omit) is showing that these separating examples are not unit tolerance graphs. In doing so, the authors introduce another interesting class of tolerance graphs.

A graph G is a *50% tolerance graph* if it has a tolerance representation $\langle \mathcal{I}, t \rangle$ so that $t_v = \frac{1}{2}|I_v|$ for all $v \in V(G)$. More generally, if G has a tolerance representation for which there is a constant c with $|I_v| - 2t_v = c$ for all $v \in V(G)$, then we say the tolerance representation has *constant cores*. Indeed these classes are equivalent.

Theorem 2.31. *The following are equivalent statements about a graph G.*

(i) *G is a unit tolerance graph.*
(ii) *G is a 50% tolerance graph.*
(iii) *G has a bounded tolerance representation with constant cores.*

The proof of (i) $\Longleftrightarrow$ (ii) appears in Langley (1993) and Bogart, Fishburn, Isaak, and Langley (1995). We include (iii) to maintain a parallel with Theorem 5.26. We defer giving the proof of Theorem 2.31 until Section 5.3 where it is a special case of the proof of Theorem 5.26.

As we discuss other types of tolerance graphs, we will revisit the question of whether the unit and proper classes are equal or not.

We conclude this section with the result due to Bogart, Jacobson, Langley, and McMorris (2001) showing that interval graphs are not just tolerance graphs, but *unit* tolerance graphs.

Theorem 2.32. *Interval graphs are unit tolerance graphs.*

The proof that interval graphs are 50% tolerance graphs is given in Corollary 6.3 as a modification of the proof of Theorem 6.2. Combining this with Theorem 2.31 gives the desired result.

Table 2.1. *Hierarchical relationships between classes of perfect graphs.*

Relationship between classes ...	... proved in
proper interval = unit interval	Roberts (1969)
unit tolerance = 50% tolerance	Theorem 2.31
bounded tolerance = parallelogram	Theorem 2.9
unit interval ⊆ unit tolerance	trivial
permutation ⊆ bounded tolerance	Theorem 2.6
permutation ⊆ trapezoid	trivial
trapezoid ⊆ cocomparability	Remark 1.10 and Theorem 1.11
trapezoid ⊆ weakly chordal	Theorem 2.21
cocomparability ⊆ co-perfectly orderable	Chvátal (1984)
tolerance ⊆ co-perfectly orderable	Theorem 2.27
tolerance ⊆ weakly chordal	Theorem 2.17
weakly chordal ⊆ perfect	Hayward (1985)
alternately orientable ⊆ perfect	Hoàng (1987)
co-perfectly orderable ⊆ perfect	Theorems 2.26 and 2.22
threshold ⊆ interval, permutation	Theorem 1.18
interval ⊆ unit tolerance	Theorem 2.32

2.7 Classes of perfect graphs

Figure 2.8 shows classes of graphs discussed in this chapter, ordered by inclusion. We include several more hierarchies of this type throughout the book because we feel they convey a great deal of information in a compact format. All of our hierarchies share three features: (1) a downward edge from class A to class B indicates that class A contains class B, (2) classes that appear in the same box are equivalent, and (3) an example appearing along an edge between two classes is a separating example for those classes.

In general, we do *not* claim that all containment relationships are necessarily given in the hierarchy, but we have tried to include as many as possible. When *all* containment relations are given, we call the hierarchy *complete*. In this case we also provide separating examples between each pair of incomparable classes.

Theorem 2.33. *The class hierarchy and separating examples illustrated in Figure 2.8 are correct. Moreover, the hierarchy is complete.*

Proof. Table 2.1 shows where the proofs of the inclusions and equivalences between classes can be found. Table 2.2 gives separating examples between incomparable classes. Finally, we prove that the examples shown along edges are separating examples.

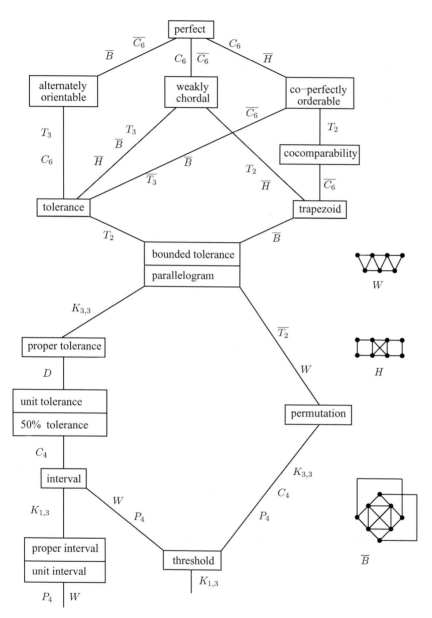

Figure 2.8. A complete hierarchy of classes of perfect graphs ordered by inclusion together with separating examples.

Table 2.2. *Separating examples between incomparable classes.*

A	B	$G_1 \in A{-}B$	$G_2 \in B{-}A$
alternately orientable	weakly chordal	C_6	$\overline{B}$
alternately orientable	co-perfectly orderable	C_6	$\overline{C_6}$
weakly chordal	co-perfectly orderable	$\overline{H}$	$\overline{C_6}$
alternately orientable	cocomp./trapezoid	T_2	$\overline{B}$
weakly chordal/tolerance	cocomparability	T_2	$\overline{C_6}$
tolerance	trapezoid	T_2	$\overline{B}$
unit/proper tolerance	permutation	W	$K_{3,3}$
unit/proper/arb. interval	permutation	W	$K_{3,3}$
unit/proper interval	threshold	P_4	$K_{1,3}$

The graph $\overline{C_6}$: The 6-cycle C_6 is a comparability graph, hence $\overline{C_6}$ is a co-comparability graph and therefore is perfect. $\overline{C_6}$ is not weakly chordal (by definition) and hence is not a trapezoid graph (Theorem 2.21). Additionally, $\overline{C_6}$ is not alternately orientable by Lemma 2.10.

The graph C_6: It is easy to check that C_6 is alternately orientable, hence it is perfect. However, C_6 is not weakly chordal by definition. In Example 2.24, we show that $\overline{C_6}$ has no admissible ordering, hence by Theorem 2.26 the graph $\overline{C_6}$ has no perfect ordering and thus C_6 is not co-perfectly orderable.

The graph T_2: A tolerance representation of the graph T_2 is shown in Figure 1.4. Hence, T_2 is a tolerance graph and thus is also weakly chordal and co-perfectly orderable. However, T_2 is not a cocomparability graph by Example 2.7, and it is therefore neither a trapezoid graph nor a bounded tolerance graph.

The graph $\overline{T_2}$: In Theorem 3.7 we will see that $\overline{T_2}$ is a bounded tolerance graph. Again, by Example 2.7, $\overline{T_2}$ is not a comparability graph, so by Theorem 1.8, $\overline{T_2}$ is not a permutation graph.

The graph C_4: A representation of C_4 as a unit tolerance graph is obtained by restricting the representation in Figure 2.7 to the intervals for vertices a, b, c, and d. It is well-known that C_4 is not an interval graph.

The graph $K_{3,3}$: The complete bipartite graph $K_{3,3}$ is shown to be a bounded tolerance graph but not a proper tolerance graph in Example 2.30. Therefore, $K_{3,3}$ is not a member of the smaller class of threshold graphs. It is both a comparability graph and a cocomparability graph, hence it is a permutation graph.

The graph $K_{1,3}$: A representation of $K_{1,3}$ as a unit tolerance graph appears in Figure 1.5. It is well-known that $K_{1,3}$ separates the classes of interval graphs and unit interval graphs (Lemma 4.5) and it not a threshold graph (Theorem 1.17).

The graph P_4: A matching diagram which shows that the graph P_4 is a permutation graph is given in Figure 1.13. It is easy to give an interval representation of P_4. However, P_4 is not a threshold graph by Theorem 1.17.

The graph W: Exercise 2.5 shows that the graph W in Figure 2.8 separates the classes indicated in Figure 2.8.

The graph D: Bogart, Fishburn, Isaak, and Langley (1995) show that the Dartmouth graph D shown in Figure 2.9 separates the classes of unit and proper tolerance graphs. Our vertex labels in Figure 2.9 match those of Bogart, Fishburn, Isaak, and Langley (1995).

The graph $\overline{B}$: The Berlin graph $\overline{B}$ (Figure 2.4) will be shown to be a trapezoid graph in Example 5.14, hence it is weakly chordal, co-perfectly orderable and perfect. It is not alternately orientable (Example 2.19) and thus is neither a tolerance graph nor a parallelogram graph. It is the incomparability graph of the order B in Figure 10.2.

The graph T_3: There are no cycles in the graph T_3 shown in Figure 3.3, hence it is weakly chordal and alternately orientable. It is not a tolerance graph by Theorem 3.7.

The graph $\overline{T_3}$: The cotree $\overline{T_3}$ is co-perfectly orderable since its complement T_3 is perfectly orderable using the following admissible ordering:

$$a \prec b_1 \prec c_1 \prec d_1 \prec b_2 \prec c_2 \prec d_2 \prec b_3 \prec c_3 \prec d_3.$$

Theorem 3.7 also shows that $\overline{T_3}$ is not a tolerance graph.

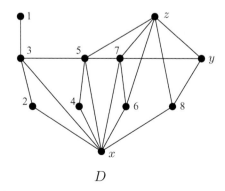

D

Figure 2.9. A proper tolerance graph that is not a unit tolerance graph.

The graph $\overline{H}$: The graph H is not perfectly orderable (Exercise 2.4). Therefore, its complement $\overline{H}$ is not co-perfectly orderable, and hence is neither a tolerance graph nor a trapezoid graph. By definition, the graph H is weakly chordal, and since the class of weakly chordal graphs is closed under taking complements, $\overline{H}$ is also weakly chordal and hence perfect. $\square$

2.8 Exercises

Exercise 2.1. Use the construction in the proof of Lemma 2.3 to convert the tolerance representation in Figure 1.4 to a regular representation of the graph T_2.

Exercise 2.2. Use the tolerance representation of $K_{3,3}$ given in Figure 2.1 and the proof of (ii) $\Longrightarrow$ (iii) in Theorem 2.6 to find linear orders L_1 and L_2 so that $K_{3,3}$ is the comparability graph of $L_1 \cap L_2$.

Exercise 2.3. Prove Theorem 2.8 algebraically, as discussed at the end of Section 2.3: (a) using right endpoints, (b) using midpoints.

Exercise 2.4. Show that the graph H shown in Figure 2.5 has no admissible ordering, and therefore is not perfectly orderable.

Exercise 2.5. Show that the graph W shown in Figure 2.8 is a unit interval graph, hence an interval graph, but is not a comparability graph, and hence not a permutation graph.

Exercise 2.6. Give a unit tolerance representation of C_4 and the associated parallelogram representation of C_4 (similar to Figure 2.3).

Exercise 2.7. "Challenge Exercise" Give a direct proof of Theorem 2.32 that every interval graph is a unit tolerance graph.

Exercise 2.8. Give a proper tolerance representation for the Dartmouth graph D in Figure 2.9.

Chapter 3
Trees, cotrees and bipartite graphs

The early papers, Golumbic and Monma (1982) and Golumbic, Monma, and Trotter (1984), leave several open questions which have shaped research in the field. The questions of characterizing the classes of tolerance graphs and bounded tolerance graphs remain open and no efficient recognition algorithms are yet known.

Theorem 2.8 gives a way to find graphs that are *not* bounded tolerance graphs, specifically by choosing graphs that are not cocomparability graphs. For example, the graph G shown in Figure 3.1 is not a bounded tolerance graph because its complement (also shown in Figure 3.1) is not transitively orientable (Exercise 1.6(c)). Yet, G is a tolerance graph as seen by the following representation: $I_a = [12, 15]$, $I_b = [3, 7]$, $I_c = [23, 27]$, $I_d = [1, 21]$, $I_e = [5, 25]$, $I_f = [10, 30]$, $t_a = \infty$, $t_b = t_c = t_d = t_f = 1$, $t_e = 17$.

In Golumbic, Monma, and Trotter (1984), the authors ask if there are other types of separating examples for the classes of tolerance graphs and bounded tolerance graphs besides graphs that are not cocomparability graphs, that is,

Question 3.1. Is there a cocomparability graph that is a tolerance graph but not a bounded tolerance graph?

This question is still open. By Theorem 2.20 and the containments in Figure 2.8, if such a graph exists, it must be a trapezoid graph.

3.1 Trees and cotrees

Independently, Felsner (1998) and Andreae, Hennig, and Parra (1993) took a first step in investigating Question 3.1, by examining graphs whose complements are trees (called *cotrees*). Clearly, cotrees are cocomparability graphs. They found that all cotrees that are tolerance graphs are also bounded tolerance

53

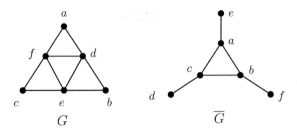

Figure 3.1. A graph G and its complement, neither of which is transitively orientable.

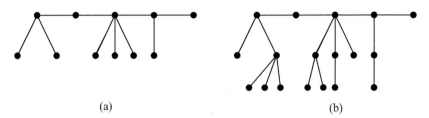

Figure 3.2. A caterpillar (a) and a caterpillar with toes (b).

graphs. Furthermore, they provide a forbidden subgraph characterization of the class of cotrees that are (bounded) tolerance graphs (our Theorem 3.7). Trees which are tolerance graphs were already characterized in Golumbic, Monma, and Trotter (1984), and, surprisingly, these two characterizations are the same. We now present these results on trees and cotrees.

We begin with two definitions. A *caterpillar* is a graph consisting of a chordless path (the spine) with any number of leaves (feet) adjacent to each vertex on the spine. A *caterpillar with toes* is a graph G' that can be obtained from a caterpillar G by attaching any number of leaves (toes) to the leaves (feet) of G. For examples of these graphs, see Figure 3.2.

Theorem 3.2. *If T is a tree, the following are equivalent.*

  (i) *T is a bounded tolerance graph.*
  (ii) *T has no subtree isomorphic to the graph T_2 in Figure 3.3.*
 (iii) *T is a caterpillar.*
 (iv) *T is an interval graph.*
  (v) *T is a permutation graph.*
 (vi) *T is a cocomparability graph.*
(vii) *T has no asteroidal triple.*

The equivalence of (ii) and (iv) appears in Lekkerkerker and Boland (1962) and the equivalence of (i) and (ii) is due to Golumbic, Monma, and Trotter

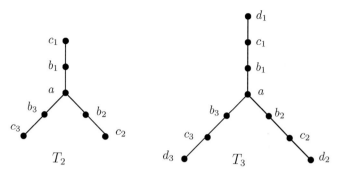

Figure 3.3. The trees T_2 and T_3.

(1984). The equivalence of caterpillars and T_2-free trees is a well-known exercise.

Proof. (i) $\Longrightarrow$ (ii): In Example 2.7 we showed that T_2 is not a cocomparability graph. Therefore, it is not a bounded tolerance graph (Theorem 2.8).

(ii) $\Longrightarrow$ (iii): Let T be a tree with no subtree isomorphic to T_2. Let P be a longest path in T. It is easy to show (by contradiction) that every other vertex of T must be a leaf adjacent to a vertex on P (Exercise 3.2).

(iii) $\Longrightarrow$ (iv): Suppose T is a caterpillar with spine vertices $\{v_1, v_2, \ldots, v_k\}$ so that $v_i v_{i+1} \in E(T)$ for $1 \leq i \leq k - 1$. Assign the interval $I_j = [j, j + 1]$ to vertex v_j and intervals $[j + \frac{1}{2}, j + \frac{1}{2}], [j + \frac{1}{3}, j + \frac{1}{3}], \ldots$ to the leaves adjacent to v_j. This gives a representation of T as an interval graph.

(iv) $\Longrightarrow$ (i): This implication follows from Theorem 2.5.

(i) $\Longrightarrow$ (vi): If T is a bounded tolerance graph, then T is a cocomparability graph by Theorem 2.8.

(vi) $\Longrightarrow$ (v): Since every tree T is a comparability graph, it follows from Theorem 1.8 that T is a permutation graph.

(v) $\Longrightarrow$ (i): This follows from Theorem 2.6.

(iv) $\Longleftrightarrow$ (vii): By Theorem 1.3, a graph is an interval graph if and only if it is chordal and has no asteroidal triple. Using this result, the forward direction follows immediately, and the converse follows since trees are chordal. $\qquad \square$

As Theorem 3.2 shows, all trees that are not bounded tolerance graphs contain the tree T_2 of Figure 3.3 as an induced subgraph. Since T_2 is not a cocomparability graph, the class of trees cannot provide an answer to Question 3.1.

The next proposition gives us a way to construct graphs that are not tolerance graphs from graphs that are not bounded tolerance graphs. Let $R_G = \{v \in V(G) \mid \mathcal{N}(v) \subseteq \mathcal{N}(y) \text{ for some } y \in V(G), y \neq v\}$. In the language of Golumbic, Monma, and Trotter (1984), vertices in the set R_G are *nonassertive* vertices.

Proposition 3.3. *Let G be a tolerance graph that is not a bounded tolerance graph. If G is induced in G' and $R_{G'} \cap V(G) = \emptyset$, then G' is not a tolerance graph.*

Proof. For a contradiction, suppose G' is a tolerance graph and fix a regular representation $\langle \mathcal{I}, t \rangle$ of it. By the hypothesis, for all $x \in V(G)$ there is no vertex $y \in V(G')$ distinct from x, with $\mathcal{N}(x) \subseteq \mathcal{N}(y)$. Thus, by Lemma 2.13, replacing t_x by $\min\{t_x, |I_x|\}$ for all $x \in V(G)$ will give a tolerance representation of G'. If we restrict this new representation to $V(G)$, we obtain a bounded representation of G, a contradiction. $\square$

Corollary 3.4. *Let G be a tolerance graph that is not a bounded tolerance graph. Then the graph G', obtained by attaching a leaf to each vertex of R_G, is not a tolerance graph.*

Proof. The set $R_{G'}$ consists of precisely those new leaves attached to the vertices of R_G. Thus, $R_{G'} \cap V(G) = \emptyset$ and the conclusion follows from Proposition 3.3. $\square$

Corollary 3.5. *The tree T_3 shown in Figure 3.3 is not a tolerance graph.*

Proof. Recall that T_2 is not a bounded tolerance graph (Theorem 3.2). We apply Corollary 3.4 with $G = T_2$. The set R_G consists of the three leaves of T_2, hence $G' = T_3$ is not a tolerance graph. $\square$

In a similar way, Proposition 3.3 can be used to show that the graphs G' in Figure 3.4 are not tolerance graphs. In each case the induced graph G is shown in bold and $R_{G'} = \emptyset$.

The following lemma allows us to build up a bounded tolerance representation of a graph one vertex at a time. When vertex a' is added in this construction, it must be adjacent to all of the existing vertices except for one, moreover, that

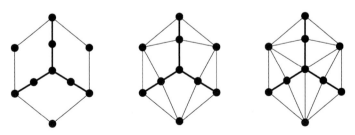

Figure 3.4. Graphs that are not tolerance graphs.

one non-neighbor must itself be adjacent to all vertices except for one. The construction is useful in building the complement $\overline{T}$ of a tree T where the vertex a' added to $\overline{T}$ corresponds to attaching a new leaf to an existing leaf of tree T. This lemma will be needed in proving (iii) $\Longrightarrow$ (v) of Theorem 3.7.

Following Andreae, Hennig, and Parra (1993), we define the *characteristic points* (or *c-points*) of a tolerance representation $\langle \mathcal{I}, t \rangle$ to be the points in the set $\{L(x), L(x) + t_x, R(x) - t_x, R(x)\}$. A *normal* representation of a tolerance graph is a regular representation in which the c-points of distinct vertices are distinct.

Lemma 3.6. *(Andreae, Hennig, and Parra, 1993) Let G be a bounded tolerance graph with normal representation* $\langle \mathcal{I}, t \rangle$. *Suppose that there exist distinct vertices* $a, b \in V(G)$ *so that* $t_a = |I_a|$ *and* $\mathcal{N}(a) = V(G) - \{a, b\}$. *Let G' be the graph that results by adding a new vertex* a' *to graph G with* $\mathcal{N}(a') = V(G) - \{a\}$. *Then G' is a bounded tolerance graph and there exists a normal representation of G' which, when restricted to G, is identical to* $\langle \mathcal{I}, t \rangle$.

Proof. Without loss of generality, we may assume $L(a) < L(b)$. If $I_b \subseteq I_a$ then $|I_a \cap I_b| = |I_b| \geq t_b$ and $ab \in E(G)$, a contradiction. Thus $R(a) < R(b)$ and, as a result, $R(a)$ is not the rightmost c-point. Let p be the leftmost c-point of $\langle \mathcal{I}, t \rangle$ which is to the right of $L(a)$, let q be the leftmost c-point of $\langle \mathcal{I}, t \rangle$ which is to the right of $R(a)$, and let s be a point to the right of all c-points of $\langle \mathcal{I}, t \rangle$. Choose new points p', q' so that $L(a) < p' < p$ and $R(a) < q' < q$. We construct an interval and tolerance for vertex a' as follows. Let $t_{a'} = q' - p'$ and let $I_{a'} = [p', s + t_{a'}]$. The c-points of a' are $p', q', s, s + t_{a'}$ which are distinct from the c-points of $\langle \mathcal{I}, t \rangle$.

It remains to show that $|I_a \cap I_{a'}| < \min\{t_a, t_{a'}\}$ and for all $x \in V(G) - \{a\}$, $|I_x \cap I_{a'}| \geq \min\{t_x, t_{a'}\}$. For the first of these, note that $I_a \cap I_{a'} = [p', R(a)]$, and we have $R(a) - p' < q' - p' = t_{a'}$ and $R(a) - p' < R(a) - L(a) = |I_a| = t_a$. To show the second, first consider $x = b$. In this case we get $|I_b \cap I_{a'}| = |I_b| \geq t_b \geq \min\{t_b, t_{a'}\}$, as desired. Now for $x \in V(G) - \{a, b\}$, we know $ax \in E$ so one of the following holds: (i) $|I_x \cap I_a| \geq t_a = |I_a|$, (ii) $(I_x \cap I_a) \supseteq [L(x), L(x) + t_x]$, or (iii) $(I_x \cap I_a) \supseteq [R(x) - t_x, R(x)]$.

In case (i) we have $L(x) < L(a) < L(a')$ and $R(x) \geq q > q' = L(a') + t_{a'}$, thus $[L(a'), L(a') + t_{a'}] \subseteq I_x \cap I_{a'}$ and $|I_x \cap I_{a'}| \geq t_{a'}$.

In case (ii), $L(a') < p \leq L(x)$ and $R(a') > s \geq L(x) + t_x$, thus $[L(x), L(x) + t_x] \subseteq (I_x \cap I_{a'})$ and $|I_x \cap I_{a'}| \geq t_x$. Finally, in case (iii), $R(x) - t_x \geq p > L(a')$ and $R(x) < s < R(a')$, so again $|I_x \cap I_{a'}| \geq t_x$. $\square$

Theorem 3.7. *If T is a tree and* $\overline{T}$ *is its complement, the following are equivalent.*

   (i) *T is a tolerance graph.*
  (ii) *T has no subtree isomorphic to the graph T_3 in Figure 3.3.*
 (iii) *T is a caterpillar with toes.*
 (iv) $\overline{T}$ *is a bounded tolerance graph.*
  (v) $\overline{T}$ *is a tolerance graph.*

    The equivalence of (i), (ii) and (iii) is due to Golumbic, Monma, and Trotter (1984) and their equivalence with (iv) and (v) is due independently to Andreae, Hennig, and Parra (1993) and Felsner (1998). The proof below of (iii) $\Longrightarrow$ (iv) and (v) $\Longrightarrow$ (ii) follows Andreae, Hennig, and Parra (1993).

Proof. (i) $\Longrightarrow$ (ii): It suffices to show that T_3 is not a tolerance graph, which we did in Corollary 3.5.

    (ii) $\Longrightarrow$ (iii): Let T be a tree with no induced graph isomorphic to T_3. Let P be a longest path in T with $V(P) = \{v_1, v_2, \ldots, v_n\}$. We consider P to be the spine of T and show that all other vertices are feet (distance 1 from P) or toes (distance 2 from P). For a contradiction, suppose there is a vertex $x \in V(T)$ whose distance to path P is 3 or more and let v_i be the unique vertex in $V(P)$ on the shortest path from x to P. If $i \in \{1, 2, 3, n - 2, n - 1, n\}$, then we get a path in T longer than P, a contradiction. Otherwise, we get an induced copy of T_3 in T.

    (iii) $\Longrightarrow$ (i): Let T be a caterpillar with toes whose set of spine vertices is $S = \{v_1, v_2, \ldots, v_m\}$ with $v_i v_{i+1} \in E(T)$ for $1 \le i \le m - 1$. Let $k = \max\{2, \deg(v_i)\}$ where the maximum is taken over all $v_i \in S$. Thus each spine vertex has at most $k - 1$ feet adjacent to it. For each $v_i \in S$, let $I_i = [2ik, (2i + 3)k]$ and $t_i = k$. Note that $|I_i \cap I_j| = k$ for $|i - j| = 1$ and $|I_i \cap I_j| = \emptyset$ for $|i - j| \ge 2$, so the intervals $\{I_i \mid 1 \le i \le m\}$ and the tolerances $t_i = k$ give a tolerance representation of the spine of T.

    The interval I_i contains the subinterval $[(2i + 1)k, (2i + 2)k]$ whose interior is disjoint from all I_j with $i \ne j$. We place the intervals for the feet attached to v_i in this subinterval, giving each foot a disjoint interval of length 1. This is possible since there are at most $k - 1$ feet attached to v_j. Each foot will have tolerance $\epsilon/2$, where ϵ is defined below.

    Let $n = \max\{\deg(w_i)\}$ where the maximum is taken over all feet w_i, and let $\epsilon = 1/n$. Finally, we assign intervals and tolerances to the toes attached to a foot vertex w_j. Since w_j has an assigned interval of length 1 and at most $n = 1/\epsilon$ toes, we can assign disjoint intervals of length ϵ to the toes adjacent to w_j and place these inside the interval for w_j. Assign infinite tolerance to each toe. It is easy to verify that this gives a tolerance representation of G.

    (iii) $\Longrightarrow$ (iv): Suppose T is a caterpillar with toes. We construct a bounded tolerance representation of $\overline{T}$. First note that if T has two toes x, y attached to foot vertex v, then $xy \in E(\overline{T})$ and $\mathcal{N}(x) = \mathcal{N}(y)$ in $\overline{T}$. Therefore, we may use $I_x = I_y$ and $t_x = t_y$ in a bounded tolerance representation of $\overline{T}$. So it suffices to consider the case in which each foot of T has at most one toe.

Let T' be the tree (caterpillar) resulting from removing the toes from T. By Theorem 3.2, T' is a permutation graph, hence $\overline{T'}$ is also a permutation graph, as we noted in Remark 1.7.

Using Theorem 2.6, fix a tolerance representation of $\overline{T'}$ in which $t_v = |I_v|$ for all $v \in V(\overline{T})$. Following the construction in the proof of (iv) $\Longrightarrow$ (ii) of Theorem 2.6, we may assume that this representation is normal.

It remains to add to this representation intervals and tolerances for the vertices that are the toes of T to arrive at a representation of $\overline{T}$. We add toes one at a time using Lemma 3.6. Let y_i be a toe of T which is attached to foot vertex x_i in T which is itself adjacent to spine vertex s in T. Apply Lemma 3.6 with $b = s, a = x_i$, and $a' = y_i$. Repeated application of Lemma 3.6 gives a bounded tolerance representation of $\overline{T}$.

(iv) $\Longrightarrow$ (v): Trivial.

(v) $\Longrightarrow$ (ii): Let $\overline{T}$ be a tolerance graph. Suppose (for a contradiction) that T_3 is an induced subgraph of T. Then $\overline{T_3}$ is an induced subgraph of $\overline{T}$, and hence a tolerance graph. Fix a regular representation $\langle \mathcal{I}, t \rangle$ of $\overline{T_3}$ and let F be the tolerance orientation associated with this representation (that is, $xy \in F \iff xy \in E(\overline{T_3})$ and $t_x < t_y$).

Figure 3.3 shows the graph T_3. We will refer to the vertices as they are labeled in this figure and remember that the edges of $\overline{T_3}$ are the ones *not* in T_3. Since F induces an acyclic orientation on the triangle induced by c_1, c_2 and c_3, we may assume, without loss of generality, that $c_1 c_2 \in F$, $c_2 c_3 \in F$, and $c_1 c_3 \in F$. By Lemma 2.15, F induces a transitive orientation on the 4-cycle induced by $\{c_2, b_2, c_3, b_3\}$, thus $c_2 b_3 \in F$. Repeating this argument, F induces a transitive orientation on the 4-cycle induced by $\{c_2, b_3, d_2, a\}$, thus $c_2 a \in F$, and F induces a transitive orientation on the 4-cycle induced by $\{c_2, b_1, d_2, a\}$, thus $c_2 b_1 \in F$. Finally, F induces a transitive orientation on the 4-cycle induced by $\{c_2, c_1, b_2, b_1\}$, thus $c_2 c_1 \in F$, a contradiction. $\square$

We saw in Theorem 3.7 that a tree is a tolerance graph precisely when its complement is a tolerance graph. However, the following example shows that it is not true in general that the class of tolerance graphs is closed under taking complements.

Example 3.8. The graph H in Figure 2.5 is a tolerance graph; a unit tolerance representation of it is given in Table 3.1. Its complement, $\overline{H}$ is not co-perfectly orderable (Exercise 2.4), hence is not a tolerance graph (Figure 2.8).

Table 3.1. *A unit tolerance representation*
of the graph H.

Vertex	Interval	Tolerance
a	[0,10]	1
b	[9,19]	4
c	[14,24]	4
d	[23,33]	1
e	[2,12]	4
f	[11,21]	1
g	[13,23]	1
h	[22,32]	4

3.2 Bipartite tolerance graphs – the bounded case

We do not know of any characterization, in general, of those graphs that are both bipartite and tolerance graphs. However, if we also require boundedness, then characterizations do exist. We will present these in this section.

A bipartite graph $G = (X, Y, E)$ is said to have the *SBS-indexing property* if there exists an ordering $\prec$ of the vertex sets X and Y, such that, for all $x_1 \prec x_2$ and $y_1 \prec y_2$, we have

$$\text{if } x_1 y_2 \in E \quad \text{and} \quad x_2 y_1 \in E,$$
$$\text{then } x_1 y_1 \in E \quad \text{and} \quad x_2 y_2 \in E,$$

that is, each pair of crossing edges must induce a $K_{2,2}$. (This property was called a strong ordering in Spinrad, Brandstädt, and Stewart (1987), but we prefer to reserve the term strong ordering for strongly chordal graphs.)

As we saw in Theorem 3.2, seven familiar classes of graphs became equivalent when restricted to trees. The analogous result for bipartite graphs is given in the next theorem.

Theorem 3.9. *Let $G = (X, Y, E)$ be a bipartite graph. The following conditions are equivalent.*

  (i) *G is a bounded tolerance graph.*
 (ii) *G is a trapezoid graph.*
(iii) *G is a cocomparability graph.*
 (iv) *G is AT-free.*
  (v) *G is a permutation graph.*

(vi) *G has the SBS-indexing property.*

(vii) $\exists \, \epsilon > 0 \text{ and } f : V(G) \to \mathbf{R} \text{ such that } xy \in E \Leftrightarrow |f(x) - f(y)| \le \epsilon \text{ for}$
$x \in X \text{ and } y \in Y.$

Proof. The implications (v) $\Rightarrow$ (i) $\Rightarrow$ (ii) $\Rightarrow$ (iii) $\Rightarrow$ (iv) hold for all graphs (Figure 2.8 and Theorem 1.13), and the implication (iv) $\Rightarrow$ (v) is mentioned on p. 93 of Brandstädt, Le, and Spinrad (1999) as following from Gallai's list of forbidden subgraphs of comparability graphs (Gallai, 1967). The equivalence (v) $\Leftrightarrow$ (vi) is due to Spinrad, Brandstädt, and Stewart (1987) and the equivalence (vi) $\Leftrightarrow$ (vii) is due to Brandstädt, Spinrad, and Stewart (1987).[1]   □

Remark 3.10. We note that interval graphs do not appear in this theorem. In fact, the bipartite interval graphs are precisely the class of disjoint caterpillars (Exercise 3.7). The cycle C_4 is a separating example between the classes of bipartite interval graphs and the equivalent classes of graphs in Theorem 3.9.

Question 3.11. The general question of characterizing bipartite tolerance graphs is still open.

3.3 Exercises

Exercise 3.1. Let T be a tree with n vertices. Give an algorithm having complexity $O(n)$ which checks whether T is a tolerance graph.

Exercise 3.2. Complete the proof of (ii) $\Longrightarrow$ (iii) in Theorem 3.2.

Exercise 3.3. Give a unit tolerance representation for the caterpillar in Figure 3.2(a). Give a constructive proof that every caterpillar is a unit tolerance graph.

Exercise 3.4. Use the proof of (iii) $\Longrightarrow$ (i) of Theorem 3.7 to give a tolerance representation of the graph shown in Figure 3.2(b).

Exercise 3.5. Use the proof of (iii) $\Longrightarrow$ (iv) of Theorem 3.7 to give a bounded tolerance representation of the *complement* of the graph shown in Figure 3.2(b).

Exercise 3.6. This exercise refers to the bipartite graph G shown in Figure 3.5.
   (a) Give an indexing of $X = \{a, b, c\}$ and $Y = \{d, e, f\}$ that satisfies the SBS-indexing property.

---

[1] Condition (vii) was at first used as a definition in Derigs, Goecke, and Schrader (1984) for what they called "bipartite tolerance graphs", however, as we see here, it is equivalent to the standard notion of bipartite bounded tolerance graphs.

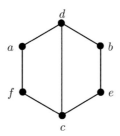

Figure 3.5. A bipartite graph satisfying the conditions of Theorem 3.9.

(b) Give a representation of G as a permutation graph.

(c) Specify an $\epsilon > 0$ and a function $f : V(G) \to \mathbf{R}$ that satisfies condition (vii) of Theorem 3.9.

Exercise 3.7. Show that the bipartite interval graphs are precisely the class of disjoint caterpillars.

Chapter 4
Interval probe graphs and sandwich problems

4.1 Physical mapping of DNA

The use of interval models in molecular biology dates back to the original studies on the linearity of genes by the well-known biologist Seymore Benzer. In Benzer (1959), interval graphs were defined in order to study overlap data on sub-elements inside the gene. The question at that time was whether the overlap data was consistent with the hypothesis that genes are linear structures with the sub-elements being intervals on that line. If perfect and complete data were to be available for of all pairs of these sub-elements, then the problem would be that of recognizing an interval graph. However, with only partial data, as is always the case in experimental genetics, the problems involve embedding the data in a larger consistent set (e.g., interval graph completion). This will generally require inferring additional intersections (e.g., selectively adding chords to a cycle), and using additional biological information or assumptions to test consistency and propose the possible linear orderings.

During the ensuing years, research in genetics has involved many combinatorial problems on intervals. One of these is DNA physical mapping, in which one wishes to find the linear order of segments (physically contiguous units) of the chromosome. It is an essential part of most sequencing, gene locating, and cloning projects. One of the main goals set for the Human Genome Project is to obtain a detailed mapping of all human chromosomes.

The elements which must be ordered are determined by experimental techniques, obtained by cutting the chromosome with various enzymes into relatively small fragments called *clones*. Different enzymes cut the chromosome in different ways, so within a large *library* of clones many clones will overlap. However, in the process, the order of the clones is lost. Physical mapping sets out to reconstruct their order, based on experimental information including clone–clone incidence.

Probes (or *markers*) are relatively short sequences of DNA (say 10–1000bp, base pairs, the "letters" of the DNA alphabet), which can be regarded as a distinguished subset of clones. In general, clones are much longer stretches of DNA and, depending on the technology of cutting the chromosome, can range in length from about 15Kbp long (*lambda* clones), to 40Kbp (*cosmid* clones), to 40Kbp–100Kbp (*P1* or *BAC* clones), to 2000Kbp–2000Kbp (yeast artificial chromosomes or *YAC*s). Probes which appear in a unique location along the DNA (e.g., a pattern which is sufficiently long that it is unlikely to occur elsewhere) are called *unique probes*, and they are the ones for which experiments are particularly designed.

One of the most common experiments is testing whether a unique probe is contained in each of the clones of the library, thus approximating a matrix of *probe–clone incidence*. If the data were to be perfect, then the problem of reconstructing the probe order could be solved using the PQ-tree algorithm in Booth and Lueker (1976) for testing the consecutive ones property on the matrix, as each clone defines a set of probes which must occur consecutively along the chromosome. In the presence of ambiguities, however, there may be some probe–clone pairs for which it is not known whether the probe is contained in the clone or not (the matrix entry is empty and may be zero or one). In this case, the problem of checking the consistency of the data and proposing a consistent solution would be to solve an *interval sandwich problem*, which we discuss in Section 4.7.

The probe–clone incidence data also provides partial information for a second matrix of *clone–clone incidence*, specifically, if two clones contain a common unique probe, then they must intersect. This is very similar to Benzer's original problem. The intersection data will certainly be incomplete, and may contain errors and ambiguity. There may be false positives, for example, where a probe may not really be unique, and false negatives, where two clones may intersect but there is no probe in the experiment which is contained in both of them. Additional intersection data can be provided by other biological techniques, but the data will still remain incomplete. Hence, we face another case of solving an interval sandwich problem.

Interval probe graphs, a generalization of interval graphs, were introduced by Zhang (1994) and used in Zhang, Schon, Fischer, Cayanis, Weiss, Kistler, and Bourne (1994) and Zhang, Ye, Liao, Russo, and Fischer (1999) to model certain problems in physical mapping of DNA when only partial data is available on the overlap of clones (i.e., the intervals). Specifically, the clones are distinguished by the experimental scientist as being either probes or non-probes, where no intersection information is known between pairs of non-probes, but complete intersection information is known between all probe/probe and all probe/non-probe pairs.

We study the interval probe graphs beginning in Section 4.2. We will prove shortly that the interval probe graphs are tolerance graphs, which is our reason for including them in this book. For further study in the area of computational biology, we refer the reader to the references in Golumbic, Kaplan, and Shamir (1994) and in Waterman (1995) and Pevzner (2000).

4.2 Interval probe graphs

An undirected graph $G = (V, E)$ is an *interval probe graph* (or, more simply, a *probe graph*) if the vertex set can be partitioned into two subsets, P (probes) and N (non-probes), where N is a stable set and there is a completion $E' \subseteq \{uv \mid u, v \in N, u \neq v\}$ such that $G' = (V, E \cup E')$ is an interval graph. Equivalently, G is an interval probe graph if we can assign an interval to each vertex such that two vertices are adjacent if and only if at least one of them is in P and their corresponding intervals intersect. We say that E' is an *interval completion* of G on N. We remark that it is possible that a graph can be a probe graph with respect to different partitions of probes and non-probes, giving very different interval completions. Clearly, *the probe graphs constitute a generalization of interval graphs* (where $N = \emptyset$). Moreover, the induced subgraph G_P must be an interval graph, although this is not sufficient to characterize probe graphs.

Example 4.1. A probe graph which is not an interval graph is shown in Figure 4.1 using the probe/non-probe partition $P = \{a, b, c, d\}$ and $N = \{x, y\}$, shown in solid and hollow circles, respectively. If the edge xy is added to the graph, we can construct an interval representation, also shown in Figure 4.1, where the intervals representing probe vertices are drawn with thick lines.

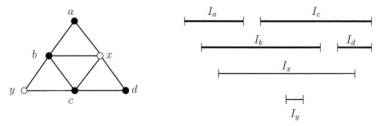

Figure 4.1. A probe graph which is not an interval graph and an interval representation if edge xy is added to the graph.

Remark 4.2. We emphasize that if there is an edge xy in a probe graph G, then $I_x \cap I_y \neq \emptyset$. But the converse holds only when at least one of x and y is a probe.

Example 4.1 also shows that probe graphs are not contained in the classes of comparability or cocomparability graphs, and they are therefore not bounded tolerance graphs. However, the following theorem is easily verified.

Theorem 4.3. *Every interval probe graph is a tolerance graph.*

Proof. Let $G = (V, E)$ be an interval probe graph, and G' an interval completion with respect to the probe/non-probe partition $V = P \cup N$. Using Lemma 1.5, fix an interval representation of G' in which all interval endpoints are distinct. Let ϵ be the smallest distance between endpoints in this representation. Assign tolerance $\epsilon > 0$ to every probe vertex and tolerance ∞ to every non-probe vertex. Since the intersection of two intervals has finite length, the non-probes form a stable set as required; and by the choice ϵ, whenever a probe interval intersects any other interval, it does so in at least length ϵ. Thus we have a tolerance representation of G.   □

The same proof shows the following.

Corollary 4.4. *Every unit probe graph is a unit tolerance graph.*

Since tolerance graphs are weakly chordal (Theorem 2.17), probe graphs are also weakly chordal, a fact that was proved independently in McMorris, Wang, and Zhang (1998). Probe graphs are not chordal, however, since the chordless cycle C_4 is easily seen to be a probe graph by assigning either pair of opposite vertices to be the non-probes. In fact, it is easy to see that the vertices of a chordless 4-cycle in an interval probe graph *must* alternate between probes and non-probes. We will use this observation repeatedly in this chapter, and return to the special role played by C_4 in Section 4.6.

4.3 The hierarchy of interval, probe, and tolerance graphs

We have seen, in the preceding section, the containment relationships between the families of graphs:

$$\text{interval} \quad \subset \quad \text{interval probe} \quad \subset \quad \text{tolerance} \qquad (4.1)$$

If we were to add to each of these classes the familiar restrictions of having either a unit or proper interval representation (as discussed in Section 2.6), then we would potentially have nine classes, taking all possible combinations $\{\text{unit}, \text{proper}, \text{arbitrary}\} \times \{\text{interval}, \text{probe}, \text{tolerance}\}$; and if we were to include bounded tolerance, that would give a tenth class. However, there are actually just eight different classes, as we will show.

This section presents the hierarchy of the eight distinct classes, together with threshold graphs and separating examples for them as illustrated in Figure 4.2. An earlier version of this section appears in Golumbic and Lipshteyn (2001).

The following classical result of Roberts (1969), which we also saw in Chapter 1, tells us that the classes of unit and proper interval graphs are equal.

Lemma 4.5. *The classes of unit interval graphs and proper interval graphs are equivalent, and furthermore, they are equivalent to the $K_{1,3}$-free interval graphs.*

Lipshteyn (2001) has shown a similar result that the classes of unit probe graphs and proper probe graphs are equal. We state this as a lemma.

Lemma 4.6. *The classes of unit probe graphs and proper probe graphs are equivalent.*

Proof. Every unit probe graph G has a vertex partition (P, N) into probes and non-probes and an edge completion E' consisting of some edges with both endpoints in N, such that the completed graph G' is a unit interval graph. Since G' is also a proper interval graph, G is a proper probe graph using the same partition and completion.

Conversely, let G be a proper probe graph with vertex partition (P, N) and edge completion E' between non-probes such that the completed graph G' is a proper interval graph. Using Lemma 4.5, G' is also a unit interval graph. Therefore, G is a unit probe graph using the same partition and edge completion. $\square$

The following new result relates threshold graphs to interval probe graphs.

Proposition 4.7. *Threshold graphs are unit probe graphs.*

Proof. Let G be a threshold graph with degree decomposition $V(G) = D_0 \cup D_1 \cup \cdots \cup D_m$, where for all distinct vertices $x \in D_i$ and $y \in D_j$, we have $xy \in E(G) \iff i + j > m$ (see Section 1.7.3). We define the interval $I_i = [L(i), L(i) + m]$ as follows. For $i = 0, 1, \ldots, \lfloor m/2 \rfloor$, define $L(i) = i$, and for $i = \lfloor m/2 \rfloor + 1, \ldots, m$, define $L(i) = 2m - i + \frac{1}{2}$, as illustrated for the case of $m = 7$ in Figure 4.3. Now take $|D_i|$ copies of interval I_i, designating those with $0 \le i \le \lfloor m/2 \rfloor$ to be non-probes and those with $\lfloor m/2 \rfloor + 1 \le i \le m$ to be probes. It is easy to verify that this is a unit probe representation of G. $\square$

We now present the complete hierarchy of these classes, as illustrated in Figure 4.2. Recall that we say a hierarchy is *complete* when *all* containment relationships are given. The examples appearing along edges not only provide

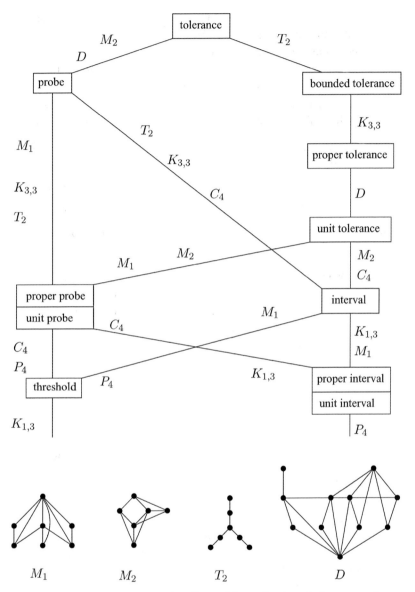

Figure 4.2. The complete hierarchy of interval, interval probe and tolerance graphs with separating examples.

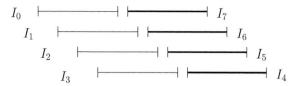

Figure 4.3. Unit intervals representing a threshold graph. Bold intervals are probes and light intervals are non-probes.

separating examples for pairs of classes where one is contained in another, but also for pairs of classes that are incomparable.

Theorem 4.8. *The class hierarchy and the separating examples illustrated in Figure 4.2 are correct. Moreover, the hierarchy is complete.*

Proof. Proposition 2.29 shows that every proper tolerance graph is a bounded tolerance graph, and Theorem 2.32 shows that every interval graph is a unit tolerance graph. Proposition 4.7 shows that threshold graphs are unit probe, and Theorem 1.18 shows that they are interval graphs. Corollary 4.4 shows unit probe graphs are unit tolerance graphs. The other edges of the hierarchy follow from the inclusions in equation (4.1) and the definitions of unit and proper representations. We now prove that the examples shown along edges of the diagram are the separating examples.

The graph $K_{1,3}$: By Lemma 4.5, $K_{1,3}$ is not a proper interval graph, however, it is an interval graph and a threshold graph. It is easy to see that $K_{1,3}$ is a proper probe graph (Exercise 4.2).

The graph $K_{3,3}$: In Example 2.30, we showed that $K_{3,3}$ is a bounded tolerance graph but not a proper tolerance graph. It also fails to be an interval graph since it contains chordless 4-cycles, but it is a probe graph, as follows: Make one side of the bipartition probes and the other side non-probes. An interval completion is obtained by adding the three edges connecting all the non-probes. It is not proper probe since it is not proper tolerance.

The graph C_4: The chordless 4-cycle C_4 is not an interval graph, and hence neither a unit interval graph nor a threshold graph. However, C_4 is a proper probe graph (Exercise 4.2) and hence also a member of the larger classes of unit tolerance graphs and probe graphs.

The graph P_4: By Theorem 1.17, the chordless 4-path P_4 is not a threshold graph. It is easy to show that P_4 is a unit interval graph, hence also interval and unit probe.

The graph M_1: Figure 4.4 shows the graph M_1 and an interval representation of it. Therefore, it is also a member of the larger classes of unit tolerance graphs and probe graphs. We will prove that M_1 is not a proper probe graph and

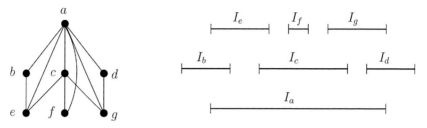

Figure 4.4. The graph M_1 and an interval graph representation of it.

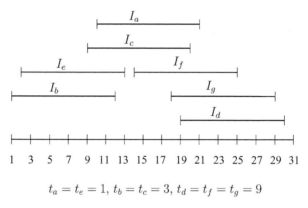

$$t_a = t_e = 1, \ t_b = t_c = 3, \ t_d = t_f = t_g = 9$$

Figure 4.5. A unit tolerance representation for the graph M_1.

therefore not a member of the smaller classes of threshold graphs and unit interval graphs. Suppose otherwise, and let G' be the completed proper interval graph with vertex partition (P, N) into probes and non-probes. Since G' is $K_{1,3}$-free, the completion must have cancelled the $K_{1,3}$ induced by $\{a, b, c, d\}$ and the $K_{1,3}$ induced by $\{c, e, f, g\}$. The only way to do this would require that at least two of $\{b, c, d\}$ are non-probes and at least two of $\{e, f, g\}$ are non-probes. But no independent set N of G can satisfy this requirement. Therefore, M_1 is not a proper probe graph.

The graph M_2: This graph is a unit tolerance graph as demonstrated by the representation in Figure 2.7, and therefore a tolerance graph. We show that M_2 is not a probe graph (and therefore not an interval graph) as follows. In order to obtain an interval completion of the graph M_2, one must add a chord to the chordless cycle $\{a, b, d, c\}$ (using vertex labels from Figure 2.7). If the added chord is bc, then a, d, e, f are probes. In this case, the chordless cycle $\{a, b, d, e\}$ has only one non-probe and has no interval completion. Similarly, if the added chord is ad, then b, c, e, f are probes. In this case, the chordless cycle $\{a, c, f, b\}$ cannot be completed. This proves that M_2 is not a probe graph.

Table 4.1. *Separating examples between incomparable classes.*

A	B	$G_1 \in A - B$	$G_2 \in B - A$
probe	unit/proper/bnd. tol.	T_2	M_2
unit/proper probe	interval	C_4	M_1
threshold	unit/proper interval	$K_{1,3}$	P_4

The graph D: Bogart *et al.* (1995) showed that this graph is a proper tolerance graph but not a unit tolerance graph. The graph D is not an interval probe graph by a simple argument using the observation that the vertices on all chordless 4-cycles would have to alternate between probes and non-probes (Exercise 4.1).

The graph T_2: Theorems 3.2 and 3.7 show that T_2 is a tolerance graph but is not a bounded tolerance graph and hence neither an interval graph nor a proper probe graph. T_2 is a probe graph, as follows. Make the three leaves and the central vertex non-probes, and make the remaining three vertices probes. An interval completion can be obtained by adding any (or all) of the three edges connecting the leaves to the central vertex.

Finally, we verify the incomparabilities between pairs of classes. We show $A \parallel B$ by exhibiting graphs $G_1 \in A - B$ and $G_2 \in B - A$ in Table 4.1. $\square$

4.4 The trees that are interval probe graphs

In this section, we present a characterization of the trees that are interval probe graphs, which is due to Li Sheng (1998,1999). We begin with a general lemma that shows the impossibility of an asteroidal triple of probe vertices.

Lemma 4.9. *Let G be an interval probe graph and let G' be an interval completion with respect to the probe partition $P \cup N$. If G has an asteroidal triple, then at least one of these three vertices must be a non-probe which has a new neighbor in G'.*

Proof. Suppose $\{c_1, c_2, c_3\}$ is an asteroidal triple of G. If the neighborhoods of all three of these vertices remained the same in G', then $\{c_1, c_2, c_3\}$ would still be an asteroidal triple in G', since all "neighborhood avoiding" paths in G would still be valid in G'. Therefore, at least one of the triple, say c_1, must have a new neighbor in G', and since only non-probes get new neighbors, we conclude that c_1 is a non-probe. $\square$

We now apply this result to the tree T_2 shown in Figure 3.3.

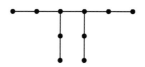

Figure 4.6. The tree Π_2.

Lemma 4.10. *The tree T_2 is an interval probe graph, and its central vertex must be a non-probe.*

Proof. We already saw, at the end of Section 4.3, that T_2 is an interval probe graph. Suppose that T_2 has an interval completion G' in which the central vertex a is a probe. (We refer to the vertices as labelled in Figure 3.3.) Using Lemma 4.9, since $\{c_1, c_2, c_3\}$ is an asteroidal triple of T_2, one of them, say c_1, is a non-probe and gets a new neighbor in G'. Now, since c_1 is a non-probe, vertex b_1 is a probe. If either $c_1 b_2 \in E(G')$ or $c_1 b_3 \in E(G')$ was a completed edge, then G' would have a chordless 4-cycle, which is not allowed in an interval completion. Therefore, either $c_1 c_2 \in E(G')$ or $c_1 c_3 \in E(G')$ must be a completed edge, but either of these would give a forbidden chordless 5-cycle in G'. Thus, c_1 has no new neighbor in G', a contradiction. $\square$

Corollary 4.11. *The tree Π_2 (in Figure 4.6) is not an interval probe graph.*

Proof. Let a_1 and a_2 denote the vertices of degree 3 in Π_2. Since they are adjacent, at least one of them would have to be a probe vertex in any interval completion. But both a_1 and a_2 are the central vertex of an induced T_2, so by Lemma 4.10, neither could be a probe. Therefore, Π_2 cannot be an interval probe graph. $\square$

We now give the characterization of trees which are interval probe graphs. Note that the graph T_3 is shown in Figure 3.3.

Theorem 4.12. *(Sheng, 1998, 1999) Let T be a tree. The following are equivalent:*

(i) *T is an interval probe graph,*
(ii) *T has no induced subgraph isomorphic to T_3 or Π_2.*

Proof. (i) $\implies$ (ii): If T is an interval probe graph, then it is also a tolerance graph (Theorem 4.3). Hence, by Theorem 3.7, T may not contain an induced T_3. Also, by Corollary 4.11, T has no induced Π_2.

(ii) $\implies$ (i): Suppose T has no induced T_3 or Π_2. We will show that T is interval probe. By Theorem 3.7, T is a caterpillar with toes. Let $[x_1, x_2, \ldots, x_k]$ be a longest path in T. If $k \leq 4$, then T contains no T_2, so T is an interval graph (our Theorem 3.2) and thus is an interval probe graph. So we may assume $k \geq 5$.

We define the *children* of x_i to be $C_i = \mathcal{N}(x_i) - \{x_{i-1}, x_{i+1}\}$ and the *grandchildren* of x_i to be $G_i = \bigcup \{\mathcal{N}(y) \mid y \in C_i\} - \{x_i\}$. We say that x_i is a *grandparent* if $G_i \neq \emptyset$.

Note that x_1 and x_k have no children and x_2 and x_{k-1} have no grandchildren since the path is longest possible. Therefore, since T is also Π_2-free, we can make the following claim:

Claim: *There are no consecutive grandparents* x_i, x_{i+1}.

We now define the partition into probes and non-probes. Let the non-probes consist of all grandparents and grandchildren, and let the probes consist of all children and all non-grandparents on the path. Formally,

$$P = \bigcup \{C_i\} \cup \{x_i \mid G_i = \emptyset\}$$

and

$$N = \bigcup \{G_i\} \cup \{x_i \mid G_i \neq \emptyset\}.$$

An interval completion is obtained by adding edges between each grandparent and all of its own grandchildren, that is,

$$E' = \{x_i z \mid G_i \neq \emptyset \quad \text{and} \quad z \in G_i\}$$

which can easily be verified (Exercise 4.3). $\qquad\qquad\qquad\qquad\square$

4.5 Partitioned interval probe graphs

The definition of probe graph does not specify a particular partition of the vertices in advance. However, in the biology applications, the partition into probes and non-probes is part of the input. Therefore, we may distinguish between the general case of probe graphs, where we must find both a partition and an interval completion for it, and the special case of *partitioned probe graphs*, where we are given a fixed partition and must only find a completion for it.

A polynomial time algorithm for the problem of recognizing partitioned probe graphs (i.e., with respect to a fixed partition) was first reported in Johnson and Spinrad (2001). Their method uses PQ-trees and constructs an interval probe model in $O(n^2)$ time. Another method, given in McConnell and Spinrad (2002), uses modular decomposition and has complexity $O(n + m\log n)$ for a graph with m edges. The minimum coloring, maximum clique and maximum stable set problems are all polynomial for partitioned probe graphs (see Exercises 4.5 and Chapter 9).

In contrast to this, however, the complexity of the general problem of recognizing probe graphs (when no partition is given) is an open problem. Similarly, the problem of recognizing tolerance graphs is open.

Interval probe graphs do not satisfy the properties of being chordal, cocomparability or having a consecutive ordering of maximal cliques, which characterize their interval completions. However, they do possess certain analogues of these properties. The next section presents one of these results, which is due to Zhang (1994) and is cited in McMorris, Wang, and Zhang (1998). [1]

4.6 The enhancement of a partitioned probe graph is chordal

Let $G = (V, E)$ be a probe graph whose vertex set is partitioned into a set P of probes and a stable set N of non-probes. The *enhanced graph* $G^* = (P \cup N, E^*)$ is the graph G together with all edges uv for which $u, v \in N$ are distinct non-probes and there exist distinct probes $x, y \in P$ for which $ux, vx, uy, vy \in E$ but $xy \notin E$. We call the edge uv an *enhanced edge*.

If G is an interval probe graph with respect to the partition $V = P \cup N$, then each enhanced edge must be in every interval completion, since no chordless 4-cycles can remain. The next lemma shows this more formally, specifically that the enhanced graph will also be an interval probe graph. Later, in Theorem 4.17, we will present a result of Zhang (1994) which shows that this enhanced graph is chordal and thus has desirable properties.

Lemma 4.13. *The graph $G = (P \cup N, E)$ is an interval probe graph with respect to this partition if and only if the enhanced graph $G^* = (P \cup N, E^*)$ has an interval completion on N.*

Proof. ($\Longleftarrow$): This direction follows immediately from the definition of a probe graph.

($\Longrightarrow$): Suppose $G = (V, E)$ is an interval probe graph where V is partitioned $V = P \cup N$ and N is the stable set on non-probes. Let $E' \subseteq \{uv \mid u, v \in N, u \neq v\}$ be any interval completion of G, that is, $G' = (V, E \cup E')$ is an interval graph. It suffices to show $E^* \subseteq E'$, for then G' will also be an interval

---

[1] Another of these analogous results, also due to Zhang (1994) and referred to in McMorris, Wang, and Zhang (1998), is a characterization of interval probe graphs in terms of having a consecutive ordering of quasi-maximal cliques.

completion of G^* on N. For a contradiction, suppose there exist $u, v \in N$ with $uv \in E^*$ and $uv \notin E \cup E'$. Since uv is an enhanced edge, there exist distinct probes, $x, y \in P$ so that $ux, uy, vx, vy \in E$ but $xy \notin E$. Note that $xy \notin E'$ since $x, y \in P$. Thus the four vertices x, y, u, v induce a chordless 4-cycle in G'. This is a contradiction since G' is an interval graph and interval graphs are chordal. □

Lemma 4.13 can be of use when we wish to determine whether a given graph $G = (V, E)$, whose vertex set is partitioned into a set P of probes and a stable set N of non-probes, is a probe graph with respect to this partition. It allows us to add edges to G in two stages and to consider G^* instead of G.[2] First we add the enhanced edges to arrive at the enhanced graph $G^* = (V, E^*)$. Then we add completion edges E^+ between other pairs of vertices in N so that the result $G' = (V, E^* \cup E^+)$ is an interval graph.

Fix an interval representation $\{I_v \mid v \in V\}$ of G'. Note that in both stages we have only added edges between pairs of non-probes, thus an edge in G' involving a probe vertex must be an original edge in G. We record this observation as a remark.

Remark 4.14. If xy is an edge of an interval completion G' of G (or equivalently, if $I_x \cap I_y \neq \emptyset$) and one of x, y is a probe vertex, then $xy \in E$.

Lemma 4.15. Suppose $I_x \subseteq I_y$ and $x \in N$. Then for $v \neq x, y$ we have $vx \in E^* \implies vy \in E^*$. That is, if non-probe x has its interval I_x contained in I_y, then any neighbor of x in G^* is also a neighbor of y in G^*.

Proof. If $vx \in E^*$ then $I_v \cap I_x \neq \emptyset$ (Remark 4.2) and thus $I_v \cap I_y \neq \emptyset$. If v or y is a probe then, by Remark 4.14, $vy \in E \subseteq E^*$. Otherwise, v and y (together with x) are non-probes. The edge $vx \in E^*$ must be an enhanced edge since N is a stable set in G. Thus there must exist probes $p, r \in P$ so that $rx, rv, px, pv \in E$ and $rp \notin E$. Since $rx \in E$, we know $I_r \cap I_x \neq \emptyset$ which implies $I_r \cap I_y \neq \emptyset$ and (since r is a probe) $ry \in E$. Similarly, $py \in E$. Therefore, the same probes, r, p, produce an enhanced edge between v and y, thus $vy \in E^*$. □

Lemma 4.16. If $x, y, z \in N$, $I_x \subseteq I_y$ and $xz \in E^*$ is an enhanced edge, then yz and xy are also enhanced edges.

Proof. Since xz is an enhanced edge, there exist probes $r, p \in P$ so that $rx, rz, px, pz \in E^*$ and $rp \notin E^*$. By Lemma 4.15, $ry, py \in E^*$, so probes r, p also produce the enhanced edges yz, xy. □

---

[2] We will see this in the proof of Theorem 4.17.

We now state the main result of this section.

Theorem 4.17. *(Zhang, 1994). If $G = (P \cup N, E)$ is a probe graph with respect to the partition $P \cup N$, then $G^* = (P \cup N, E^*)$ is chordal.*

Proof. Let $G = (V, E)$ be a probe graph whose vertex set is partitioned into a set P of probes and a stable set N of non-probes. Let $G^* = (V, E^*)$ be its enhanced graph. For a contradiction, assume G^* is not chordal. Let C be a shortest chordless cycle induced in G^*. Label the vertex set $V(C) = \{v_1, v_2, \ldots, v_n\}$ so that $E(C) = \{v_i v_{i+1} \mid 1 \leq i < n\} \cup \{v_1 v_n\}$. Since we have assumed G^* is not chordal, we know $n \geq 4$.

Lemma 4.13 tells us that for G^* there is a completion set E^+ of edges uv between distinct vertices in N so that $G' = (V, E^* \cup E^+)$ is an interval graph. Fix an interval representation $\{I_x \mid x \in V\}$ of G'. To avoid double subscripts we will denote the interval I_{v_j} by I_j for $j = 1, 2, 3, \ldots, n$. If necessary, permute the labels of vertices in C so that I_2 has the smallest left endpoint of all $\{I_j \mid 1 \leq j \leq n\}$ and the left endpoints of intervals I_1, I_2, I_3 appear in the order I_2, I_3, I_1. We know $v_j v_{j+1} \in E^*$ for $1 \leq j < n$ and $v_1 v_n \in E^*$ and we record the following remark.

Remark 4.18. $I_j \cap I_{j+1} \neq \emptyset$ for $1 \leq j < n$ and $I_1 \cap I_n \neq \emptyset$.

There are two cases which we consider separately: (1) $I_3 \subseteq I_2$ and (2) $(I_1 \cap I_2) \subseteq (I_2 \cap I_3)$.

Case 1: $I_3 \subseteq I_2$.

By Remark 4.18, $I_4 \cap I_3 \neq \emptyset$, thus $I_4 \cap I_2 \neq \emptyset$ and $v_2 v_4 \in E(G')$. Since $v_2 v_4 \notin E^*$ (it would be a chord in C) we must have $v_2 v_4 \in E^+$ and thus $v_2, v_4 \in N$. Now consider whether or not v_3 is a probe. If v_3 is a non-probe, then by Lemma 4.15 (with $x = v_3$, $y = v_2$, $v = v_4$) we obtain the contradiction $v_2 v_4 \in E^*$ (a chord in C). Thus v_3 must be a probe vertex. In summary, we are working under the assumptions $I_3 \subseteq I_2$, $v_2, v_4 \in N$, and $v_3 \in P$. We next consider two subcases.

Subcase 1a: v_1 is a probe.

If $v_1 v_4 \in E^*$ then there is an enhanced edge between v_2 and v_4 (using probes $r = v_1$, $p = v_3$), which would be a chord in C, a contradiction. Otherwise, $v_1 v_4 \notin E^*$ (which means $n \geq 5$) and by Remark 4.14, $I_1 \cap I_4 = \emptyset$. This (together with Remark 4.18 and our assumption about locations of intervals) restricts I_4 to be a subset of I_2. Now apply Lemma 4.15 with $x = v_4$, $y = v_2$, and $v = v_5$ to conclude $v_2 v_5 \in E^*$, a contradiction.

Subcase 1b: v_1 is a non-probe.

The edge $v_1 v_2 \in E^*$ must be an enhanced edge (since $v_1, v_2 \in N$). So there exist probes $r, p \in P$ with $r v_1, r v_2, p v_1, p v_2 \in E^*$ and $rp \notin E^*$. If $I_4 \nsubseteq I_2$

then our assumptions about locations of intervals imply that any interval which intersects both I_1 and I_2 will also intersect I_4. Thus rv_4, $pv_4 \in E^*$ and there is an enhanced edge between v_2 and v_4, a contradiction.

Alternatively, $I_4 \subseteq I_2$. If $v_1 v_4 \in E^*$, then it is an enhanced edge and we may apply Lemma 4.16 with $x = v_4$, $y = v_2$, $z = v_1$ to conclude $v_2 v_4 \in E^*$, a contradiction. Thus $v_1 v_4 \notin E^*$ and $n \geq 5$. Vertex v_5 must be a non-probe, or else there would be a chord $v_2 v_5 \in E^*$ because $I_4 \cap I_5 \neq \emptyset$ would imply $I_2 \cap I_5 \neq \emptyset$. Now apply Lemma 4.16 with $x = v_4$, $y = v_2$, $z = v_5$ to conclude $v_2 v_5 \in E^*$, a contradiction.

Case 2: $(I_1 \cap I_2) \subseteq (I_2 \cap I_3)$.
By Remark 4.18, $I_1 \cap I_2 \neq \emptyset$ thus $I_1 \cap I_3 \neq \emptyset$ and $v_1 v_3 \in E^+$. Since $v_1 v_3$ is a chord of C, $v_1 v_3 \notin E^*$, so v_1 and v_3 are both non-probes. We consider whether or not v_2 is a probe.

Subcase 2a: v_2 is a probe.
First consider the possibility in which the right endpoint of I_1 is greater than the right endpoint of I_3 and thus $I_3 \subseteq I_1 \cup I_2$. By Remark 4.18, $I_3 \cap I_4 \neq \emptyset$ and so $(I_1 \cup I_2) \cap I_4 \neq \emptyset$. However, $I_2 \cap I_4 = \emptyset$ since v_2 is a probe and $v_2 v_4 \notin E$ (Remark 4.14). Thus $I_1 \cap I_4 \neq \emptyset$. If v_4 is a probe then $v_1 v_3$ is an enhanced edge (using probes v_2, v_4), a contradiction. If v_4 is a non-probe, then the probes that produce $v_3 v_4$ as an enhanced edge also produce the enhanced edge $v_1 v_3$, a contradiction.

If, instead, the right endpoint of I_1 is less than the right endpoint of I_3, the argument is similar (using v_n in place of v_4).

Subcase 2b: v_2 is a non-probe.
Now $v_1 v_2$ is an enhanced edge. The restrictions imposed on locations of I_1, I_2, I_3 in this case imply that any interval that intersects both I_1 and I_2 will also intersect I_3. Thus any two probes that produce the enhanced edge $v_1 v_2$ also produce an enhanced edge between v_1 and v_3, a contradiction. $\square$

4.7 The Interval Graph Sandwich Problem

The *graph sandwich problem* for a graph property Π asks the following. Given a graph $G = (V, E)$ and a specified subset E_0 of nonedges, can G be augmented by adding some of the (*optional*) edges from E_0 to obtain a new graph G' having the property Π? We denote by $E_f = \overline{E \cup E_0}$ the *forbidden* edges, i.e., those which may not be added to form the sandwich graph G'. We use the notation $< V, E, E_0, E_f >$ for an instance of the graph sandwich problem.

The Interval Graph Sandwich Problem (IGSP), presented in this section, was shown to be NP-complete in Golumbic and Shamir (1993), and a variety of other graph sandwich problems have been studied in Golumbic, Kaplan, and Shamir (1995). The graph sandwich problem is polynomial for split graphs, threshold graphs, cographs and k-trees for fixed k and is NP-complete for comparability graphs, permutation graphs, k-trees for general k, chordal graphs, unit interval graphs and others. A modified version of the figure from Golumbic, Kaplan, and Shamir (1995) showing the computational complexity of the graph sandwich problem on many classes of graphs is reproduced in Figure 4.7. The complexity of the sandwich problem for tolerance graphs remains open.

Formally, the Interval Graph Sandwich Problem can be stated as follows:

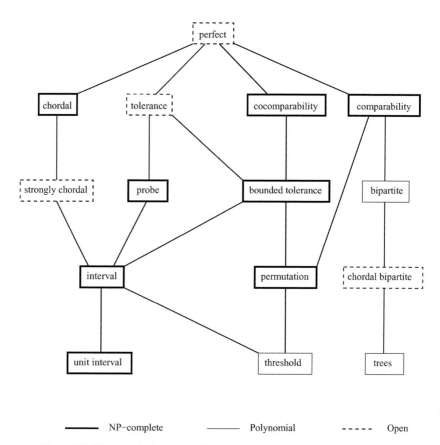

Figure 4.7. The complexity status of the sandwich problem for some graph classes. Additional classes appear in Golumbic, Kaplan, and Shamir (1995).

Interval Graph Sandwich Problem (IGSP)

Input: A graph $G = (V, E)$ and a subset $E_0 \subseteq \overline{E}$ of nonedges (which may optionally be added to G).

Question: Does there exist a subset $E' \subseteq E_0$ such that $G' = (V, E \cup E')$ is an interval graph?

Theorem 4.19. *(Golumbic and Shamir, 1993) The Interval Graph Sandwich Problem is NP-complete.*

We present here the proof from Golumbic, Kaplan, and Shamir (1994). In Section 4.8, we will use this theorem to prove a new result that the Probe Graph Sandwich Problem is NP-complete.

Proof. The IGSP is in the class NP, since we can nondeterministically generate a subset $E' \subseteq E_0$ and test whether the graph G' is an interval graph in polynomial time. To prove that the IGSP is NP-hard, we follow Golumbic, Kaplan, and Shamir (1994) by giving a reduction from the Betweenness Problem, stated below, which was shown to be NP-complete by Opatrný (1979) and which appears in Garey and Johnson (1979).[3]

Betweenness problem

Input: A set $S = \{a_1, \ldots, a_n\}$ of elements and a collection $T = \{T_1, \ldots, T_m\}$ of ordered triples of distinct elements of S, where $T_i = (a_{i_1}, a_{i_2}, a_{i_3}), i = 1, \ldots, m$.

Question: Does there exist a one-to-one function $f : S \to \{1, 2, \ldots, n\}$ such that either $f(a_{i_1}) < f(a_{i_2}) < f(a_{i_3})$ or $f(a_{i_3}) < f(a_{i_2}) < f(a_{i_1})$ for $i = 1, \ldots, m$?

A simple example of the following reduction is given in Exercise 4.4.

Reduction from betweenness to IGSP: Given an instance of the Betweenness Problem, we construct an instance of Interval Graph Sandwich Problem as follows. We create one vertex v_i for each element $a_i \in S$, and two vertices u_j, w_j for each triple T_j. Let $V = \{v_1, \ldots, v_n\} \cup \{u_j, w_j \mid j = 1, \ldots, m\}$. There are $4m$ (required) edges, namely, $E = \{v_{j_1} u_j, u_j v_{j_2}, v_{j_2} w_j, w_j v_{j_3} \mid j = 1, \ldots, m\}$. The forbidden edges are $E_f = \{v_i v_j \mid i \neq j\} \cup \{v_{j_1} w_j, u_j w_j, u_j v_{j_3} \mid j = 1, \ldots, m\}$. All other edges are optional, i.e., $E_0 = \overline{E \cup E_f}$. Clearly, this reduction is polynomial.

---

[3] This proof is illustrative of a useful technique of exploiting the unique way of representing P_5. Almost the same reduction from Betweenness is used in Golumbic, Kaplan, and Shamir (1995) to show NP-hardness of the sandwich versions for permutation graphs and cocomparability graphs. It can be adapted for parallelogram graphs and trapezoid graphs as well (see Exercise 4.7).

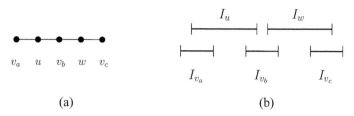

(a)   (b)

Figure 4.8. A 5-path and an interval representation of it.

Thus, to each triple T_j there corresponds a copy of the chordless 5-path P_5 induced by the vertices $\{v_{j_1}, u_j, v_{j_2}, w_j, v_{j_3}\}$ which must remain in the interval sandwich solution. Furthermore, the set of all v_i must remain a stable set. We make the key observation that any interval representation of a 5-path on vertices $\{v_a, u, v_b, w, v_c\}$ (shown in Figure 4.8(a)) must appear as in Figure 4.8(b) or its mirror image, either $I_{v_a} \prec I_{v_b} \prec I_{v_c}$ or $I_{v_c} \prec I_{v_b} \prec I_{v_a}$.

Suppose there is an ordering f of the set S which satisfies the betweenness conditions. Create an interval realization for an interval sandwich G' as follows. Draw n disjoint intervals corresponding to the v_i in the same order that the a_i appear in f. That is, $I_{v_i} \prec I_{v_j} \iff f(a_i) < f(a_j)$. Since the order conditions on triples are satisfied, additional intervals corresponding to u_j and w_j can be added for each triple T_j to create an induced 5-path on the vertices $\{v_{j_1}, u_j, v_{j_2}, w_j, v_{j_3}\}$. The resulting interval graph G' on V from this realization is a solution to the sandwich problem.

Conversely, suppose we are given a sandwich solution G' with interval realization $\mathcal{I}$. Since the set $\{v_1, \ldots, v_n\}$ induces a stable set in G', a total ordering is generated by the order of the corresponding n (pairwise disjoint) intervals from $\mathcal{I}$. Construct a function f according to this order. By the key observation above, every triple satisfies the betweenness condition. $\square$

4.8 The NP-completeness of the Interval Probe Graph Sandwich Problem

Recall from the generic definition of a sandwich problem in Section 4.7 that the Probe Graph Sandwich Problem asks, for a given graph $G = (V, E)$ and a specified subset E_0 of (*optional*) nonedges, whether G can be augmented by adding some subset $E' \subseteq E_0$ to the edge set E to obtain a new graph $G' = (V, E \cup E')$ which is a probe graph. In this section, we prove that the sandwich problem for probe graphs is NP-complete. Formally, the problem can be stated as follows:

Probe Graph Sandwich Problem (PGSP)

Input: A graph $G = (V, E)$ and a subset $E_0 \subseteq \overline{E}$ of nonedges (which may optionally be added to G).

Question: Does there exist a subset $E' \subseteq E_0$ and a partition of the vertices into two sets P and N, where N is a stable set of $G' = (V, E \cup E')$, and an interval completion $E'' \subseteq \{uv \mid u, v \in N, u \neq v\}$? In other words, $G'' = (V, E \cup E' \cup E'')$ is an interval graph.

Theorem 4.20. *The Probe Graph Sandwich Problem is NP-complete.*

Proof. The PGSP is in the class NP since we can nondeterministically choose a partition and generate arbitrary subsets E', E'' for this partition, and then test whether G'' is an interval graph in polynomial time. Next we prove that the PGSP is NP-hard by giving a reduction from the IGSP.

Let $\mathcal{G} = \langle V, E, E_0, E_f \rangle$ be an instance of the IGSP, where $G = (V, E)$ is the graph of required edges, E_0 are the optional edges and $E_f = \overline{E \cup E_0}$ are the forbidden edges. From $\mathcal{G}$, we define an instance of the Probe Graph Sandwich Problem $\mathcal{H} = \langle V \cup U, E \cup F, E_0 \cup F_0, E_f \cup F_f \rangle$ as follows.

For each forbidden edge $e = uv \in E_f$, we create a new pair of vertices u_e, v_e, and we add edges $uu_e, u_e v_e, v_e v$, forbid uv_e and vu_e, but make optional in $\mathcal{H}$ all other adjacencies involving u_e, v_e. Formally,

$U = \{u_e \mid e \in E_f\} \cup \{v_e \mid e \in E_f\}$,

$F = \{uu_e, u_e v_e, v_e v \mid e \in E_f\}$,

$F_f = \{uv_e, vu_e \mid e \in E_f\}$,

$F_0 = \{u_e u_{e'}, u_e v_{e'}, v_e u_{e'}, v_e v_{e'} \mid e' \in E_f, \ e \neq e'\} \cup \{u_e x, v_e x \mid$
$\quad x \in V, \ x \notin e\}$.

Clearly $\mathcal{H}$ can be constructed from $\mathcal{G}$ in polynomial time, so the theorem will follow from proving the following claim.

Claim: $\mathcal{G}$ has an interval sandwich solution if and only if $\mathcal{H}$ has a probe sandwich solution.

Suppose that $G' = (V, E \cup E')$ is a solution for $\mathcal{G}$, that is, G' is an interval graph with $E' \subseteq E_0$. Let $\{I_v \mid v \in V\}$ be an interval representation for G'. For each $e = uv \in E_f$, we must have $I_u \cap I_v = \emptyset$, so we may augment the representation by adding intervals I_{u_e} and I_{v_e} in the gap between I_u and I_v (as illustrated in Figure 4.9) to form a chordless path with edges $uu_e, u_e v_e, v_e v$. This new interval representation $\{I_v \mid v \in V\} \cup \{I_{u_e}, I_{v_e} \mid e = uv \in E_f\}$ is easily seen to give a solution to $\mathcal{H}$ where the probe/non-probe partition is $P = V \cup U$ and $N = \emptyset$.

Conversely, suppose that $H' = (V \cup U, E \cup F \cup E' \cup F')$ is a probe graph solution for $\mathcal{H}$, where $E' \subseteq E_0$ and $F' \subseteq F_0$. Let the probe/non-probe partition

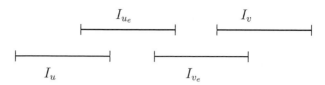

Figure 4.9. Augmenting the interval representation of G'.

of H' be $P \cup N$, where $E'' \subseteq \{yy' \mid y, y' \in N,\ y \neq y'\}$ denotes the interval completion, i.e., $H'' = (V \cup U, E \cup F \cup E' \cup F' \cup E'')$ is an interval graph. We next show that no forbidden edge $e = uv \in E_f$ was added by E''. If $uv \in E''$, then both $u, v \in N$, implying that both $u_e, v_e \in P$ since uu_e and $v_e v$ are required edges in H' and N is a stable set. But adding the edge uv would result in a chordless 4-cycle $[u_e, v_e, v, u]$ contradicting the fact that H'' is an interval graph. Therefore, H'' contains no forbidden edges from E_f. From this it follows that the induced subgraph H''_V is an interval sandwich solution for $\mathcal{G}$.   □

4.9 Exercises

Exercise 4.1. Prove that the graph D shown in Figure 2.9 and the graph H shown in Figure 2.5 are not interval probe graphs. Use the observation that the vertices on all chordless 4-cycles would have to alternate between probes and non-probes.

Exercise 4.2. Give proper probe representations of $K_{1,3}$ and C_4.

Exercise 4.3. Verify the assertion at the end of the proof of Theorem 4.12.

Exercise 4.4. Let $S = \{a_1, a_2, a_3, a_4\}$ and $T = \{T_1, T_2\}$, where $T_1 = (a_2, a_1, a_4)$ and $T_2 = (a_3, a_2, a_1)$.
  (a) Find a solution to this betweenness problem.
  (b) Construct the associated interval graph sandwich problem as in the reduction from betweenness to IGSP in the proof of Theorem 4.19.
  (c) Give the resulting solution to this interval graph sandwich problem.

Exercise 4.5. Coloring partitioned interval probe representations
  Consider the following algorithm applied to a set of intervals, partitioned into probes and non-probes.
  Sort the intervals according to their left endpoints.
  Sweep across the representation from left to right. (An interval is *active* from the time its left endpoint is scanned until its right endpoint is scanned.)

When a new probe is encountered, assign the lowest numbered available color, and when it is finished, its color becomes available again.

When a non-probe is encountered, it is assigned the same color as all the currently active non-probe intervals, should they exist, and otherwise it is assigned the lowest available color. When the last of the current non-probes is finished, its color becomes available again. (Note that there may be non-probes assigned different colors.)

(a) Add to the intervals of Figure 4.5 the additional intervals $I_h = [3, 7]$ and $I_i = [4, 8]$, and draw the interval probe graph with probe set $P = \{a, b, c, d, e\}$ and non-probe set $N = \{f, g, h, i\}$.

(b) Apply this coloring algorithm to the intervals. The solution should use four colors.

(c) Prove that the algorithm finds a minimum coloring by using the observation that a maximum clique can be found just at the point where the highest numbered color k is first used.

(d) Explain why (c) proves that the interval probe graphs are perfect.

(e) Determine the complexity of this coloring algorithm.

Exercise 4.6. Let G be an interval probe graph with n vertices.

(a) Prove that the number of maximal cliques is at most $O(n^2)$. (Hint: see Exercise 1.15.)

(b) Find an example which achieves this bound.

Exercise 4.7. In Golumbic, Kaplan, and Shamir (1995) the authors show that the permutation graph sandwich problem is NP-complete using a reduction similar to the one we have seen in the proof of Theorem 4.19, exploiting the unique way a chordless 5-path is represented in a permutation diagram.

Adapt this technique to prove that the sandwich problem is NP-complete for parallelogram graphs (and hence for bounded tolerance graphs).

Chapter 5

Bitolerance and the ordered sets perspective

5.1 The concept of a bounded tolerance order

A set of real intervals $\{I_v \mid v \in V\}$ can be viewed as a representation of the interval graph $G = (V, E)$ where $xy \in E \iff I_x \cap I_y \neq \emptyset$. It can also be interpreted as representing an interval order $P = (V, \prec)$ where $x \prec y$ if and only if I_x is completely to the left of I_y (which we denote $I_x \ll I_y$). The graph G and the order P are related in that G is the incomparability graph of P, that is, $xy \in E(G) \iff x \parallel y$ in P. Thus, results about interval graphs have counterparts in the world of ordered sets. For example, note the similarity in the characterization theorems below.

Theorem 5.1. *(Gilmore and Hoffman, 1964) A graph G is an interval graph if and only if it is a cocomparability graph with no induced C_4.*

Theorem 5.2. *(Fishburn, 1970) An order P is an interval order if and only if it has no induced $\mathbf{2} + \mathbf{2}$.*

The graph C_4 is the incomparability graph of the order $\mathbf{2} + \mathbf{2}$, so the same 4-element structure is forbidden in both theorems. Theorem 5.1 has the extra condition that G be a cocomparability graph. This is not needed in Theorem 5.2 since the relation in any ordered set is transitive, and thus the incomparability graph of any ordered set is always a cocomparability graph.

In the next several chapters, we study tolerance from the perspective of orders. In order to do this, it is necessary that the representations we use yield cocomparability graphs. Tolerance graphs are not necessarily cocomparability graphs. The graph, T_2 in Figures 2.2 and 3.3 is a tolerance graph by Theorem 3.7 but is not a cocomparability graph, as seen in Example 2.7. However, bounded tolerance graphs are cocomparability graphs (Theorem 2.8) and give rise to the first family we study, the bounded tolerance orders.

84

Definition 5.3. An order $P = (V, \prec)$ is a *bounded tolerance order* if each element $v \in V$ can be assigned a real interval I_v and a tolerance t_v with $0 < t_v \leq |I_v|$ so that $x \prec y$ if and only if the center of I_x is less than the center of I_y and $|I_x \cap I_y| < \min\{t_x, t_y\}$.

The collection $\langle \mathcal{I}, t \rangle$ of intervals $\mathcal{I} = \{I_v \mid v \in V\}$ and tolerances $t = \{t_v \mid v \in V\}$ is called a *bounded tolerance representation* of P. The collection $\langle \mathcal{I}, t \rangle$ is also a bounded tolerance representation of the incomparability graph G of P because $xy \in E(G) \iff |I_x \cap I_y| \geq \min\{t_x, t_y\} \iff x \parallel y$. This leads to the following remark.

Remark 5.4. If P is a bounded tolerance order, the incomparability graph G is a bounded tolerance graph. If G is a bounded tolerance graph, then there exists a bounded tolerance order P whose incomparability graph is G.

In Theorem 7.8 we will see that if G is a bounded tolerance graph, *every* order P whose incomparability graph is G is a bounded tolerance order.

In the next section, we give another definition of bounded tolerance orders as a special case of bounded bitolerance orders. Proposition 5.16 shows the two definitions are equivalent.

5.2 Classes of bounded bitolerance orders

Bogart and Trenk (1994) define a broader class of tolerance orders by assigning two tolerances to each vertex, a left tolerance and a potentially different right tolerance.

Definition 5.5. (Bitolerance Order): An ordered set $P = (V, \prec)$ is a *bounded bitolerance order* if it has a representation $\langle \mathcal{I}, p, q \rangle$ as follows. Each $v \in V$ is assigned a real interval $I_v = [L(v), R(v)]$ and two additional tolerant points $p(v), q(v) \in I_v$ satisfying $p(v) \neq L(v)$ and $q(v) \neq R(v)$ so that $x \prec y \iff R(x) < p(y)$ and $q(x) < L(y)$. The collection $\langle \mathcal{I}, p, q \rangle$, where $\mathcal{I} = \{I_v \mid v \in V\}$, $p = \{p(v) \mid v \in V\}$ and $q = \{q(v) \mid v \in V\}$ is called a *bounded bitolerance representation* of P.

Figure 5.1 shows the ordered set **3** + **2** together with a bounded bitolerance representation of it. In this and subsequent bounded bitolerance representations, we depict the tolerant point $p(v)$ by a rectangle and the tolerant point $q(v)$ by an oval. In a bounded bitolerance representation, the *left tolerance* of v (denoted $t_l(v)$) is $p(v) - L(v)$, and the *right tolerance* of v (denoted $t_r(v)$) is $R(v) - q(v)$.

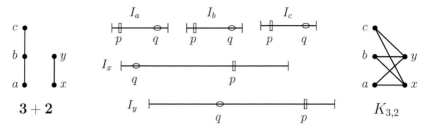

Figure 5.1. A bounded bitolerance representation of the order $\mathbf{3 + 2}$ and the graph $K_{3,2}$.

The restrictions on $p(v)$ and $q(v)$ ensure that these quantities are positive and at most $|I_v|$. The center point of interval I_v is denoted $c(v)$.

Any collection $\langle \mathcal{I}, p, q \rangle$ of intervals and tolerant points satisfying $p(v), q(v) \in I_v = [L(v), R(v)]$, $p(v) \neq L(v)$ and $q(v) \neq R(v)$, gives rise to a relation $\prec$ defined by $x \prec y$ if and only if $R(x) < p(y)$ and $q(x) < L(y)$. This relation is easily seen to be transitive and irreflexive, thus $P = (V, \prec)$ is an ordered set. The collection $\langle \mathcal{I}, p, q \rangle$ can also be viewed as a representation of a graph G which is the incomparability graph of the order P. A graph G with such a representation is called a *bounded bitolerance graph*. For example, the complete bipartite graph $K_{3,2}$ is a bounded bitolerance graph and a representation for it is given in Figure 5.1.

We note the following for future reference.

Remark 5.6. Every bounded bitolerance graph is a cocomparability graph.

5.2.1 Three types of restrictions

If further restrictions are placed on the intervals in $\mathcal{I}$ or the points in p and q, we get the following subclasses of bounded bitolerance orders.

Restrictions on intervals I_v

Definition 5.7. (Unit): P is a *unit bitolerance order* if it has a bounded bitolerance representation $\langle \mathcal{I}, p, q \rangle$ in which $|I_x| = |I_y|$ for all $x, y \in V$.

Definition 5.8. (Proper): P is a *proper bitolerance order* if it has a bounded bitolerance representation $\langle \mathcal{I}, p, q \rangle$ in which I_x is not properly contained in I_y for all $x, y \in V$.

Restrictions on tolerant points $p(v)$, $q(v)$

Definition 5.9. (Point-core): P is a *point-core bitolerance order* if it has a bounded bitolerance representation $\langle \mathcal{I}, p, q \rangle$ in which $p(v) = q(v)$ for all

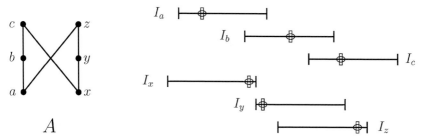

Figure 5.2. A unit point-core bitolerance representation of order A.

$v \in V$. In this case, we let $f(v) = p(v) = q(v)$, call this point the *splitting point* of I_v, and denote the representation by $\langle \mathcal{I}, f \rangle$.

Definition 5.10. (Totally bounded): P is a *totally bounded bitolerance order* if it has a bounded bitolerance representation $\langle \mathcal{I}, p, q \rangle$ in which $p(v) \le q(v)$ for all $v \in V$.

Restrictions on left and right tolerance

Definition 5.11. (Tolerance): P is a *bounded tolerance order* if it has a bounded bitolerance representation $\langle \mathcal{I}, p, q \rangle$ in which $t_l(v) = t_r(v)$ for all $v \in V$. In this case we write $t_v = t_l(v) = t_r(v)$.

Analogous to the totally bounded restriction is one in which $q(v) \le p(v)$ for all v. However, this class turns out to be equivalent to the class of bounded bitolerance orders, as we show in Proposition 10.2.

Figure 5.2 shows an order A and a unit point-core bitolerance representation of it. Point-core bitolerance orders appear in the literature under several different names. They are called *point-core orders* in Langley (1993), *orders with Fishburn representations* in Bogart and Isaak (1998), and *split interval orders* in Fishburn and Trotter (1999). The class of bitolerance orders with both the unit and point-core restrictions are called *split semiorders* in Fishburn and Trotter (1999).

Additional classes of bounded bitolerance orders can be formed by combining some of the restrictions above. For example, unit totally bounded tolerance orders are those in which $|I_x| = |I_y|$, $p(x) \le q(x)$, and $t_l(x) = t_r(x)$ for all $x, y \in V$. When one combines the "point-core" restriction in Definition 5.9 with the "tolerance" restriction in Definition 5.11, the result is a bounded bitolerance representation in which $t_l(v) = t_r(v) = \frac{1}{2}|I_v|$ for all $v \in V$. This class is known as the *50% tolerance orders*, the analog of 50% tolerance graphs discussed in Section 2.6.

Remark 5.12. In all, there are 18 possible ways to combine these restrictions and obtain classes of bounded bitolerance orders. These are discussed in Chapter 10 and their hierarchy is shown in Figure 10.1.

Each of these 18 classes has a graph theoretic counterpart, just as bounded bitolerance graphs are the counterpart of bounded bitolerance orders. Any bounded bitolerance representation can be interpreted as representing an order P or its incomparability graph G. Thus we may define 18 classes of bounded bitolerance graphs as being the incomparability graphs of the analogous classes of bounded bitolerance orders.

Remark 5.13. If P is a bounded bitolerance order, then its incomparabilty graph is a bounded bitolerance graph. If G is a bounded bitolerance graph, then there exists a bounded bitolerance order P whose incomparability graph is G. Moreover, the above hold if "bounded bitolerance" is replaced by any of the 18 classes that appear in Figure 10.1.

The next two examples illustrate the utility of being able to translate order theoretic results to graph theory.

Example 5.14. The graph $\overline{B}$ in Figure 2.4 is a bounded bitolerance graph.

Proof. Table 10.3 gives a (totally) bounded bitolerance representation of the order B whose incomparability graph is $\overline{B}$.   □

Example 5.15. The graph $\overline{C_{2n}}$ is not a bounded bitolerance graph for $n \geq 3$.

Proof. Suppose $\overline{C_{2n}}$ were a bounded bitolerance graph for some $k \geq 3$. By Remark 5.13, there exists a bounded bitolerance order P whose incomparability graph is $\overline{C_{2n}}$. Thus, the comparability graph of P is the chordless cycle C_{2n}. However, C_{2n} has only one transitive orientation (up to isomorphism) so P must be the n-crown. In Example 10.6, we will show the n-crown is not a bounded bitolerance order for $n \geq 3$.   □

5.2.2 Equivalent definitions of bounded tolerance

In this section we prove that the two definitions given for bounded tolerance orders are equivalent.

Proposition 5.16. *The definitions of bounded tolerance orders given in Definitions 5.3 and 5.11 are equivalent.*

Proof. Suppose $P = (V, \prec)$ is a bounded tolerance order according to Definition 5.3 and fix a representation $\langle \mathcal{I}, t \rangle$ of P so that $x \prec y \iff c(x) < c(y)$

and $|I_x \cap I_y| < \min\{t_x, t_y\}$. Let $I_v = [L(v), R(v)]$, $p(v) = L(v) + t_v$, $q(v) = R(v) - t_v$ and note that $p(v), q(v) \in I_v$, and $p(v) \neq L(v)$ and $q(v) \neq R(v)$ since $t_v > 0$. The collection $\langle \mathcal{I}, p, q \rangle$ represents an order $Q = (V, \prec')$ using Definition 5.11, where $x \prec' y \iff R(x) < p(y)$ and $q(x) < L(y)$. Since the transformation between the representations $\langle \mathcal{I}, t \rangle$ and $\langle \mathcal{I}, p, q \rangle$ is reversible, it remains to show that $x \prec y \iff x \prec' y$.

Suppose $x \prec y$. Hence, $c(x) < c(y)$ and $|I_x \cap I_y| < \min\{t_x, t_y\}$. If $I_x \subseteq I_y$ or $I_y \subseteq I_x$ then $|I_x \cap I_y| = \min\{|I_x|, |I_y|\} \geq \min\{t_x, t_y\}$, a contradiction. Thus neither of I_x, I_y is contained in the other. Combining this with $c(x) < c(y)$ we obtain $I_x \cap I_y = [L(y), R(x)]$. Therefore, $R(x) - L(y) = |I_x \cap I_y| < t_y = p(y) - L(y)$, so $R(x) < p(y)$. Similarly, $R(x) - L(y) = |I_x \cap I_y| < t_x = R(x) - q(x)$, so $q(x) < L(y)$. Together these imply $x \prec' y$.

Conversely, suppose $x \prec' y$. Thus $R(x) < p(y)$ and $q(x) < L(y)$. Since $R(x) < p(y) \leq R(y)$ and $L(x) \leq q(x) < L(y)$, we know

$$c(x) = \tfrac{1}{2}(L(x) + R(x)) < \tfrac{1}{2}(L(y) + R(y)) = c(y).$$

To show $x \prec y$, it remains to prove $|I_x \cap I_y| < \min\{t_x, t_y\}$. Again, $I_x \cap I_y = [L(y), R(x)]$. Hence $|I_x \cap I_y| = R(x) - L(y) < R(x) - q(x) = t_x$, and $|I_x \cap I_y| = R(x) - L(y) < p(y) - L(y) = t_y$. $\qquad \square$

This proof can be easily adapted to yield the following analogous result for bounded bitolerance orders by replacing t_x by $t_r(x)$ and replacing t_y by $t_l(y)$.

Proposition 5.17. *Let P be a bounded bitolerance order with representation $\langle \mathcal{I}, p, q \rangle$ in which the center points $\{c(v) \mid v \in V\}$ of the intervals in $\mathcal{I}$ are distinct. Then $R(x) < p(y)$ and $q(x) < L(y)$ if and only if $c(x) < c(y)$ and $|I_x \cap I_y| < \min\{t_r(x), t_l(y)\}$.*

5.2.3 Distinct endpoints and tolerant points

In many papers on tolerance graphs and orders it is noted without proof that one can find a representation with distinct endpoints, tolerant points and tolerances. This fact is not immediately obvious when additional conditions such as unit or proper are assumed. Indeed, in Fishburn and Trotter (1999) there is a lengthy proof of two variants on this comment, and we gave a different variant in Lemma 1.5. In this section, we present a general lemma from Isaak, Nyman, and Trenk (2001) about distinctness of endpoints, tolerant points and tolerances.

Lemma 5.18. *Let $\mathcal{C}$ be any of the 18 classes of bounded bitolerance orders referred to in Remark 5.12. If $P = (V, \prec)$ is a member of $\mathcal{C}$, then P has a* *representation as a member of $\mathcal{C}$ in which all endpoints, tolerant points and tolerances are distinct.*

Proof. First we focus on achieving distinct endpoints and tolerant points, and later we make further modifications to achieve distinct tolerances as well.

Given a bounded bitolerance representation $\langle \mathcal{I}, p, q \rangle$ of P, let ϵ be the smallest positive distance between two distinct points in $\{L(x), p(x), q(x), R(x) | x \in V\}$. Assume that $V = \{1, 2, \ldots, |V|\}$. Form a new representation with $L'(x) = L(x) - \epsilon/10 + \epsilon/10^{x+2}$, $p'(x) = p(x) - \epsilon/10^2 + \epsilon/10^{x+2}$, $q'(x) = q(x) + \epsilon/10^2 + \epsilon/10^{x+2}$ and $R'(x) = R(x) + \epsilon/10 + \epsilon/10^{x+2}$. The following are straightforward to check from the definitions of $L'(x), p'(x), q'(x), R'(x)$ and the use of terms involving ϵ: the prime representation has all endpoints and tolerant points distinct, if the original representation was proper or unit or totally bounded or satisfied the 'tolerance' property then so is the prime representation. It remains to check that the prime representation also represents P. If $p(y) - R(x) > 0$ then by the choice of ϵ, $p(y) - R(x) > \epsilon$ and thus $p'(y) - R'(x) = (p(y) - \epsilon/10^2 + \epsilon/10^{y+2}) - (R(x) + \epsilon/10 + \epsilon/10^{x+2}) > 0$. If $p(y) - R(x) \le 0$ then $p'(y) - R'(x) = (p(y) - \epsilon/10^2 + \epsilon/10^{y+2}) - (R(x) + \epsilon/10 + \epsilon/10^{x+2}) \le 0$ since $(-\epsilon/10^2 + \epsilon/10^{y+2}) - (\epsilon/10 + \epsilon/10^{x+2}) \le 0$. So $R(x) < p(y) \Leftrightarrow R'(x) < p'(y)$. Similarly $q(x) < L(y) \Longleftrightarrow q'(x) < L'(y)$.

For point-core representations do as above but omit the terms $\epsilon/10^2$ from $p'(x)$ and $q'(x)$ so that $p(x) = q(x) \Rightarrow p'(x) = q'(x)$. Finally, suppose P is a unit or proper interval order with a representation in which $v \in V$ is assigned the interval $I_v = [L(v), R(v)]$. Define $L'(x)$ and $R'(x)$ as above. The prime representation has all endpoints distinct and it is straightforward to check that the unit or proper property is maintained. In a manner similar to above we can check that $R(x) < L(y) \Leftrightarrow R'(x) < L'(y)$ and hence the prime intervals also represent P.

Note that if two tolerances are equal the transformation above maintains this. If in addition we want distinct tolerances, consider the following additional changes. Let δ be the smallest positive distance between two distinct points in $\{L'(x), p'(x), q'(x), R'(x) | x \in V\}$. Let $L''(x) = L'(x) - \delta/10^x$ and $R''(x) = R'(x) + \delta/10^x$ and let $p''(x) = p'(x)$ and $q''(x) = q'(x)$. This preserves the distinct representation and yields distinct tolerances. However, it does not preserve the unit property. For unit orders let $p'''(x) = p'(x) - \delta/10^x$ and $q'''(x) = q'(x) + \delta/10^x$ and let $L'''(x) = L'(x)$ and $R'''(x) = R'(x)$. This preserves the distinct representation and yields distinct tolerances. However, it does not preserve the point-core property. All of the classes we have considered fit into at least one of the representations above except for unit point-core tolerance representations which by definition have all tolerances equal. □

5.3 Geometric interpretations

In this section, we show that bounded bitolerance graphs and orders can be represented using trapezoids. This is a generalization of Theorem 2.9 where we showed that bounded tolerance graphs can be represented as intersection graphs of parallelograms. The trapezoid representations can facilitate our proofs about bounded bitolerance orders and provide more insight than strictly algebraic proofs.

Trapezoid graphs were introduced by Dagan, Golumbic, and Pinter (1988) in connection with VLSI design, and independently by Corneil and Kamula (1987). We begin by repeating the definition of a *trapezoid representation* from Section 1.6. Fix two horizontal lines L_1 and L_2 with L_1 above L_2. Given a set V, assign a trapezoid T_v to each $v \in V$ where the parallel sides of T_v lie along L_1 and L_2. If $T_x \cap T_y = \emptyset$, then either T_x is to the left of T_y and we write $T_x \ll T_y$, or T_y is to the left of T_x and we write $T_y \ll T_x$. The collection $\mathcal{T} = \{T_v | v \in V\}$ can also be interpreted as a *trapezoid representation* for the graph $G = (V, E)$ with $xy \in E$ if and only if $T_x \cap T_y \neq \emptyset$. A graph with a trapezoid representation is called a *trapezoid graph*. The collection $\mathcal{T} = \{T_v | v \in V\}$ is also a trapezoid representation for the ordered set $P = (V, \prec)$ with $x \prec y \iff T_x \ll T_y$. An order with a trapezoid representation is called a *trapezoid order*. The graph G is the incomparability graph of the order P since $T_x \cap T_y = \emptyset \iff (T_x \ll T_y)$ or $(T_y \ll T_x)$.

Remark 5.19. If P is a trapezoid order then its incomparability graph G is a trapezoid graph. If G is a trapezoid graph then there exists a trapezoid order P whose incomparability graph is G.

By combining the results of Theorems 5.24 and 7.1 we will see that if G is a trapezoid order, then in fact *every* order P whose incomparability graph is G is a trapezoid order.

In Section 1.6 we defined parallelogram graphs. A *parallelogram order* is an ordered set that has a parallelogram representation, that is, a trapezoid representation in which each trapezoid is actually a parallelogram. We state the analog of Theorem 2.9 in terms of orders. Its proof is similar to that of Theorem 2.9.

Lemma 5.20. *An order is a bounded tolerance order if and only if it is a parallelogram order.*

As in the case of parallelograms, we allow degenerate trapezoids, that is, the sides along L_1 and L_2 may be points in which case the resulting trapezoid is a triangle or a line.

Recall from Chapter 1 that the interval dimension of an ordered set P is the least number of interval orders whose intersection is P. In their paper introducing trapezoid graphs, Dagan, Golumbic, and Pinter (1988) observed a connection with interval dimension. In a trapezoid representation, each trapezoid naturally gives rise to two intervals, one for each of its parallel sides, and thus a trapezoid representation leads to two interval orders. The next proposition makes this more precise.

Lemma 5.21. *(Dagan, Golumbic, and Pinter, 1988) An order P is a trapezoid order if and only if P has interval dimension at most 2.*

Proof. Let P be a trapezoid order with representation $T = \{T_v | v \in V\}$. Let $I_v = [a(v), c(v)] = T_v \cap L_1$ and $J_v = [b(v), d(v)] = T_v \cap L_2$. The sets of intervals $\mathcal{I} = \{I_v \mid v \in V\}$ and $\mathcal{J} = \{J_v \mid v \in V\}$ are interval representations of interval orders we will call $P_1 = (V, \prec_1)$ and $P_2 = (V, \prec_2)$, respectively. Moreover, the transformation is completely reversible. It remains to show that $P = P_1 \cap P_2$, which follows because $x \prec y \iff T_x \ll T_y \iff (I_x \ll I_y$ and $J_x \ll J_y) \iff (x \prec_1 y$ and $x \prec_2 y)$. $\qquad\square$

We call the diagonal from the top left corner of T_v to the bottom right corner the *left diagonal*, denoted D_v. The opposite diagonal, the *right diagonal*, is denoted D'_v. A *left-leaning* trapezoid representation is one in which the non-horizontal sides of each trapezoid representation have negative slope. Similarly, in a *right-leaning* representation, the non-horizontal sides have positive slope. In proving results in this section, it will be useful to make transformations between different types of trapezoid representations. The next lemma and remark make that possible.

Lemma 5.22. *Every trapezoid order has a left-leaning representation.*

Proof. We transform a trapezoid representation $T = \{T_v | v \in V\}$ of P into a left-leaning representation $\{T'_v \mid v \in V\}$ of P by sliding all points on line L_1 to the left by a sufficiently large constant k. More precisely, trapezoid T_v with horizontal sides $[a(v), c(v)]$ along L_1 and $[b(v), d(v)]$ along L_2 becomes trapezoid T'_v with sides $[a(v) - k, c(v) - k]$ along L_1 and $[b(v), d(v)]$ along L_2. Note that $T_x \ll T_y \iff T'_x \ll T'_y$. If k is sufficiently large, the new representation $\{T'_v \mid v \in V\}$ is left-leaning. $\qquad\square$

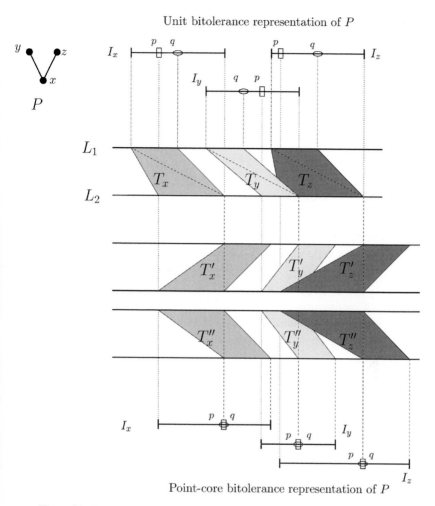

Figure 5.3. The top portion shows an order P together with unit bitolerance and trapezoid representations of it. The rest of the figure illustrates the correspondence between unit bitolerance and point-core bitolerance representations.

Remark 5.23. A left-leaning trapezoid representation of P can be transformed into to a right-leaning trapezoid representation of P (and vice versa) by reflecting the representation about the line L_2. This transformation also has the effect of reversing the roles of the left and right diagonals of each trapezoid.

The top portion of Figure 5.3 illustrates the correspondence between bounded bitolerance representations and left-leaning trapezoid representations which is proven in the next theorem. The bounded bitolerance representation in Figure 5.3 happens to be a unit bitolerance representation, but this extra condition is not used in the transformation to a trapezoid representation. The connection between bounded bitolerance orders and trapezoid orders was first observed in Langley (1993).

Theorem 5.24. *The following are equivalent statements about an ordered set P.*

(i) *P is a bounded bitolerance order.*
(ii) *P is a trapezoid order.*
(iii) *P has interval dimension at most 2.*

Proof. The equivalence (ii) $\Longleftrightarrow$ (iii) is proved in Lemma 5.21.

(i) $\Longleftrightarrow$ (ii): Given a bounded bitolerance representation $\langle \mathcal{I}, p, q \rangle$ of order $P = (V, \prec)$ with $I_v = [L(v), R(v)]$, we associate a trapezoid T_v with each element $v \in V$ as follows. The trapezoid T_v has top edge $[L(v), q(v)]$ along L_1 and bottom edge $[p(v), R(v)]$ along L_2. Let $Q = (V, \prec')$ be the trapezoid order with representation $\mathcal{T} = \{T_v | v \in V\}$ and note that this representation is left-leaning by the definition of a bounded bitolerance representation. This construction is reversible since all trapezoid orders have left-leaning representations (Lemma 5.22). Thus it remains to show that $P = Q$, which follows because

$$x \prec' y \iff T_x \ll T_y$$
$$\iff (T_x \cap L_1) \ll (T_y \cap L_1) \text{ and } (T_x \cap L_2) \ll (T_y \cap L_2)$$
$$\iff q(x) < L(y) \text{ and } R(x) < p(y)$$
$$\iff x \prec y. \qquad \square$$

Combining Theorem 5.24 with Remarks 5.13 and 5.19 yields the following.

Theorem 5.25. *A graph G is a trapezoid graph if and only if G is a bounded bitolerance graph.*

The construction in the proof of Theorem 5.24 establishes a correspondence between left-leaning trapezoid representations $\mathcal{T}$ and bounded bitolerance representations $\langle \mathcal{I}, p, q \rangle$. We say that $\mathcal{T}$ is the (left-leaning) trapezoid representation *associated* with the bounded bitolerance representation $\langle \mathcal{I}, p, q \rangle$ and vice versa.

A bounded bitolerance representation $\langle \mathcal{I}, p, q \rangle$ of $P = (V, \prec)$ has *constant cores* if $q(v) - p(v)$ is a constant for all $v \in V$. This constant may be positive (giving a totally bounded representation), negative, or zero (giving a point-core representation). The constant also determines the slope of the right diagonals of the trapezoids in the trapezoid representation associated with $\langle \mathcal{I}, p, q \rangle$. This definition of constant cores for bitolerance representations reverts to the definition in Section 2.6 in the case of a tolerance representation since in that instance,

$$|I_v| - 2t_v = (R(v) - L(v)) - (p(v) - L(v)) - (R(v) - q(v)) = q(v) - p(v).$$

Our next theorem is a bitolerance analog of Theorem 2.31. Our proof is geometric and is based on the proof in Bogart, Fishburn, Isaak, and Langley (1995) that the classes of unit and 50% tolerance graphs are equal. An algebraic proof of (i) $\Longleftrightarrow$ (ii) in Theorem 5.26 appears in Bogart and Isaak (1998), where point-core bitolerance representations are called *Fishburn representations*. An example which illustrates the correspondence between unit bitolerance representations and point-core bitolerance representations is given in Figure 5.3.

Theorem 5.26. *(Langley, 1993) The following are equivalent statements about an order P.*

(i) *P is a unit bitolerance order.*
(ii) *P is a point-core bitolerance order.*
(iii) *P has a bitolerance representation with constant cores.*

Proof. (i) $\Longrightarrow$ (ii): Suppose $P = (V, \prec)$ is a unit bitolerance order and fix a representation $\langle \mathcal{I}, p, q \rangle$ of it where $I_v = [L(v), R(v)]$. Let $\mathcal{T} = \{T_v | v \in V\}$ be the associated trapezoid representation of P. Note that this representation has the property that the left diagonal D_v has the same slope for each $v \in V$ since the upper left corner of T_v is $L(v)$ and the lower right corner is $R(v)$ and all intervals I_v are unit length. Create a new set of trapezoids $\mathcal{T}' = \{T_v' \mid v \in V\}$ by sliding all points on the line L_1 to the right by a constant k until the diagonals simultaneously become vertical. Note that $T_x \ll T_y$ if and only if $T_x' \ll T_y'$ so $\mathcal{T}'$ is also a trapezoid representation of P.

The representation $\mathcal{T}'$ is a right-leaning trapezoid representation of P and we may transform it to a left-leaning trapezoid representation $\mathcal{T}''$ of P by reflecting it around the line L_2, as discussed in Remark 5.23. The bounded bitolerance representation associated with $\mathcal{T}''$ is a point-core bitolerance representation.

(ii) $\Longrightarrow$ (iii): This follows immediately from the definitions of point-core and constant core, the former being a special case of the latter.

(iii) $\implies$ (i): Let $\langle \mathcal{I}, p, q \rangle$ be a constant core bounded bitolerance representation of P where $I_v = [L(v), R(v)]$. Thus there is a constant k for which $q(v) - p(v) = k$ for all $v \in V$. Let $\mathcal{T} = \{T_v | v \in V\}$ be the associated trapezoid representation of P and note that the constant core condition translates into the condition that the right diagonals of each trapezoid each have the same slope. As in Remark 5.23, reflect each trapezoid around the line L_2 to get a new trapezoid representation $\mathcal{T}'$ of P which has the property that left diagonals of each trapezoid have the same slope. Finally, slide all points on the line L_1 to the left (as in the proof of Lemma 5.22) until the non-horizontal sides of each trapezoid have negative slope. The result is a left-leaning trapezoid representation $\mathcal{T}''$ of P in which all the left diagonals have the same slope and hence the same length. The bounded bitolerance representation associated with $\mathcal{T}''$ is a unit bitolerance representation. $\square$

The proof of Theorem 2.31 can be obtained by modifying the proof of Theorem 5.26. The modification is achieved by changing all occurrences of "bitolerance" to "tolerance", all occurrences of "trapezoid" to "parallelogram", all occurrences of "order" to "graph" and observing that the results in Lemma 5.22 and Remark 5.23 also apply to parallelogram graphs. As noted earlier, the class of point-core tolerance orders is the same as 50% tolerance orders, and we study these further in the next chapter.

5.4 Exercises

Exercise 5.1. Prove Proposition 5.17.

Exercise 5.2. (a) What bounded tolerance order P is represented in Figure 1.5?

(b) Use the construction in the proof of Lemma 5.18 to obtain a bounded tolerance representation of this order P in which all endpoints, tolerant points and tolerances are distinct.

Exercise 5.3. (a) What order P is represented by the trapezoid representation in Figure 1.14?

(b) Verify that the incomparability graph of this order P is the graph G from Exercise 1.9.

(c) Use the construction in the proof of Lemma 5.21 to find interval orders P_1 and P_2 such that $P = P_1 \cap P_2$.

(d) Use the construction in the proof of Theorem 5.24 to give a bounded bitolerance representation of P.

Exercise 5.4. (a) A unit tolerance representation for the order $3 + 2$ is given in the proof of Proposition 10.9. Use the proof of (i) $\implies$ (ii) in Theorem 5.24 to transform this unit tolerance representation to a parallelogram representation of $3 + 2$.

(b) Use the result of (a) and the proof of (i) $\implies$ (ii) in Theorem 5.26 to construct a point-core tolerance representation of $3 + 2$ (i.e., a 50% tolerance representation).

Chapter 6
Unit and 50% tolerance orders

In this chapter, we focus on the classes of unit and 50% tolerance orders. Recall that a *unit tolerance order* is a bounded tolerance order which has a representation in which all intervals have the same length. A *50% tolerance order* is a bounded tolerance order $P = (V, \prec)$ which has a representation $\langle \mathcal{I}, t \rangle$ in which $t_v = \frac{1}{2}|I_v|$ for each $v \in V$. These classes are equal, which we record below.

Theorem 6.1. *The following are equivalent statements about an order* P.

(i) *P is a unit tolerance order.*
(ii) *P is a 50% tolerance order.*
(iii) *P has a tolerance representation with constant cores.*
(iv) *P is a point-core tolerance order.*

The same result is stated in Theorem 2.31 in terms of graphs and without the fourth condition. Its proof is discussed at the end of Section 5.3. The fourth condition is easily seen to be equivalent to the second condition using the definitions of those classes.

6.1 Unit tolerance orders with six or fewer elements

The main result of this section is a characterization of unit tolerance orders with six or fewer elements. This result also will be used in Section 6.3 in characterizing width 2 orders which are unit tolerance orders.

Theorem 6.2. *(Bogart, Jacobson, Langley, and McMorris, 2001) Every interval order is a 50% tolerance order. Furthermore, if P is an interval order then P has a 50% tolerance representation in which comparable elements get disjoint intervals.*

98

Proof. Fix an interval representation $\mathcal{I} = \{I_v \mid v \in V\}$ of an interval order $P = (V, \prec)$, where $I_v = [L(v), R(v)]$. By Lemma 5.18, we may assume that all interval endpoints are distinct. Label the set $V = \{v_1, v_2, \ldots, v_n\}$ so that $R(v_1) < R(v_2) < R(v_3) < \cdots < R(v_n)$.

Algorithm for converting an interval representation to a 50% tolerance representation

for $i := 1, 2, \ldots, n$
  $x := v_i$
  $m_x := |I_x|$
  for $y := v_i, v_{i+1}, \ldots, v_n$
    (i) If $I_x \cap I_y \neq \emptyset$, replace I_y by $[L(y), R(y) + m_x]$.
    (ii) If $I_x \ll I_y$, replace I_y by $[L(y) + m_x, R(y) + m_x]$.

An application of the algorithm to a unit representation of the order $\mathbf{2} + \mathbf{1}$ is shown in Figure 6.1.

We will show that this algorithm transforms $\mathcal{I}$ into a 50% tolerance representation of P. First consider $a, b \in V$ which are comparable in P. Without loss of generality, we may assume $a \prec b$ and thus $I_a \ll I_b$ in the original representation $\mathcal{I}$. Each time $R(a) \to R(a) + m_x$ we have $R(x) < R(a) < L(b)$ so $L(b) \to L(b) + m_x$ using (ii). Thus we will have $I_a \ll I_b$ in the final representation. This proves that comparable elements get disjoint intervals in the final representation.

Now consider the remaining case in which $a \parallel b$ in P. Without loss of generality, assume $R(a) < R(b)$, thus $L(b) \leq R(a) < R(b)$ in the original representation. We will show that the final interval for b contains the center of the final interval for a. Whenever $L(b) \to L(b) + m_x$ we have $R(x) < L(b) \leq R(a)$ so we also have $R(a) \to R(a) + m_x$ using (i) or (ii). Thus $L(b) \leq R(a)$ at all times. Each time $R(a) \to R(a) + m_x$, we also have $R(b) \to R(b) + m_x$, and thus $R(a) \leq R(b)$ at all times. When $x = a$, (i) applies for both $y = a$ and $y = b$ and the new interval for b contains the current point $R(a)$, and $R(a)$ becomes the center point of the new interval for I_a. After this, the interval I_a remains unchanged. The left endpoint of I_b also remains the same since only (i) will apply. Thus at the end we will have $c(a) \in I_b$. $\qquad\square$

The representation produced by the algorithm is simultaneously a 50% tolerance representation and an interval representation of order P. It has the additional property that for any pair of incomparable elements, the interval with the larger right endpoint contains the center of the other interval (see Figure 6.1). Other authors have studied instances of representations giving rise to the same

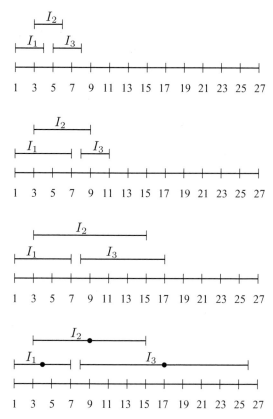

Figure 6.1. An interval representation of the order **2 + 1** converted to a 50% tolerance representation using the algorithm in the proof of Theorem 6.2.

order *P* when interpreted in two different ways (for example, see Tanenbaum, 1996, 1999).

Corollary 6.3. *Every interval graph is a* 50% *tolerance graph. Furthermore, if G is an interval graph, then G has a* 50% *tolerance representation in which nonadjecent vertices get disjoint intervals.*

The proof of Corollary 6.3 can be obtained by modifying the proof of Theorem 6.2 as follows: change all occurrences of "order" to "graph", of "*P*" to "*G*", of "comparable" to "nonadjacent" and "incomparable" to "adjacent".

We now present the main result of this section.

Theorem 6.4. *(Bogart, Jacobson, Langley, and McMorris, 2001) All orders on five or fewer elements are unit tolerance orders. The only 6-element orders that are not unit tolerance orders are shown in Figure 6.2.*

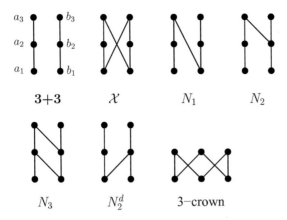

Figure 6.2. The 6-element orders which are not unit tolerance orders.

We will give the proof that these 6-element orders are not unit tolerance orders (omitting one lemma) and sketch the proof that they are the only forbidden 6-element orders for unit tolerance orders.

Sketch of Proof. Recall that the dual of an order is obtained by reversing all comparabilities, and this is equivalent to reflecting its Hasse diagram about a horizontal axis. The 3-crown is not a unit tolerance order because its incomparability graph is $\overline{C_6}$ which is not a tolerance graph by Corollary 2.16.

For the other six orders in Figure 6.2, label the elements $a_1, a_2, a_3, b_1, b_2, b_3$ as shown for the $\mathbf{3+3}$. Each of these has a $\mathbf{2+2}$ whose only comparabilities are $x \prec y$ and $w \prec z$ and a fifth element a with $y \prec a$ and $z \parallel a$. In Bogart, Jacobson, Langley, and McMorris (2001), the authors prove that this configuration of five elements forces the centers of the intervals in a 50% tolerance representation to lie in the order $c(w) < c(x) < c(y) < c(z)$. Similarly, if the fifth element b satisfies $b \prec x$ and $b \parallel w$, then the same conclusion holds (Exercise 6.1). We omit the proof of this lemma.

For the $\mathbf{3+3}$, $\mathcal{X}$ and N_1, let $x = a_1$, $y = a_2$, $w = b_1$, and $z = b_2$. Using $a = a_3$ forces $c(b_1) < c(a_1) < c(a_2) < c(b_2)$, and using $a = b_3$ forces $c(a_1) < c(b_1) < c(b_2) < c(a_2)$, a contradiction. For the N_2 and N_3, let $x = a_1$, $y = a_2$, $w = b_2$, and $z = b_3$. Using $a = a_3$ forces $c(b_2) < c(a_1) < c(a_2) < c(b_3)$, and using $b = b_1$, the dual configuration forces $c(a_1) < c(b_2) < c(b_3) < c(a_2)$, a contradiction. Since N_2 is not a 50% tolerance order, neither is its dual N_2^d. This shows that the orders in Figure 6.2 are not 50% tolerance orders, and hence, by Theorem 6.1, they are not unit tolerance orders.

For the converse, let P be a 6-element ordered set which is not a 50% tolerance order. By Theorem 6.2, we know that P is not an interval order and hence P contains an induced $\mathbf{2} + \mathbf{2}$. This accounts for four of the six elements of P. The proof proceeds by considering all possible ways of adding two additional elements. In each case (and there are many), either an order from Figure 6.2 is formed, or a 50% tolerance representation is given (see Bogart, Jacobson, Langley, and McMorris, 2001). Thus we may conclude that the orders in Figure 6.2 are the only 6-element orders which are not unit tolerance orders.

Finally, we consider orders with five or fewer elements. Let $P = (X, \prec)$ be a unit tolerance order with $|X| \leq 5$. Add $6 - |X|$ isolated elements to P to get the order $Q = (X', \prec')$ with $|X'| = 6$. None of the orders shown in Figure 6.2 have isolated elements, so by the previous paragraph, Q is a unit tolerance order. Hence the induced suborder P is also a unit tolerance order.   □

For completeness, we include the graph theory analog of Theorem 6.4.

Corollary 6.5. *All cocomparability graphs with five or fewer vertices are unit tolerance graphs. The only 6-element cocomparability graphs which are not unit tolerance graphs are given in Figure 6.3.*

Proof. The graphs in Figure 6.3 are the incomparability graphs of the orders in Figure 6.2. If G is a 6-vertex cocomparability graph that does not appear in Figure 6.3, then G is the incomparability graph of a 6-element order P that does not appear in Figure 6.2. By Theorem 6.4, P has a unit tolerance representation and this is also a unit tolerance representation of its incomparability graph G. The same argument shows that cocomparability graphs with five or fewer vertices are unit tolerance graphs.

Finally we show that the graphs in Figure 6.3 are not unit tolerance graphs. For each graph G in Figure 6.3, its complement $\overline{G}$ has a unique transitive orientation (up to isomorphism and reversing all arc directions) which appears

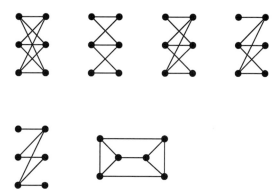

Figure 6.3. The cocomparability graphs which are not unit tolerance graphs.

as an order P in Figure 6.2 (Exercise 6.2). Any unit tolerance representation of the graph G would also give a unit tolerance representation of the order P, contradicting Theorem 6.4. Thus G is not a unit tolerance order.   □

6.2 Unit vs. proper for bounded bitolerance orders

In this section we return to the unit vs. proper question. As mentioned in Section 2.6, the classes of unit and proper interval graphs are equal (Roberts, 1969), but the classes of unit and proper tolerance graphs (and orders) are unequal (Bogart, Fishburn, Isaak, and Langley, 1995). Here we present a proof adapted from Bogart and Isaak (1998) to show that the classes of unit and proper bitolerance orders are equal.

Each of the unit vs. proper questions discussed so far has involved a third condition equivalent to the "unit" class. In the case of interval orders, P is a unit interval order if and only if it is an interval order with no induced $\mathbf{3 + 1}$. In the case of tolerance orders, P is a unit tolerance order if and only if it is a 50% tolerance order (Theorem 6.1). The bitolerance case is no different, and the third condition will be introduced in Lemmas 6.7 and 6.8. A shorter proof of the equivalence of unit and proper bitolerance orders follows from the proof of Theorem 10.3, but this does not show the additional third equivalence.

Given a proper bitolerance representation $\langle \mathcal{I}, p, q \rangle$ of $P = (V, \prec)$ in which the center points $\{c(v) \mid v \in V\}$ are distinct, the relation $\prec_c$ defined by $x \prec_c y$ if and only if $c(x) < c(y)$ is a linear extension of P by Proposition 5.17. The linear order $(V, \prec_c)$ is called the *central extension* of P associated with the representation $\langle \mathcal{I}, p, q \rangle$. The next lemma shows that it will always be possible to find a central extension of a proper bitolerance order.

Lemma 6.6. *Every proper bitolerance order has a proper bitolerance representation in which the centers of intervals are distinct.*

Proof. Fix a proper bitolerance representation $\langle \mathcal{I}, p, q \rangle$ of $P = (V, \prec)$. Let $S = \{L(v), R(v), p(v), q(v), c(v) \mid v \in V\}$, an analog of the c-points of a tolerance representation introduced in Section 3.1. Let ϵ be the smallest positive number that occurs as a difference between two elements of S. Suppose x and y are distinct elements of V with $c(x) = c(y)$. Since the representation is proper, this means $I_x = I_y$. We form a new representation of P by replacing the interval and tolerant points for x by $I'_x = [L(x) + \epsilon/2, R(x) + \epsilon/2]$, $p'(x) = p(x)$ and $q'(x) = q(x) + \epsilon/2$, and for each $z \in V$ with $q(z) = L(x)$, replace $q(z)$ by $q'(z) = q(z) + \epsilon/2$.

One can check that the new representation is a proper bitolerance representation of P that has one fewer pair of elements sharing a center point. If necessary, compute a new ϵ and repeat the process until all center points are distinct. □

Lemma 6.7. *(Bogart and Isaak, 1998) Let $P = (V, \prec)$ be a proper bitolerance order and $N = (V', \prec)$ an induced suborder with $V' = \{i, j, a, b\}$ and whose only comparabilities are $i \prec a$, $b \prec j$, and $b \prec a$. Then for any proper bitolerance representation of P, in which the centers of the intervals are distinct, the associated central extension $\prec_c$ satisfies either (i) $b \prec_c i \prec_c a$ or (ii) $b \prec_c j \prec_c a$ (or both).*

Proof. For a contradiction, assume there is a proper representation $\langle \mathcal{I}, p, q \rangle$ of P whose central extension $\prec_c$ of P violates both (i) and (ii). Then $i \prec_c b \prec_c a \prec_c j$, that is, the centers satisfy $c(i) < c(b) < c(a) < c(j)$. Since the representation is proper, the left and right endpoints of each interval also appear in this order, thus $R(i) < R(b)$ and $L(a) < L(j)$. Now $b \prec j$ so $R(b) < p(j)$ and thus $R(i) < R(b) < p(j)$. Similarly, $i \prec a$ so $q(i) < L(a) < L(j)$. However, $R(i) < p(j)$ and $q(i) < L(j)$ imply $i \prec j$, a contradiction. □

Lemma 6.8. *(Bogart and Isaak, 1998) Let $P = (V, \prec)$ be a proper bitolerance order and let $Q = (V', \prec)$ be an induced $\mathbf{2} + \mathbf{2}$ with $V' = \{i, j, a, b\}$ and whose only comparabilities are $i \prec a$, and $b \prec j$. Then for any proper bitolerance representation of P, in which the centers of the intervals are distinct, the associated central extension $\prec_c$ of P satisfies (i) $i \prec_c b \prec_c j \prec_c a$ or (ii) $b \prec_c i \prec_c a \prec_c j$.*

Proof. Fix a proper bitolerance representation $\langle \mathcal{I}, p, q \rangle$ of P where $I_v = [L(v), R(v)]$ and the centers of the intervals are distinct. Let $\prec_c$ be the associated central extension. Thus $i \prec_c a$ and $b \prec_c j$ and without loss of generality we may assume $i \prec_c b$. It remains to show $j \prec_c a$, so for a contradiction, assume $a \prec_c j$.

Since the representation is proper, the left and right endpoints of each interval are also ordered by $\prec_c$, thus $R(i) < R(b)$ and $L(a) < L(j)$. Since $b \prec j$, we have $R(b) < p(j)$, thus $R(i) < R(b) < p(j)$. Likewise, $i \prec a$ so $q(i) < L(a)$ and thus $q(i) < L(a) < L(j)$. Combining these yields $i \prec j$, a contradiction. □

Theorem 6.9. *(Bogart and Isaak, 1998) The following are equivalent statements about an ordered set $P = (V, \prec)$.*

(i) *P is a unit bitolerance order.*
(ii) *P is a proper bitolerance order.*

(iii) *P has a linear extension* $\prec_L$ *so that*

(a) *if* $N = (V', \prec)$ *is any induced suborder with* $V' = \{i, j, a, b\}$ *and whose only comparabilities are* $i \prec a$, $b \prec j$, *and* $b \prec a$, *then either* (i) $b \prec_L i \prec_L a$ *or* (ii) $b \prec_L j \prec_L a$ *(or both), and*

(b) *if* $Q = (V', \prec)$ *is any induced* $\mathbf{2} + \mathbf{2}$ *with* $V' = \{i, j, a, b\}$ *and whose only comparabilities are* $i \prec a$, *and* $b \prec j$, *then either* (i) $i \prec_L b \prec_L j \prec_L a$ *or* (ii) $b \prec_L i \prec_L a \prec_L j$.

Proof. (i) $\Longrightarrow$ (ii): This follows directly from the definitions of unit and proper bitolerance orders.

(ii) $\Longrightarrow$ (iii): By Lemma 6.6 we may fix a proper bitolerance representation of P in which the centers of intervals are distinct. Let $\prec_L$ be the associated central extension of P. The conclusion follows from Lemmas 6.7 and 6.8.

(iii) $\Longrightarrow$ (i): Let $P = (V, \prec)$ be an ordered set satisfying the conditions of (iii) for linear extension $\prec_L$. Index the elements $V = \{v_1, v_2, \ldots, v_n\}$ according to $\prec_L$ so that $v_i \prec_L v_j \iff i < j$. We use this linear extension to construct a unit bitolerance representation $\langle \mathcal{I}, p, q \rangle$ for P, where $I_v = [L(v), R(v)]$.

- Assign endpoints of I_{v_i}:
  - Let $L(v_i) = i$ for $i : 1 \leq i \leq n$.
  - Let $R(v_i) = n + i$ for $i : 1 \leq i \leq n$.
  Note that each interval I_{v_i} has length n, so the representation will satisfy the "unit" requirement.
- Assign right tolerant points:
  - If v_i is a maximal element in P, let $q(v_i) = n + 1/2$.
  - Otherwise, let a be the smallest index so that $v_i \prec v_a$ and set $q(v_i) = a - 1/2$.
- Assign left tolerant points:
  - If v_i is a minimal element in P, let $p(v_i) = n + 1/2$.
  - Otherwise, let b be the largest index so that $v_b \prec v_i$ and set $p(v_i) = n + b + 1/2$.

First we show that $q(v_i) \in I_{v_i}$ for each $v_i \in V$. If $q(v_i) = n + 1/2$, then $L(v_i) \leq n < q(v_i) < n + 1 \leq R(v_i)$. Otherwise, $q(v_i) = a - 1/2$ where $i < a$. Since i and a are integers, $i \leq a - 1$ and $L(v_i) = i \leq a - 1 < q(v_i) < a \leq n < R(v_i)$. Thus in either case, $q(v_i) \in I_{v_i}$. Similarly $p(v_i) \in I_{v_i}$ for each $i : 1 \leq i \leq n$.

Therefore, the intervals $I_{v_i} = [L(v_i), R(v_i)]$ and the tolerance points $p(v_i), q(v_i)$ give a unit bitolerance representation of an order $Q = (V, \prec')$. It remains to show that $Q = P$, that is, $v_i \prec v_j \iff v_i \prec' v_j$. Without loss of generality, assume $i < j$.

Case 1: $v_i \prec v_j$.

We must show $v_i \prec' v_j$, that is, $q(v_i) < L(v_j)$ and $R(v_i) < p(v_j)$. Since v_i is not maximal, by definition we have $q(v_i) = a - 1/2$, and by our choice of a, we know $a \leq j$. Thus $q(v_i) = a - 1/2 < a \leq j = L(v_j)$.

Likewise, since v_j is not minimal, by definition we have $p(v_j) = n + b + 1/2$, and by our choice of b, we know $i \leq b$. Thus $p(v_j) = n + b + 1/2 > n + b \geq n + i = R(v_i)$.

Case 2: $v_i \parallel v_j$ in P.

Since $p(v_i) < R(v_i) < R(v_j)$, we know $v_j \not\prec' v_i$. Thus it remains to show $v_i \not\prec' v_j$. For a contradiction, assume $v_i \prec' v_j$, thus $R(v_i) < p(v_j)$ and $q(v_i) < L(v_j)$.

Since $q(v_i) < L(v_j) \leq n$, we know $q(v_i) \neq n + 1/2$ and thus v_i is not maximal in P. Therefore, $q(v_i) = a - 1/2$ where a is the smallest index for which $v_i \prec v_a$.

Claim 1: $a < j$.

We have $a - 1/2 = q(v_i) < L(v_j) = j$ which means $a \leq j$ since a and j are integers. Note that $a \neq j$ since $v_i \prec v_a$ and $v_i \parallel v_j$. This proves Claim 1.

Similarly, since $p(v_j) > R(v_i) \geq n + 1$, we know $p(v_j) \neq n + 1/2$ and thus v_j is not minimal in P. Therefore, $p(v_j) = n + b + 1/2$ where b is the largest index so that $v_b \prec v_j$.

Claim 2: $b > i$.

We have $n + b + 1/2 = p(v_j) > R(v_i) = n + i$, which means $b \geq i$ since b and i are integers. Again, $b \neq i$ since $v_b \prec v_j$ and $v_i \parallel v_j$. This proves Claim 2.

Additionally, $a \neq b$ since otherwise we would have $v_i \prec v_a = v_b \prec v_j$, contradicting $v_i \parallel v_j$. Therefore, $V' = \{v_i, v_j, v_a, v_b\}$ consists of four distinct elements in P with $v_i \parallel v_j$, $v_i \prec v_a$ and $v_b \prec v_j$. The only other comparability possible among these elements is $v_b \prec v_a$.

If $v_b \prec v_a$ then the induced order $(V', \prec)$ is the order N. By hypothesis (a), $v_b \prec_L v_i \prec_L v_a$ or $v_b \prec_L v_j \prec_L v_a$, which by our indexing means $b < i < a$ or $b < j < a$. The first of these contradicts $b > i$ and the second contradicts $a < j$.

If $v_b \parallel v_a$ in P then the induced order $(V', \prec)$ is a $\mathbf{2 + 2}$. By hypothesis (b), either $v_i \prec_L v_b \prec_L v_j \prec_L v_a$ or $v_b \prec_L v_i \prec_L v_a \prec_L v_j$. Thus $i < b < j < a$ or $b < i < a < j$ by definition of L. The first of these contradicts $a < j$ and the second contradicts $b > i$. $\qquad\square$

6.3 Width 2 bounded tolerance orders

In this section we focus attention on orders of width 2. We begin with some definitions. The *width* of an ordered set P is the size of the largest antichain in P. A *chain cover* of P is a collection of chains in P whose union gives all of P. Since any subset of a chain is also a chain, we may assume that the chains in a chain cover partition the elements of P. The *size* of a chain cover is the number of chains in the cover. A famous theorem of Dilworth (see Bogart, 2000; Trotter, 1992) states that for any order P, the width of P equals the minimum size of a chain cover of P. Using this theorem, an order $P = (X, \prec)$ has width 2 if the ground set X can be partitioned nontrivially as $X = A \cup B$ where $(A, \prec)$, $(B, \prec)$ are chains.

If $P = (X, \prec)$ is a width 2 order, then its incomparability graph $G = (X, E)$ is bipartite since the chains A, B that partition X in P become independent sets in G. Conversely, if G is a cocomparability graph that is bipartite, then any transitive orientation of $\overline{G}$ gives a width 2 order. In Theorem 3.9 we showed that all cocomparability graphs which are bipartite are permutation graphs (and bounded tolerance graphs and trapezoid graphs). The order-theoretic equivalent is the following.

Theorem 6.10. *Any finite order of width 2 has dimension 2 (and hence is a bounded tolerance order and a trapezoid order).*

Proof. This theorem follows from the more general inequality $\dim(P) \leq width(P)$ due to Dilworth, which can be found in Bogart (2000) and Trotter (1992). An order of dimension 2 admits a representation as a permutation diagram which is a special case of a parallelogram (resp. trapezoid) diagram. $\square$

While Theorem 6.10 shows that all width 2 orders are bounded tolerance orders, they are *not* all unit tolerance orders. In Bogart, Jacobson, Langley, and McMorris (2001), the authors characterize those width 2 orders which are unit tolerance orders.

Theorem 6.11. *(Bogart, Jacobson, Langley, and McMorris, 2001) A width 2 order P is a unit tolerance order if and only if P does not contain any of the orders $\mathbf{3} + \mathbf{3}$, $\mathcal{X}$, N_1, N_2, N_3, N_2^d shown in Figure 6.2.*

We show that none of the orders in Figure 6.2 are unit tolerance orders in Theorem 6.4. The proof of the other direction appears in Bogart, Jacobson, Langley, and McMorris (2001) and is omitted here.

6.4 Exercises

Exercise 6.1. The following result is proven in Bogart, Jacobson, Langley, and McMorris (2001): given an order P consisting of a $2+2$ whose only comparabilities are $x \prec y$ and $w \prec z$, together with a fifth element a with $y \prec a$ and $z \parallel a$, the centers of the intervals in any 50% tolerance representation of P must satisfy $c(w) < c(x) < c(y) < c(z)$. Use this result and duals to prove that if a fifth element b satisfies $b \prec x$ and $b \parallel w$, then the same conclusion holds.

Exercise 6.2. For each graph G in Figure 6.3, show that its complement $\overline{G}$ has a unique transitive orientation (up to isomorphism and reversing all arc directions) which appears as an order P in Figure 6.2.

Exercise 6.3. Using the construction from the proof of Theorem 6.2 (and Theorem 5.26) transform the interval representation in Figure 1.1 into a unit tolerance representation.

Exercise 6.4. (a) What is the central extension associated with the unit (point-core) bitolerance representation of the order A in Figure 5.2? Show that this extension satisfies condition (iii) of Theorem 6.9.

(b) Use the central extension from part (a) and the proof of (iii) $\Longrightarrow$ (i) of Theorem 6.9 to construct a (different) unit bitolerance representation of order A.

Chapter 7
Comparability invariance results

Any transitive orientation of the edges of a comparability graph $G = (V, E)$ gives an ordered set $P = (V, \prec)$, and we say that G is the *comparability graph* of P. A graph can have many different transitive orientations, so there may be many different orders with the same comparability graph. In Figure 7.1, orders P, Q, and R (and their duals) all have the comparability graph G shown, and they represent all six transitive orientations of G. Determining the number of transitive orientations of a comparability graph was studied by Shevrin and Filippov (1970) and Golumbic (1977) (see also Section 5.3 of Golumbic, 1980).

Interval orders illustrate an interesting invariance property. *If G has a transitive orientation F which gives an interval order P, then every transitive orientation of G gives an interval order.* This can be seen as follows. Since P has an interval representation, this same representation demonstrates that G is an interval graph. Suppose F' is another transitive orientation of G whose ordered set P' is not an interval order. Then P' must contain a $2 + 2$ (Theorem 1.6) in which case G contains an induced C_4, a contradiction (Theorem 1.3).

In this chapter, we investigate a variety of order-theoretic properties and parameters which exhibit this kind of invariance. We present a standard technique for proving invariance based on a theorem of Gallai, and illustrate its use on the dimension of an order. We then turn our attention to tolerance properties.

7.1 Comparability invariance

A parameter of an ordered set is said to be a *comparability invariant* if all orders with a given comparability graph have the same value of that parameter. Likewise, a property of an ordered set is said to be a *comparability invariant* if either all orders with a given comparability graph have that property, or none have that property. For example, the property of having a unique maximal

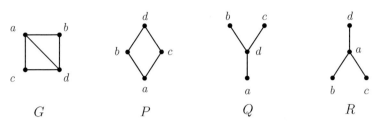

Figure 7.1. A graph G and three ordered sets each having comparability graph G.

element is *not* a comparability invariant, as seen by the example in Figure 7.1 where P and R have a unique maximal element, but Q does not. The parameter *dimension* is a well-known comparability invariant, that is, $\dim(P) = \dim(Q)$ whenever P and Q have the same comparability graph. We present this result in Theorem 7.7. More recently, Habib, Kelly, and Möhring (1991) showed that *interval dimension* is also a comparability invariant. As a consequence of this and Theorem 5.24, the property of being a bounded bitolerance order is a comparability invariant. We record this below.

Theorem 7.1. *Let P and Q be ordered sets with the same comparability graph. Then P is a bounded bitolerance order if and only if Q is a bounded bitolerance order.*

It is a good exercise to prove Theorem 7.1 by modifying the proof of Theorem 7.8 (Exercise 7.1).

Recall from Remark 5.13 that if G is a bounded bitolerance graph, then *there exists* a bounded bitolerance order P for which G is the incomparability graph of P. Once we know that membership in the class of bounded bitolerance orders is a comparability invariant (Theorem 7.1), we can make the stronger statement that *every* order P whose incomparability graph is G is a bounded bitolerance order.

We can make similar claims for other classes of bounded bitolerance graphs (discussed in Remarks 5.12 and 5.13) once we know membership in those classes is a comparability invariant. The property of a graph G belonging to the graph class $\mathcal{S}$ is shown to be a comparability invariant for $\mathcal{S} = \{$bounded tolerance orders$\}$, $\mathcal{S} = \{$unit bitolerance orders$\}$, and $\mathcal{S} = \{$unit tolerance orders$\}$ in Bogart, Isaak, Laison, and Trenk (2001) (our Theorems 7.8, 7.12, and 7.13). Other than these classes and bounded bitolerance orders (Theorem 7.1), there are no other classes of tolerance orders for which membership is known to be a comparability invariant, and no classes of tolerance orders for which membership is known *not* to be a comparability invariant.

We begin by presenting the standard technique for proving that an ordered set property is a comparability invariant, following Trotter (1992).

7.2 Autonomous sets and Gallai's Theorem

Given a graph $G = (V, E)$, a set $A \subseteq V$ is called an *autonomous set* if every vertex in $V \setminus A$ is either adjacent to all of the vertices in A or to none of the vertices in A. For the graph G in Figure 7.1, there are many autonomous sets, including $\{b, c, d\}, \{b, c\}, \{a, d\}, \{a, b, c, d\}$, and some that are not autonomous such as $\{b, d\}$. Autonomous sets play a key role in relating ordered sets that have the same comparability graph.

Let $P = (V, \prec_1)$ and $Q = (V, \prec_2)$ be ordered sets with the same comparability graph G. We say that Q is obtained from P by an *elementary reversal* if there is a set $A \subseteq V$ that is autonomous in G and satisfies the following.

(1) A is not a stable set of G.
(2) If x, y are not both in A, then $x \prec_1 y \iff x \prec_2 y$.
(3) If $x, y \in A$, then $x \prec_1 y \iff y \prec_2 x$.

In this process, Q is obtained from P (and vice versa) by reversing the comparabilities in A. For example, in Figure 7.1, Q is obtained from P by reversing the autonomous set $\{b, c, d\}$, and R is obtained from Q by first reversing the autonomous set $\{a, d\}$, and then reversing the autonomous set $\{a, b, c, d\}$.

Autonomous sets that participate in elementary reversals are called *order autonomous sets* in Kelly (1986). They satisfy an additional property which we record as Remark 7.2. We leave the proof of this remark as an exercise.

Remark 7.2. If $Q = (V, \prec_2)$ is obtained from $P = (V, \prec_1)$ by an elementary reversal using the order autonomous set A, then the sets $Pred(A) = \{v \in V \setminus A \mid v \prec_1 a \text{ for all } a \in A\}$, $Succ(A) = \{w \in V \setminus A \mid a \prec_1 w \text{ for all } a \in A\}$, and $Inc(A) = \{z \in V \setminus A \mid z \parallel a \text{ for all } a \in A\}$ partition $V \setminus A$.

By condition (2) in the definition of an elementary reversal, we could also use the relation $\prec_2$ of Q in defining the sets $Pred(A)$, $Succ(A)$, and $Inc(A)$ and get the same sets.

Note that the set $\{a, d\}$ is an autonomous set in graph G of Figure 7.1 but not an order autonomous set in the order P of Figure 7.1.

If A is an order autonomous set and $a \in A$ is incomparable to every other element of A, then $A' = A \setminus \{a\}$ is another order autonomous set. Furthermore, Q can be obtained from P by an elementary reversal of A if and only if Q can be obtained from P by an elementary reversal of A'. When all such elements are removed from an order autonomous set, the resulting set will not be empty by the first condition of our definition. We record this as a remark.

Remark 7.3. If one ordered set is obtained from another by an elementary reversal, this can be achieved using an order autonomous set A in which every element of A is comparable to another element of A.

The following theorem of Gallai (1967) (which appears in Trotter, 1992, pp. 61–62) shows that we can move between any two orders with the same comparability graph by a sequence of elementary reversals. It allows us to show a property is a comparability invariant by considering pairs of orders for which one can be obtained from the other by an elementary reversal.

Theorem 7.4 (*Gallai*). *Let $G = (V, E)$ be the comparability graph associated with distinct ordered sets $P = (V, \prec_P)$ and $Q = (V, \prec_Q)$. Then there exists a sequence of ordered sets $P_0, P_1, \ldots, P_m$ so that $P_0 = P$, $P_m = Q$ and P_{i+1} is obtained from P_i by an elementary reversal for $i = 0, 1, \ldots, m - 1$.*

A corollary of Theorem 7.4 which will be useful to us later in this chapter is given below.

Corollary 7.5. *Let P and Q be finite ordered sets with the same comparability graph and let S be a class of orders. To prove that $P \in S \iff Q \in S$, it suffices to prove $P \in S \Rightarrow Q \in S$ where Q can be obtained from P by an elementary reversal.*

Proof. By Theorem 7.4 we need only prove $P \in S \iff Q \in S$ in the case that Q can be obtained from P by an elementary reversal. However, since the process of obtaining one order from another by an elementary reversal is symmetric, the result follows. □

7.3 Dimension is a comparability invariant

In this section we show that the comparability invariance of dimension is a consequence of Gallai's Theorem, as pointed out in Trotter (1992).

If $P = (V, \prec)$ is an ordered set and A is an order autonomous set of P, then we can form a new order P_A by deleting all elements of A except for one. Formally, choose any $a \in A$ and let $P_A = ((V \setminus A) \cup \{a\}, \prec)$. By Remark 7.2, the order P_A does not depend on which representative of A is chosen.

The next lemma is illustrated in an example in Exercise 7.3.

Lemma 7.6. *If A is an order autonomous set of $P = (V, A)$, then $\dim(P) = \max(\dim(P_A), \dim(A, \prec))$.*

Proof. The inequality $\dim(P) \geq \max\{\dim(P_A), \dim(A, \prec)\}$ follows because P_A and $(A, \prec)$ are suborders of P, hence they each have dimension at most $\dim(P)$.

We now show the reverse inequality. Let $L_1, L_2, \ldots, L_t$ be a linear realizer of P_A and let $L'_1, L'_2, \ldots, L'_s$ be a linear realizer of $(A, \prec)$. If $t < s$ let $L_k = L_t$ for $k = t+1, t+2, \ldots, s$, and if $s < t$ let $L'_j = L'_s$ for $j = s+1, s+2, \ldots, t$. Now replace the element a in L_i by the linear order L'_i for each $i \leq \max\{s, t\}$. This gives a realizer of P with $\max\{s, t\}$ elements. $\qquad\square$

Theorem 7.7. *Dimension is a comparability invariant.*

Proof. By Theorem 7.4, it suffices to show that $\dim(P) = \dim(Q)$ whenever Q can be obtained from P by an elementary reversal. Let $Q = (V, \prec_2)$ be obtained from $P = (V, \prec_1)$ by an elementary reversal of autonomous set A. Condition (2) of the definition of an elementary reversal implies that $P_A = Q_A$. Condition (3) of the same definition implies that $(A, \prec_1)$ is the dual of $(A, \prec_2)$. Thus $\dim(A, \prec_1) = \dim(A, \prec_2)$. Now $\dim(P) = \dim(Q)$ follows from Lemma 7.6. $\qquad\square$

7.4 Bounded tolerance orders

In this section we prove the following result from Bogart, Isaak, Laison, and Trenk (2001), which is implied (but not stated) in the work of Habib, Kelly, and Möhring (1992).

Theorem 7.8. *Let P and Q be ordered sets with the same comparability graph. Then P is a bounded tolerance order if and only if Q is a bounded tolerance order.*

Recall from Lemma 5.20 that bounded tolerance orders are equivalent to parallelogram orders. We follow Bogart, Isaak, Laison, and Trenk (2001) and prove Theorem 7.8 using parallelogram representations. The proof uses the notion of scaling down some (but not all) of the parallelograms in a representation and is based on a construction in Habib, Kelly, and Möhring (1992).

Given a parallelogram representation $\{P_v \mid v \in V\}$ of an order P and a subset $W \subseteq V$, we can scale down the parallelograms in the set $\{P_w \mid w \in W\}$ as follows. Fix a sufficiently large number M and translate the parallelograms in $\{P_w \mid w \in W\}$ horizontally so that they fit between the lines $x = 0$ and $x = M$. To *scale down* the representation so that it fits between the lines $x = 0$ and $x = m$, map the point (a, b) of parallelogram P_w to the point $(\frac{am}{M}, b)$. Each parallelogram P'_w in the resulting set of parallelograms $\{P'_w \mid w \in W\}$

still has sides along L_1 and L_2 and can be translated horizontally to fit in any rectangle of width m with sides along L_1 and L_2. Note that for all w and y in W we have $P_w \ll P_y \iff P'_w \ll P'_y$. However, scaling down the parallelograms in W can change their comparability with other parallelograms representing members of $V \setminus W$, and this will be taken into consideration in our proof of Theorem 7.8.

We are now ready to present this proof.

Proof of Theorem 7.8. By Corollary 7.5, it suffices to prove the following. If P is a bounded tolerance order and Q can be obtained from P by an elementary reversal, then Q is a bounded tolerance order.

Using Lemma 5.20, we may fix a parallelogram representation of $P = (V, \prec)$. Recall that each parallelogram in the representation has one side along the horizontal line L_1 and the opposite along the parallel line L_2. Let Q be the order obtained from P by an elementary reversal using the order autonomous set A. Since A is not a stable set, there exist $x, y \in A$ with $x \prec y$. Therefore, in the parallelogram representation of P we have $P_x \ll P_y$.

Add the appropriate constant to each point on line L_1 so that parallelogram P_x becomes a rectangle (geometrically this is equivalent to moving the line L_1 to the left or right until P_x becomes a rectangle). This provides another parallelogram representation $\{P'_v \mid v \in V\}$ of P in which, lying strictly between P'_x and P'_y, there is a rectangular gap R of width $\epsilon > 0$.

As discussed above, we may scale down and translate the parallelogram representation of A in the horizontal direction so that it fits inside R (but each parallelogram still has sides on L_1 and L_2). Reflect these parallelograms about the vertical line bisecting R and denote by P''_v the new parallelogram assigned to $v \in V$. Let $P_1 = (V, \prec_1)$ be the ordered set with this parallelogram representation. Our goal is to show that $P_1 = Q$.

The reflection serves to reverse all the comparabilities between elements of A as desired. It remains to show that the reflection leaves all other comparabilities and incomparabilities of P intact. Since $P''_v = P'_v$ for all $v \in V \setminus A$, we need only consider pairs of elements where one element is in A and the other is not. By Remark 7.2, we know $V \setminus A = Pred(A) \cup Succ(A) \cup Inc(A)$.

For all $u \in Pred(A)$ we have $u \prec x$ so $P''_u = P'_u \ll P'_x \ll R$ and since the parallelograms representing elements of A are located inside R we have $u \prec_1 a$ for all $a \in A$. Similarly, for all $w \in Succ(A)$ we have $y \prec w$ so $R \ll P'_y \ll P'_w = P''_w$ and thus $a \prec_1 w$ for all $a \in A$. Finally, for all $z \in Inc(A)$, the parallelogram $P''_z = P'_z$ intersects both P'_x and P'_y and thus it intersects every line segment which lies entirely between P'_x and P'_y and has one endpoint on L_1 and the other on L_2. The left edge of P''_a is such a line segment for each $a \in A$.

Thus $P_z'' \cap P_a'' \neq \emptyset$ for all $a \in A$, so $z \parallel a$ in P_1 for all $a \in A$. Therefore, the new set of parallelograms gives a parallelogram representation of Q as desired. By Lemma 5.20, Q is a bounded tolerance order. □

7.5 Unit bitolerance and unit tolerance orders

In this section, we present results analogous to Theorem 7.8 for unit bitolerance orders and unit tolerance orders. We introduce the ideas common to both proofs first. As in the proof of Theorem 7.8, we will need the notion of scaling down. However, scaling down some (but not all) of the intervals in a unit bitolerance representation will not result in another unit bitolerance representation since this would violate the "unit" property. Instead we use a different representation of unit bitolerance orders which can be scaled down.

In Theorem 5.26 we proved that unit bitolerance orders are equivalent to point-core bitolerance orders. Recall from Definition 5.9 that a point-core bitolerance representation of $P = (V, \prec)$ consists of an assignment to each $v \in V$ an interval $I_v = [L(v), R(v)]$ and a splitting point $f(v)$ in the open interval $(L(v), R(v))$. We denote the representation by $\langle \mathcal{I}, f \rangle$ where $\mathcal{I} = \{I_v \mid v \in V\}$ and $\mathcal{F} = \{f(v) \mid v \in V\}$. In a point-core bitolerance representation of $P = (V, \prec)$ we have $x \prec y \iff R(x) < f(y)$ and $f(x) < L(y)$. This leads us to make the following remark which we will need in the proof of Theorem 7.12. The same idea will be used in the proofs of Theorem 10.3 and Theorem 13.38.

Remark 7.9 (Beads on a wire). If $P = (V, \prec)$ is a point-core bitolerance order with representation $\langle \mathcal{I}, f \rangle$, then the relation between two elements x and y is completely determined by the order of the points $L(x), f(x), R(x), L(y), f(y), R(y)$. This allows us to convert one point-core bitolerance representation of an order P into another by perturbing endpoints and splitting points, as long as we do not change the order of these points.

The name "beads on a wire", suggested to us by Kathryn Nyman, comes from visualizing the endpoints and splitting points as beads which can be slid freely along a wire as long as their order is preserved.

Given a point-core bitolerance representation $\langle \mathcal{I}, f \rangle$ of $P = (V, \prec)$ and a subset $W \subseteq V$, we may scale down the intervals and splitting points assigned to elements of W as follows. Translate the intervals in $\{I_w \mid w \in W\}$ and the splitting points $\{f(w) \mid w \in W\}$ horizontally so that they fit in the interval $[0, M]$ for a sufficiently large number M. To *scale down* the representation of elements of W so that it fits in $[0, m]$, map the intervals and splitting points by $a \to \frac{am}{M}$ so the interval $I_w = [a, b]$ maps to $I'_w = [\frac{am}{M}, \frac{bm}{M}]$ and the splitting point $f(w)$ maps to $\frac{f(w)m}{M}$. It is easy to check that these new intervals and splitting points give another point-core bitolerance representation of $(W, \prec)$. We may translate the new representation of W so that it fits in any interval of width m.

Similarly, by Theorem 6.1, unit tolerance orders are equivalent to 50% tolerance orders. Recall that the latter class is defined as bounded tolerance orders $P = (V, \prec)$ with a representation $\langle \mathcal{I}, t \rangle$ in which $t_v = \frac{1}{2}|I_v|$ for all $v \in V$. Thus a 50% tolerance representation is a point-core bitolerance representation in which the splitting point $f(v)$ lies at the center of the interval I_v for all $v \in V$. As in a point-core bitolerance representation, $x \prec y \iff f(x) < L(y)$ and $R(x) < f(y)$.

Intervals and splitting points in a 50% tolerance order can be scaled down in the same way we scaled down intervals and splitting points in a point-core bitolerance representation. If $f(w)$ lies at the center of interval $I_w = [a, b]$, then the new splitting point $f'(w) = \frac{f(w)m}{M}$ lies at the center of the new interval $I'_w = [\frac{am}{M}, \frac{bm}{M}]$, and thus the scaled down representation is still a 50% tolerance representation.

The following lemma appears in Bogart, Isaak, Laison, and Trenk (2001) and is based on ideas from Habib, Kelly, and Möhring (1992).

Lemma 7.10. *Let $P = (V, \prec)$ be a point-core bitolerance order with a representation in which element v is assigned interval $I_v = [L(v), R(v)]$ and splitting point $f(v)$. Let Q be obtained from P by an elementary reversal using the order autonomous set A. If there exist $x, y \in A$ with $x \prec y$ and $R(x) < L(y)$, then Q is a point-core bitolerance order. Moreover, if P is a 50% tolerance order, then so is Q.*

Proof. Let J be the interval $[R(x), L(y)]$, representing the gap between I_x and I_y. Each $v \in Pred(A)$ has $v \prec x$ and thus $R(v) < f(x) < R(x)$. Similarly, each $w \in Succ(A)$ has $y \prec w$ and thus $L(y) < f(y) < L(w)$. For each $z \in Inc(A)$ the interval I_z must intersect I_x and I_y, thus $L(z) \leq R(x)$ and $R(z) \geq L(y)$. This means that all intervals assigned to elements in $Pred(A)$ are completely to the left of J, all intervals assigned to elements in $Succ(A)$ are completely to the right of J, and all intervals assigned to elements in $Inc(A)$ completely contain J.

As discussed above, scale down those intervals representing elements of A so that the entire representation of A fits inside J. Next, reflect the intervals representing elements of A about the midpoint of J. The reflection serves to reverse all comparabilities in A while keeping the intervals assigned to elements in A entirely inside J. This has the desired effect of leaving all other comparabilities and incomparabilities in P intact. Thus the new set of intervals and splitting points provides a point-core bitolerance representation of Q. To justify the final sentence of the lemma, note that if the original representation of P was a 50% tolerance representation, then so is the final representation. ☐

Lemma 7.11. *Let $P = (V, \prec)$ be a point-core bitolerance order with a representation in which element $v \in V$ is assigned the interval $I_v = [L(v), R(v)]$ and the splitting point $f(v)$. Let Q be obtained from P by an elementary reversal of the autonomous set A. Further, suppose that $R(x) \geq L(y)$ for all $x, y \in A$ with $x \prec y$. Then*

  (i) *there exists an interval S with $S \subseteq I_a$ for all $a \in A$,*
  (ii) *for every $v \in Pred(A)$ the interval I_v is completely to the left of S,*
  (iii) *for every $w \in Succ(A)$ the interval I_w is completely to the right of S, and*
  (iv) *for every $z \in Inc(A)$ we have either $S \subseteq I_z$ or $f(z) \in S$.*

Proof. By Lemma 5.18, we may assume that the endpoints of the intervals in $\{I_v \mid v \in V\}$ are distinct. By the hypothesis that $R(x) \geq L(y)$ for all $x, y \in A$ with $x \prec y$, we have $I_u \cap I_v \neq \emptyset$ for any pair u, v of comparable elements in A. Also the intervals representing any pair of incomparable elements in A certainly have nonempty intersection. Thus $I_u \cap I_v \neq \emptyset$ for every pair $u, v \in A$. By the Helly property of intervals, there is a common intersection point for all the intervals assigned to elements of A. Since we have assumed interval endpoints are distinct, we know there exists an interval $S = [s_1, s_2]$ with $S \subseteq I_a$ for all $a \in A$. This establishes (i).

By Remark 7.3, we may assume that every element of A is comparable with another element of A. In particular, this means that for each $a \in A$, we have $f(a) \notin S$. If there were an element $a \in A$ for which $L(a) = s_1$ and $R(a) = s_2$ then $f(a) \in S$, contradicting our last assertion. By taking S to have maximum possible size, we may assume that there exist distinct $x, y \in A$ with $R(x) = s_2$ and $L(y) = s_1$. Furthermore, since $f(x) \notin S$, we have $f(x) < s_1 = L(y)$, and since $f(y) \notin S$, we have $f(y) > s_2 = R(x)$, so $x \prec y$. Every $v \in Pred(A)$ satisfies $v \prec x$ and thus $R(v) < f(x) < s_1$. So I_v is completely to the left of S for all $v \in Pred(A)$, proving (ii). Every $w \in Succ(A)$ satisfies $y \prec w$ and thus $s_2 < f(y) < L(w)$. So I_w is completely to the right of S for all $w \in Succ(A)$, proving (iii).

Finally we show (iv). Assume $z \in Inc(A)$ and $S \nsubseteq I_z$. We need to show $f(z) \in S$, so for a contradiction we first assume $f(z) < s_1$. In this case, $f(z) < s_1 = L(y)$ but since $z \parallel y$ we must have $R(z) > f(y)$. However,

$f(z) < s_1 < s_2 < f(y) < R(z)$, so $S \subseteq I_z$, contradicting our original assumption. We get a similar contradiction if we assume $f(z) > s_2$. Thus $f(z) \in S$ and this establishes (iv). $\qquad\square$

We now present the comparability invariance result for unit bitolerance orders from Bogart, Isaak, Laison, and Trenk (2001).

Theorem 7.12. *Let P and Q be finite ordered sets with the same comparability graph. Then P is a unit bitolerance order if and only if Q is a unit bitolerance order.*

Proof. By Corollary 7.5, it suffices to prove the following. If P is a unit bitolerance order and Q can be obtained from P by an elementary reversal, then Q is a unit bitolerance order.

Using Theorem 5.26, fix a point-core bitolerance representation of $P = (V, \prec)$ in which $v \in V$ is assigned the interval $I_v = [L(v), R(v)]$ and splitting point $f(v)$ with $L(v) < f(v) < R(v)$. By Lemma 5.18, we may assume that the endpoints of these intervals and the splitting points are distinct. Let $Q = (V, \prec')$ be the order which is obtained from P by an elementary reversal using the order autonomous set A. By Remark 7.2, the sets $Pred(A)$, $Succ(A)$, and $Inc(A)$ partition $V \setminus A$.

Case 1: There exist $x, y \in A$ with $x \prec y$ and $R(x) < L(y)$.

By Lemma 7.10, Q is a point-core bitolerance order, and therefore by Theorem 5.26, Q is a unit bitolerance order as desired.

Case 2: For all $x, y \in A$ with $x \prec y$ we have $R(x) \geq L(y)$.

In this case, Lemma 7.11 applies, so we know (i) there exists a real interval $S = [s_1, s_2]$ with $S \subseteq I_a$ for all $a \in A$, (ii) for every $v \in Pred(A)$ the interval I_v is completely to the left of S, (iii) for every $w \in Succ(A)$ the interval I_w is completely to the right of S, and (iv) for every $z \in Inc(A)$ we have either $S \subseteq I_z$ or $f(z) \in S$.

Now choose a point $h \in S$ which is different from all splitting points in the representation of P. Reflect each interval assigned to an element of A about h and denote the resulting interval for a by $I'_a = [L'(a), R'(a)]$ and the new splitting point by $f'(a)$. Since $S \subseteq I_a$ we have $L'(a) < h < R'(a)$ for each $a \in A$. The reflection serves to reverse all comparabilities in A. However, this reflection may affect other comparabilities between elements in A and elements in $V \setminus A$, and so we will make a further adjustment to the intervals in A. Our goal is to create new intervals $\{I''_a \mid a \in A\}$ so that each I''_a contains S and is contained in an interval slightly larger than S.

Choose $\epsilon > 0$ sufficiently small so that there are no endpoints of intervals or center points within ϵ of s_1 and s_2. Thus the intervals I_v which contain S also contain the larger interval $[s_1 - \epsilon, s_2 + \epsilon]$.

Consider the set $\{L'(a) \mid a \in A\} \cup \{f'(a) \mid a \in A\} \cup \{R'(a) \mid a \in A\}$ and create a new representation by sliding these points so that (i) the ordering of these points is maintained, (ii) any point less than h ends up in the interval $[s_1 - \epsilon, s_1]$, and (iii) any point greater than h ends up in the interval $[s_2, s_2 + \epsilon]$. By (i) and Remark 7.5, this will not disturb the comparabilities among elements in A. The new interval assigned to $a \in A$, denoted by I_a'', will contain S for each $a \in A$, and is contained in the slightly larger interval $[s_1 - \epsilon, s_2 + \epsilon]$.

Now for each $v \in Pred(A)$ and each $a \in A$ we have $I_v \ll I_a''$, thus $v \prec a$ in the new representation. For each $w \in Succ(A)$ we have $I_a'' \ll I_w$ for each $a \in A$, thus $a \prec w$ in the new representation. For each $z \in Inc(A)$ we have either $I_a'' \subseteq [s_1 - \epsilon, s_2 + \epsilon] \subseteq I_z$ or $f(z) \in S \subseteq I_a''$ for each $a \in A$. In either case, $z \parallel a$ in the new representation. So the original intervals and splitting points for elements in $V \setminus A$ together with the new intervals and splitting points for elements of A gives a point-core bitolerance representation of Q. By Theorem 5.26, Q is a unit bitolerance order. $\square$

Finally we present the comparability invariance result for unit tolerance orders from Bogart, Isaak, Laison, and Trenk (2001).

Theorem 7.13. *Let P and Q be finite ordered sets with the same comparability graph. Then P is a unit tolerance order if and only if Q is a unit tolerance order.*

Proof. By Corollary 7.5, it suffices to prove the following: if P is a unit tolerance order and Q can be obtained from P by an elementary reversal, then Q is a unit tolerance order.

We proceed by induction. The theorem is easy to check for orders with three or fewer elements. Assume the result is true for orders with fewer than n elements, and let $P = (V, \prec)$ be a unit tolerance order with $|V| = n$. Let Q be the order which is obtained from P by an elementary reversal using the order autonomous set A.

Using Theorem 6.1, fix a 50% tolerance representation of P in which $v \in V$ is assigned interval $I_v = [L(v), R(v)]$ with splitting point $f(v) = \frac{1}{2}(L(v) + R(v))$ and tolerance $t_v = f(v) - L(v) = R(v) - f(v) = \frac{1}{2}|I_v|$. By Lemma 5.18, we may assume that the endpoints of these intervals are distinct.

By the definition of a 50% tolerance representation, $x \prec y \iff f(x) < L(y)$ and $R(x) < f(y)$, and by Remark 7.2, the sets $Pred(A)$, $Succ(A)$, and $Inc(A)$ partition $V \setminus A$.

Case 1: There exist x, $y \in A$ with $x \prec y$ and $R(x) < L(y)$.

By Lemma 7.10, Q is a 50% tolerance order, and therefore a unit tolerance order by Theorem 6.1.

Case 2: For all x, $y \in A$ with $x \prec y$ we have $R(x) \geq L(y)$.

In this case, Lemma 7.11 applies, so we know (i) there exists a real interval $S = [s_1, s_2]$ with $S \subseteq I_a$ for all $a \in A$, (ii) for every $v \in Pred(A)$ the interval I_v is completely to the left of S, (iii) for every $w \in Succ(A)$ the interval I_w is completely to the right of S, and (iv) for every $z \in Inc(A)$ we have either $S \subseteq I_z$ or $f(z) \in S$.

Partition $Inc(A)$ as $I_C(A) \cup I_N(A)$ where $I_C(A) = \{z \in Inc(A) \mid S \subseteq I_z\}$ is the set of elements incomparable to A whose intervals "cover" S and $I_N(A) = \{u \in Inc(A) \mid S \nsubseteq I_u\}$ is the set of elements incomparable to A whose intervals do *not* "cover" S. By condition (iv) we know $f(u) \in S$ for all $u \in I_N(A)$. If $I_N(A) = \emptyset$ then the argument in the second paragraph of the proof of Lemma 7.10 applies and we conclude that Q is a unit tolerance order.

Otherwise $I_N(A) \neq \emptyset$. As noted above, $f(u) \in S$, that is, $s_1 \leq f(u) \leq s_2$ for all $u \in I_N(A)$. Here our proof diverges from that of Theorem 7.12 since we can not slide the splitting points $f(a)$ without disturbing the property that they lie in the centers of their respective intervals.

Claim: The set $A \cup I_N(A)$ is an order autonomous set of P.

Proof of Claim. To prove the claim it suffices to show that for any $u \in I_N(A)$ and any $v \in V \setminus (A \cup I_N(A))$, the relation between u and v is the same as the relation between a and v for any $a \in A$. Thus, for $z \in I_C(A)$, $v \in Pred(A)$ and $w \in Succ(A)$, we will show

(a) $u \parallel z$,
(b) $v \prec u$,
(c) $u \prec w$.

Fix elements $u \in I_N(A)$, $z \in I_C(A)$, $v \in Pred(A)$, $w \in Succ(A)$, and in addition fix elements x, $y \in A$ with $x \prec y$. To prove (a) we note that $f(u) \in S \subseteq I_z$, so $u \parallel z$.

We next prove (b). By the definition of $S = [s_1, s_2]$ we know $L(x)$, $L(y) \leq s_1$ and $s_2 \leq R(x)$, $R(y)$ (see Figure 7.2). Since $x \prec y$ we have $f(x) < L(y) \leq s_1$ and $s_2 \leq R(x) < f(y)$. Also, since $v \in Pred(A)$ and $x \in A$, we have $v \prec x$ so

$$R(v) < f(x) < L(y) \leq s_1 < f(u). \tag{7.1}$$

We wish to show $f(v) < L(u)$, which together with $R(v) < f(u)$ from (7.1) would imply $v \prec u$ and prove (b). Suppose for a contradiction that $L(u) \leq$

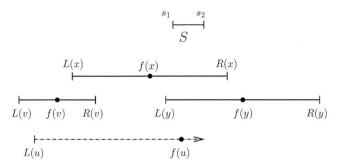

Figure 7.2. A figure to accompany the proof of part (b) of the claim.

$f(v)$ (as shown by the dashed line of I_u in Figure 7.2). Since $v \prec x$ we have $f(v) < L(x)$ so $L(u) < L(x)$. By (7.1) we have $f(x) < f(u)$. Because we have a 50% tolerance representation, the splitting points $f(x)$ and $f(u)$ lie at the centers of their respective intervals, so $\frac{1}{2}(L(x) + R(x)) = f(x) < f(u) = \frac{1}{2}(L(u) + R(u)) < \frac{1}{2}(L(x) + R(u))$, and thus $R(u) > R(x)$. But then $L(u) < L(y) \le s_1$ and $s_2 \le R(x) < R(u)$, which means $S \subset I_u$, contradicting the fact that $u \in I_N(A)$. This completes the proof of (b). A similar argument shows (c) and finishes the proof of the claim.

Case 2a: $V \neq A \cup I_N(A)$

In this case, the order P_1, induced in P by the elements in $A \cup I_N(A)$ is a 50% tolerance order with fewer than n elements. Furthermore, A is an order autonomous set in P_1. By the induction hypothesis we may fix a 50% tolerance representation of the order resulting from P_1 by reversing all comparabilities in A. As discussed at the beginning of this section, scale down and translate this representation so that it fits entirely in S and place it there. This representation captures the comparabilities and incomparabilities between elements of $A \cup I_N(A)$ in Q. We leave the intervals representing elements of $V \setminus (A \cup I_N(A))$ intact, so our representation gives the correct order relations between elements of $V \setminus (A \cup I_N(A))$ in Q. By (ii), (iii) of Lemma 7.11, the claim, and the definition of $I_C(A)$, our representation also realizes the comparabilities and incomparabilities between elements of $A \cup I_N(A)$ and elements of $V \setminus (A \cup I_N(A))$ in Q. Therefore, Q is a 50% tolerance order and by Theorem 6.1, Q is a unit tolerance order.

Case 2b: $V = A \cup I_N(A)$

In this case, reflect each interval in A about the midpoint of S in order to reverse the comparabilities in A. Let I'_a be the new interval assigned to $a \in A$. We leave the intervals representing elements in $V \setminus A = I_N(A)$ intact, so our representation correctly realizes the order relation between elements of $I_N(A)$ in Q. It remains to consider the order relations between an element of A and an element of $I_N(A)$. Recall that $S \subseteq I_a$ for all $a \in A$ by condition (i) of case 2. Since I'_a results from reflecting I_a about the midpoint of S, we also have $S \subseteq I'_a$ for all $a \in A$. By condition (iv) of case 2, we have $f(u) \in S$ for all $u \in I_N(A)$, thus $u \parallel a$ for all $a \in A$ as desired. Thus the new representation is a 50% tolerance representation of Q when $V = A \cup I_N(A)$. By Theorem 6.1, Q is a unit tolerance order. $\qquad\square$

7.6 Exercises

Exercise 7.1. Prove Theorem 7.1 by modifying the proof of Theorem 7.8.

Exercise 7.2. Prove Remark 7.2.

Exercise 7.3. This exercise refers to the order P shown in Figure 7.3 and the order autonomous set $A = \{u, v, w, x\}$ of P.

(a) Find a minimum size linear realizer of P_A.
(b) Find a minimum size linear realizer of $(A, \prec)$.
(c) Use these to find a minimum size linear realizer of P as in the proof of Lemma 7.6.

Exercise 7.4. Let Q be the order obtained from the order P in Figure 7.3 by an elementary reversal of the autonomous set $A = \{u, v, w, x\}$.

Draw Q, then find a minimum size linear realizer of Q using the result of Exercise 7.3.

Exercise 7.5. Give a point-core bitolerance representation of the order P in Figure 7.3 that satisfies the hypothesis of Lemma 7.10 for the autonomous set

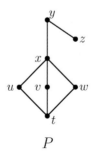

P

Figure 7.3. The order P used in several of the exercises.

$A = \{u, v, w, x\}$. Find an interval J as in the proof of Lemma 7.10, and verify that all intervals assigned to elements in $Pred(A)$ are completely to the left of J, all intervals assigned to elements in $Succ(A)$ are completely to the right of J, and all intervals assigned to elements in $Inc(A)$ completely contain J.

Exercise 7.6. Give a point-core bitolerance representation of the order P in Figure 7.3 that satisfies the hypothesis of Lemma 7.11 for the autonomous set $A = \{u, v, w, x\}$. Verify that the conclusion of Lemma 7.11 holds in this case.

Chapter 8

Recognition of bounded bitolerance orders and trapezoid graphs

In this chapter we present the theoretical basis for recognizing orders of interval dimension at most 2, which we have seen are equivalent to trapezoid orders and bounded bitolerance orders (Theorem 5.24). A polynomial time recognition algorithm for trapezoid orders also provides a method for recognizing trapezoid graphs in polynomial time, as described in the following remark.

Remark 8.1 (Trapezoid graph recognition). If a graph G is a trapezoid graph, its complement $\overline{G}$ must be a cocomparability graph. So first, find a transitive orientation F of the complement $\overline{G}$. If none exists, then G is not a trapezoid graph. Otherwise, test to see if the order given by F is a trapezoid order. By Theorems 5.24 and 7.1, being a trapezoid order is a comparability invariant. Hence, G is a trapezoid graph if and only if F exists and gives a trapezoid order.

Three groups of authors solved the recognition problem for trapezoid orders independently. The fastest algorithm is due to Ma and Spinrad (1994) with a running time of $O(n^2)$, where n is the number of elements in the order. This approach reduces the interval dimension 2 question to a problem of covering an associated bipartite graph by chain graphs (i.e., graphs with no induced $2K_2$).

We will follow the approaches of Felsner, Habib, and Möhring (1994) and Langley (1995) which are purely order theoretic and involve an interesting auxiliary order $B(P)$. First we outline this approach. Given an order $P = (X, \prec)$ we will construct an order $B(P)$ (denoted $\mathcal{PS}$ in Langley, 1995) consisting of at most $2n$ subsets of X, ordered by set inclusion, where $n = |X|$. The goal is to show $\mathrm{idim}(P) = \dim(B(P))$ and then use the fact that recognizing orders of dimension 2 can be accomplished in time $O(n^2)$ (Spinrad, 1985). The inequality $\mathrm{idim}(P) \leq \dim(B(P))$ is shown in Proposition 8.4 by converting a linear realizer of $B(P)$ into an interval realizer of P of the same size.

The reverse inequality $\mathrm{idim}(P) \geq \dim(B(P))$ is more complicated, and here we follow the approach of Felsner, Habib, and Möhring (1994). We let

$k = \mathrm{idim}(P)$ and fix an interval realizer of P of size k. We construct an order $\mathcal{B}(\mathcal{I}^*)$ based on a geometric representation of P in which each element $x \in X$ is assigned a box in $\mathbf{R}^k$. It is easy to show that $\dim(\mathcal{B}(\mathcal{I}^*)) \leq k$ (Proposition 8.6). The proof is completed by proving the isomorphism $B(P) \cong \mathcal{B}(\mathcal{I}^*)$ (Theorem 8.7).

8.1 Preliminaries

8.1.1 Dimension and realizers

Recall from Chapter 1 that a collection $\{L_1, L_2, \ldots, L_t\}$ of linear extensions of poset P is a *realizer* of P if $P = L_1 \cap L_2 \cap \cdots \cap L_t$. The *dimension* of P (denoted $\dim(P)$) is the minimum positive integer t for which P has a t-element linear realizer.

A poset $P = (X, \prec)$ is said to be *embedded* in $\mathbf{R}^t$ if each $x \in X$ can be assigned a t-tuple $(x_1, x_2, \ldots, x_t)$ such that $x \prec y$ if and only if $x_i \leq y_i$ for each i and $\exists j$ such that $x_j < y_j$. The term *dimension* comes from embedding an order P in $\mathbf{R}^n$ for sufficiently large n. For completeness, we include the following theorem of Dushnik and Miller and its proof.

Theorem 8.2. *(Dushnik and Miller, 1941) Let $P = (X, \prec)$ be an ordered set. Then $\dim(P)$ is the smallest t for which P can be embedded in $\mathbf{R}^t$.*

Proof. Given a realizer $\{L_1, L_2, \ldots, L_t\}$ of $P = (X, \prec)$, we can embed P in $\mathbf{R}^t$ by assigning to $x \in X$ the t-tuple $(x_1, x_2, \ldots, x_t)$ where x_i is the height of x in L_i. Then $x \prec y$ if and only if $x_i < y_i$ for each i, by definition of a realizer.

Conversely, suppose P is embedded in $\mathbf{R}^t$ and $x \in X$ is assigned the t-tuple $(x_1, x_2, \ldots, x_t)$. We assume there are no identical t-tuples in the embedding (although the proof can easily be modified so this assumption is not needed). Let L_i be the linear extension of P in which $x \prec_i y$ if and only if either $x_i < y_i$ or $[x_i = y_i$ and $x_j < y_j]$, where j is the lowest coordinate in which the tuples differ. That is, L_i is obtained by projecting the t-tuples onto the ith coordinate axis (and breaking ties according to the first place where their coordinates differ). Then $\{L_1, L_2, \ldots, L_t\}$ is a realizer of P. $\square$

8.1.2 Predecessor and successor sets

Fix an order $P = (X, \prec)$ and let $n = |X|$. For $x \in X$, the *predecessor set* of x in P is $Pred_P(x) = \{z \in X \mid z \prec x\}$ and the *successor set* of x in P is $Succ_P(x) = \{y \in X \mid x \prec y\}$. Since we will only refer to predecessors and successors in P,

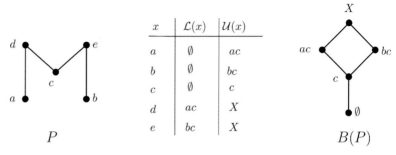

x	$\mathcal{L}(x)$	$\mathcal{U}(x)$
a	$\emptyset$	ac
b	$\emptyset$	bc
c	$\emptyset$	c
d	ac	X
e	bc	X

Figure 8.1. An order P and the resulting order $B(P)$.

we abbreviate $Pred_P(x)$ by $Pred(x)$ and $Succ_P(x)$ by $Succ(x)$. In this section, it will be useful for us to consider open interval representations of interval orders.

We now define $B(P)$. For each $x \in X$, let $\mathcal{L}(x) = Pred(x)$ and $\mathcal{U}(x) = \bigcap_{z \in Succ(x)} Pred(z)$ which we take to be all of X if $Succ(x) = \emptyset$. The order $B(P)$ consists of the set $\mathcal{M} = \{\mathcal{L}(x), \mathcal{U}(x) \mid x \in X\}$ ordered by set inclusion. Figure 8.1 gives an example of an order P, a chart showing the values of $\mathcal{L}(x), \mathcal{U}(x)$ and the resulting order $B(P) = (\mathcal{M}, \subset)$.

Lemma 8.3. *Let $P = (X, \prec)$ be an order, and $\mathcal{L}(x), \mathcal{U}(x)$ be defined as above. Then for all $x, y \in X$ we have*

(i) $x \in \mathcal{U}(x)$,
(ii) $\mathcal{L}(x) \subset \mathcal{U}(x)$,
(iii) *if $x \prec y$ then $\mathcal{U}(x) \subseteq \mathcal{L}(y)$.*

Proof. (i) For all $z \in Succ(x)$ we have $x \in Pred(z)$ so $x \in \mathcal{U}(x)$.

(ii) If $y \in \mathcal{L}(x)$ then $y \prec x$. Thus for all $z \in Succ(x)$, we have $y \in Pred(z)$ because $y \prec x \prec z$. Thus $y \in \bigcap_{z \in Succ(x)} Pred(z) = \mathcal{U}(x)$.

(iii) If $x \prec y$ then $y \in Succ(x)$ so $\mathcal{U}(x) = \bigcap_{z \in Succ(x)} Pred(z) \subseteq Pred(y) = \mathcal{L}(y)$. □

Proposition 8.4. *If P is an ordered set, then* $\mathrm{idim}(P) \leq \dim(B(P))$.

Proof. First we show how to convert a linear extension L of $B(P)$ into an interval extension I_L of P. We complete the proof by showing that if $L_1, L_2, \ldots, L_k$ is a linear realizer of $B(P) = (\mathcal{M}, \subset)$, then $I_{L_1}, I_{L_2} \ldots, I_{L_k}$ is an interval realizer of $P = (X, \prec)$.

Let $L_t = M_1 \prec M_2 \prec \cdots \prec M_r$ be a linear extension of $B(P)$ where $\mathcal{M} = \{M_1, M_2, \ldots, M_r\}$. To each element $x \in X$, assign the open interval $I_x^t = (i, j)$ where $M_i = \mathcal{L}(x)$ and $M_j = \mathcal{U}(x)$. These intervals are nontrivial since $\mathcal{L}(x) \subseteq \mathcal{U}(x)$ (by Lemma 8.3) and $\mathcal{L}(x) \neq \mathcal{U}(x)$ because $x \in \mathcal{U}(x)$ but

$x \notin \mathcal{L}(x)$. The interval order I_{L_t}, represented by the set of intervals $\{I_x^t \mid x \in X\}$, is an extension of P since by Lemma 8.3, $x \prec y$ implies $\mathcal{U}(x) \subseteq \mathcal{L}(y)$ and thus $I_x^t \ll I_y^t$.

Now suppose $L_1, L_2, \ldots, L_k$ is a linear realizer of $B(P)$. Since $I_{L_1}, I_{L_2}, \ldots, I_{L_k}$ are interval extensions of P, it suffices to show that each incomparable pair in P is incomparable in $I_{L_1} \cap I_{L_2} \cap \cdots \cap I_{L_k}$. Suppose $x \parallel y$ in P. Then $x \in \mathcal{U}(x)$ but $x \notin \mathcal{L}(y)$, so $\mathcal{U}(x) \nsubseteq \mathcal{L}(y)$. Since $L_1, L_2, \ldots, L_k$ is a realizer of $B(P)$, there exists a j for which $\mathcal{L}(y) \prec \mathcal{U}(x)$ in L_j, and thus $x \not\prec y$ in I_{L_j}. A symmetric argument shows that there is an ℓ for which $y \not\prec x$ in I_{L_ℓ}, thus $x \parallel y$ in $I_{L_1} \cap I_{L_2} \cap \cdots \cap I_{L_k}$ as desired. $\qquad\square$

Example 8.5. The order $B(P)$ in Figure 8.1 has dimension 2 with linear realizer $L_1 : \emptyset \subseteq c \subseteq ac \subseteq bc \subseteq X$ and $L_2 : \emptyset \subseteq c \subseteq bc \subseteq ac \subseteq X$. Applying the construction used in the proof of Proposition 8.4, we get the interval realizer $\{I_{L_1}, I_{L_2}\}$ of P, with the following assignments.

$$I_a^1 = (1,3) \quad I_b^1 = (1,4) \quad I_c^1 = (1,2) \quad I_d^1 = (3,5) \quad I_e^1 = (4,5)$$
$$I_a^2 = (1,4) \quad I_b^2 = (1,3) \quad I_c^2 = (1,2) \quad I_d^2 = (4,5) \quad I_e^2 = (3,5)$$

In Figure 8.2 we use a different interval realizer of P to better illustrate the process of P-normalization which is discussed in the next section.

8.2 The order $\mathcal{B}(\mathcal{I})$ of extreme corners

Let $P = (X, \prec)$ be an order with $n = |X|$ and fix an interval realizer $\mathcal{I} = \{I_1, I_2, \ldots, I_k\}$ of P. Furthermore, fix an open interval representation of each I_j where (a_x^j, b_x^j) is the interval assigned to x in I_j. Each $x \in X$ has the open box $\prod_{j=1}^{k} (a_x^j, b_x^j)$ in $\mathbf{R}^k$ associated with it and this collection of boxes is called a *box embedding* of P in $\mathbf{R}^k$. The box associated with x is completely determined by its *extreme lower corner* $\ell_x = (a_x^1, a_x^2, \ldots, a_x^k)$ and its *extreme upper corner* $u_x = (b_x^1, b_x^2, \ldots, b_x^k)$. We write $u_x \le \ell_y$ to mean that each component of u_x is at most as large as the corresponding component of ℓ_y. Since $\mathcal{I}$ is an interval realizer of P, we have $x \prec y$ in $P \iff x \prec y$ in each $I_j \iff b_x^j \le a_y^j$ for all $j \iff u_x \le \ell_y$.

Note that different interval realizers of P, and even different representations of these interval orders, may yield different box embeddings of P. Given a box embedding of P, we can recover the representations of an interval realizer of P by projecting downward in each coordinate direction. Figure 8.2 shows the order P from Figure 8.1 together with an interval order realizer of P and the resulting box embedding of P. In the box embedding, circles are used to mark the extreme lower corners and extreme upper corners.

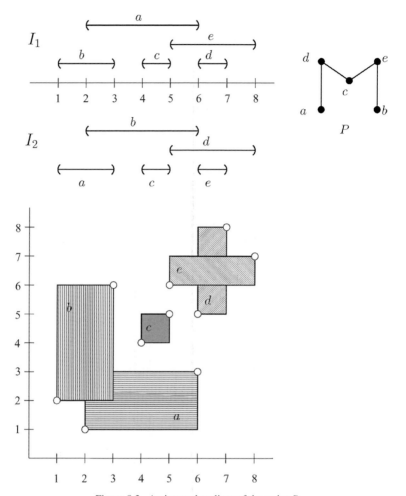

Figure 8.2. An interval realizer of the order P.

In addition to providing a geometric representation of P, a box embedding of $P = (X, \prec)$ also gives rise to a second ordered set. The order $\mathcal{B}(\mathcal{I})$ consists of the set $\{\ell_x, u_x \mid x \in X\}$ of *extreme corners* ordered componentwise, that is, one element is below another in $\mathcal{B}(\mathcal{I})$ precisely when each component of the first is less than or equal to the corresponding component of the second. The order of extreme corners $\mathcal{B}(\mathcal{I})$ has at most $2n$ elements, as does $B(P)$. Figure 8.3 shows the order $\mathcal{B}(\mathcal{I})$ associated with the box embedding of P shown in Figure 8.2.

Clearly, the order $\mathcal{B}(\mathcal{I})$ in Figure 8.3 is not isomorphic to the order $B(P)$ of Figure 8.1. To achieve an isomorphism between $B(P)$ and an order of extreme corners, we must modify the interval representations of the interval orders in $\mathcal{I}$.

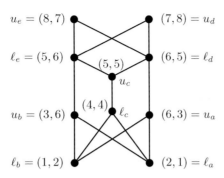

Figure 8.3. The order of extreme corners $\mathcal{B}(\mathcal{I})$.

Given an interval representation of $I_j \in \mathcal{I}$ in which element x is assigned open interval (a_x^j, b_x^j), we normalize it to obtain the P-normalization as follows.

First decrease left endpoints as much as possible while still respecting the comparabilities in P and the outer limits of the original representation. Specifically, let

$$a_x^{j*} = \begin{cases} \max\{b_z^j \mid z \in Pred(x) \text{ in } P\} & \text{if } x \text{ is not minimal in } P \\ \min\{a_z^j \mid z \in X\} & \text{if } x \text{ is minimal in } P. \end{cases}$$

Then increase right endpoints as much as possible while still respecting the comparabilities in P and the outer limits of the original representation. Thus, let

$$b_x^{j*} = \begin{cases} \min\{a_z^{j*} \mid z \in Succ(x) \text{ in } P\} & \text{if } x \text{ is not maximal in } P \\ \max\{b_z^j \mid z \in X\} & \text{if } x \text{ is maximal in } P. \end{cases}$$

Let I^* be the resulting P-normalization of I and $\mathcal{I}^* = \{I_1^*, I_2^*, \ldots, I_k^*\}$. Figure 8.4 shows the box embedding resulting from the P-normalization of each interval order in Figure 8.2. By construction, the order I_j^* may be different from I_j. In our example, $b \prec c$ in I_1 but $b \parallel c$ in I_1^*. In general, any comparability in P will persist in I_j^*, and no new comparabilities are introduced in the transformation $I_j \to I_j^*$. Thus the new set $\mathcal{I}^*$ will also be an interval realizer of P.

Figure 8.5 shows the order of extreme corners of $\mathcal{B}(\mathcal{I}^*)$. Note that it is isomorphic to $B(P)$ via the isomorphism $\ell_x^* \to \mathcal{L}(x)$, $u_x^* \to \mathcal{U}(x)$. In the next section, we show this isomorphism is present in general. We conclude this section with an easy result about $\dim(\mathcal{B}(\mathcal{I}^*))$.

Proposition 8.6. *Let P be an ordered set and $k = \mathrm{idim}(P)$. If $\mathcal{I} = \{I_1, I_2, \ldots, I_k\}$ is an interval realizer of P, and $\mathcal{I}^* = \{I_1^*, I_2^*, \ldots, I_k^*\}$ is its P-normalization, then $\dim(\mathcal{B}(\mathcal{I}^*)) \le k = \mathrm{idim}(P)$.*

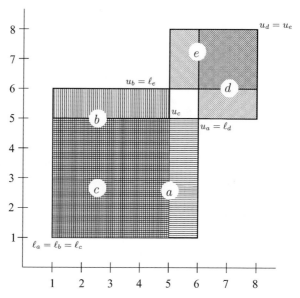

Figure 8.4. The box embedding of P corresponding to $\mathcal{I}^*$.

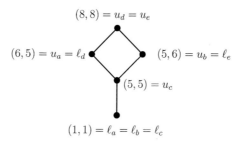

Figure 8.5. The order of extreme corners $\mathcal{B}(\mathcal{I}^*)$.

Proof. The box embedding of P corresponding to the interval representations of $I_1^*, I_2^*, \ldots, I_k^*$ can also be viewed as an embedding of $\mathcal{B}(\mathcal{I}^*)$ in $\mathbf{R}^k$. By Theorem 8.2, $\dim(\mathcal{B}(\mathcal{I}^*)) \leq k$. $\square$

8.3 The isomorphism between $B(P)$ and $\mathcal{B}(\mathcal{I}^*)$

The remaining step in proving $\mathrm{idim}(P) = \dim(B(P))$ is showing the isomorphism between $B(P)$ and $\mathcal{B}(\mathcal{I}^*)$ which we do in the following theorem.

Theorem 8.7. *Let* $P = (X, \prec)$ *be an ordered set. Let* $\mathcal{I} = \{I_1, I_2, \ldots, I_k\}$ *be any interval realizer of P, and* $\mathcal{I}^* = \{I_1^*, I_2^*, \ldots, I_k^*\}$ *its P-normalization. Then* $B(P) \cong \mathcal{B}(\mathcal{I}^*)$.

Proof. Recall that $B(P)$ is the set $\{\mathcal{L}(x), \mathcal{U}(x) \mid x \in X\}$ ordered by set inclusion. $\mathcal{B}(\mathcal{I}^*)$ is the set $\{\ell_x^*, u_x^* \mid x \in X\}$ of extreme corners of the box embedding of P associated with $\mathcal{I}^*$, ordered componentwise. Thus $\ell_x^* = (a_x^{1*}, a_x^{2*}, \ldots, a_x^{k*})$ and $u_x^* = (b_x^{1*}, b_x^{2*}, \ldots, b_x^{k*})$ where (a_x^{j*}, b_x^{j*}) is the interval assigned to x in I_j^*.

The following four statements, which we will prove below, show that the mapping $\mathcal{L}(x) \to \ell_x^*, \mathcal{U}(x) \to u_x^*$ is an isomorphism from $B(P)$ to $\mathcal{B}(\mathcal{I}^*)$.

(a) $\mathcal{U}(x) \subseteq \mathcal{L}(y) \iff u_x^* \leq \ell_y^*$.
(b) $\mathcal{L}(x) \subseteq \mathcal{L}(y) \iff \ell_x^* \leq \ell_y^*$.
(c) $\mathcal{U}(x) \subseteq \mathcal{U}(y) \iff u_x^* \leq u_y^*$.
(d) $\mathcal{L}(x) \subseteq \mathcal{U}(y) \iff \ell_x^* \leq u_y^*$.

(a) Suppose $\mathcal{U}(x) \subseteq \mathcal{L}(y)$. Since $x \in \mathcal{U}(x)$ we have $x \in \mathcal{L}(y)$ and thus $x \prec y$ in P. Since $\mathcal{I}^*$ is an interval realizer of P, we know $x \prec y$ in each I_j^*. This means $b_x^{j*} \leq a_y^{j*}$ for all j and thus $u_x^* \leq \ell_y^*$.

Conversely, suppose $u_x^* \leq \ell_y^*$. Using the box embedding representation of P associated with $\mathcal{I}^*$ we conclude $x \prec y$ in P. Since $y \in Succ(x)$ we get $\mathcal{U}(x) = \bigcap_{z \in Succ(x)} Pred(z) \subseteq Pred(y) = \mathcal{L}(y)$.

(b) Suppose $\mathcal{L}(x) \subseteq \mathcal{L}(y)$, thus $Pred(x) \subseteq Pred(y)$. If x is minimal in P, then by definition, $a_x^{j*} = \min\{a_z^j \mid z \text{ in } X\} \leq a_y^{j*}$ for all j. Otherwise, $a_x^{j*} = \max\{b_z^j \mid z \in Pred(x)\} \leq \max\{b_z^j \mid z \in Pred(y)\} = a_y^{j*}$ for all j. Thus $\ell_x^* \leq \ell_y^*$.

Conversely, suppose $\ell_x^* \leq \ell_y^*$, that is, $a_x^{j*} \leq a_y^{j*}$ for all j. For any $z \in Pred(x)$ we have $z \prec x$ in each $\mathcal{I}_j^*$ and thus $b_z^{j*} \leq a_x^{j*} \leq a_y^{j*}$ for each j. Hence $z \prec y$ in each I_j^* and therefore $z \prec y$ in P. This proves $Pred(x) \subseteq Pred(y)$.

(c) Suppose $\mathcal{U}(x) \subseteq \mathcal{U}(y)$. First we show $Succ(y) \subseteq Succ(x)$. Let $z \in Succ(y)$. Since $x \in \mathcal{U}(x) \subseteq \mathcal{U}(y)$ we know $x \in Pred(z)$ and thus $z \in Succ(x)$ as desired. Now the proof is analogous to (b).

Conversely, assume $u_x^* \leq u_y^*$. Using an argument analogous to part (b) we obtain $Succ(y) \subseteq Succ(x)$. Thus $\mathcal{U}(x) = \bigcap_{z \in Succ(x)} Pred(z) \subseteq \bigcap_{z \in Succ(y)} Pred(z) = \mathcal{U}(y)$ as desired.

(d) Suppose $\mathcal{L}(x) \subseteq \mathcal{U}(y)$. If x is minimal in P then $a_x^{j*} = \min\{a_z^j \mid z \in X\} \leq b_y^{j*}$ for each j. If y is maximal in P then $b_y^{j*} = \max\{b_z^j \mid z \in X\} \geq a_x^{j*}$ for each j. Otherwise, there exists $v \in Pred(x)$ and $w \in Succ(y)$ with $a_x^{j*} = b_v^j$ and $b_y^{j*} = a_w^{j*}$. Since $v \in \mathcal{L}(x) \subseteq \mathcal{U}(y) = \bigcap_{z \in Succ(y)} Pred(z)$, we know $v \in$

$Pred(w)$, that is, $v \prec w$ in P. Therefore, $v \prec w$ in each $\mathcal{I}_j^*$ and thus $a_x^{j*} = b_v^j \leq b_v^{j*} \leq a_w^{j*} = b_y^{j*}$ for each j. This means $\ell_x^* \leq u_y^*$, as desired.

Conversely, suppose $\ell_x^* \leq u_y^*$. Then $a_x^{j*} \leq b_y^{j*}$ for all j. For any $z \in Succ(y)$ we have $a_x^{j*} \leq b_y^{j*} \leq a_z^{j*}$ for each j. Let $w \in \mathcal{L}(x) = Pred(x)$. Since $\mathcal{I}^*$ is an interval realizer of P, $b_w^{j*} \leq a_x^{j*}$ for all j. Similarly, for any $z \in Succ(y)$ we have $b_y^{j*} \leq a_z^{j*}$. Thus $b_w^{j*} \leq a_x^{j*} \leq b_y^{j*} \leq a_z^{j*}$ for all j. Since $w \prec z$ for all $w \in \mathcal{L}(x)$ and all $z \in Succ(y)$ we get $w \in \bigcap_{z \in Succ(y)} Pred(z) = \mathcal{U}(y)$, thus $\mathcal{L}(x) \subseteq \mathcal{U}(y)$. This completes the proof of the theorem. □

Now we combine the results of Theorem 8.7 with Propositions 8.4 and 8.6 to conclude that we may determine the interval dimension of an order P by computing the dimension of $B(P)$. Thus, we obtain the main result of this section.

Theorem 8.8. *For any ordered set P we have* $\mathrm{idim}(P) = \dim(B(P))$.

Proof. From Theorem 8.7 we know that if P is an ordered set, $\mathcal{I}$ is an interval realizer of P, and $\mathcal{I}^*$ is its P-normalization, then $B(P) \cong B(\mathcal{I}^*)$ and thus $\dim(B(P)) = \dim(B(\mathcal{I}^*))$. Using the inequalities from Propositions 8.4 and 8.6 we get, $\mathrm{idim}(P) \leq \dim(B(P)) = \dim(B(\mathcal{I}^*)) \leq \mathrm{idim}(P)$. So equality holds throughout and $\mathrm{idim}(P) = \dim(B(P))$. □

8.4 The recognition algorithm and its complexity

The question of determining whether an order P is a bounded bitolerance order (or equivalently, a trapezoid order) reduces to determining if $\mathrm{idim}(P) \leq 2$ by Theorem 5.24. By Theorem 8.8, the problem reduces further to determining if $\dim(B(P)) \leq 2$. This method consists of two parts: first compute $Q = B(P)$ from P, and second determine whether $\dim(Q) \leq 2$. Note that if P has n elements then Q has at most $2n$ elements. The calculation of $Q = B(P)$ is straightforward in time $O(n^3)$ (Langley, 1995) and with more care can be accomplished in time $O(n^\alpha)$ where this denotes the time complexity for matrix multiplication, (Felsner, Habib, and Möhring, 1994).

The problem of determining whether $\dim(Q) \leq k$ is NP-complete in general (Yannakakis, 1982). However, orders of dimension at most 2 are characterized in Dushnik and Miller (1941) as those orders Q for which the incomparability graph of Q is transitively orientable. This result leads to efficient algorithms to determine if $\dim(Q) \leq 2$ which, in the affirmative case, produce a linear realizer of Q of size two (see Golumbic, 1980). The complexity of these algorithms varies from $O(n^3)$ for the straightforward algorithms based on transitive

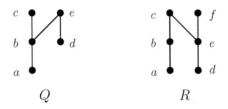

Figure 8.6. The orders Q and R used in the exercises.

orientation to $O(n + e)$ for the fastest algorithms based on recognizing permutation graphs (McConnell and Spinrad, 1997, 1999).

Using these methods, the straightforward approach for recognizing bounded bitolerance orders has time complexity $O(n^3)$ and the faster approaches $O(n^\alpha)$. The proof of Proposition 8.4 gives a construction for achieving an interval realizer of P given a linear realizer of $B(P) = Q$. Thus, we have an efficient algorithm for recognizing orders P of interval dimension at most 2 and, in the affirmative case, producing two interval orders whose intersection is P. The proofs of Lemma 5.21 and Theorem 5.24 give constructions for producing a trapezoid representation and a bounded bitolerance representation of P in this case.

The fastest known algorithm for determining whether an order has interval dimension at most 2 is due to Ma and Spinrad (1994) with a running time of $O(n^2)$. Therefore, *the class of bounded bitolerance orders (or equivalently trapezoid orders) can be recognized in time $O(n^2)$.*

By Remark 8.1, trapezoid graphs can be recognized within these same complexity bounds.

8.5 Exercises

Exercise 8.1. Give an embedding in $\mathbf{R}^2$ of the order shown in Figure 7.3.

Exercise 8.2. Consider the following embedding of a six element order P in $\mathbf{R}^3$: $(7, 4, 7)$, $(1, 4, 5)$, $(3, 5, 1)$, $(1, 8, 6)$, $(3, 3, 3)$, $(2, 4, 3)$. Use the proof of Theorem 8.2 to construct a linear realizer of P of size 3.

Exercise 8.3. This exercise refers to the order $Q = (X, \prec)$ in Figure 8.6.
   (a) Make a chart showing $\mathcal{L}(x)$ and $\mathcal{U}(x)$ for each $x \in X$.
   (b) Draw the Hasse diagram for $B(Q)$.
   (c) Verify that $\mathrm{idim}(Q) = \dim(B(Q))$.
   (d) Use the construction in Proposition 8.4 to find a minimum size interval realizer of Q.

Exercise 8.4. This exercise refers to the order $R = (X, \prec)$ in Figure 8.6.

(a) Repeat parts (a) – (d) of Exercise 8.3 for the order R.

(b) Find $\mathcal{I}$, the R-normalization of the interval representation of R in part (a).

(c) Draw the box embedding corresponding to $\mathcal{I}$ from part (b) (analogous to Figure 8.4).

(d) Give the Hasse diagram for $\mathcal{B}(\mathcal{I}^*)$ (analogous to Figure 8.5) and verify that it is isomorphic to $B(R)$.

Exercise 8.5. Using the constructions provided in this chapter, implement the algorithm to test whether an order P is a trapezoid order, and in the affirmative case, produce a trapezoid representation of it.

Chapter 9

Algorithms on tolerance graphs

Interval relations play a significant role in many resource allocation, temporal reasoning, biological and scheduling problems. We saw this in Sections 1.1 and 4.1 in our motivating examples for interval graphs, tolerance graphs and interval probe graphs. Intervals can represent events in time, which may conflict or may be compatible. They can represent certain tasks to be performed according to a timetable which must be assigned distinct processors or people. Or they may represent fragments of DNA, which are compatible or incompatible.

For many optimization problems, such as graph coloring or finding maximum stable sets, there are efficient algorithms that give solutions when the set of graphs under consideration is restricted to a structured family. Many applications reduce to solving optimization problems on such families of graphs. Indeed, at the very beginning of this book, a 4-coloring of the tolerance graph in Figure 1.3 provided an assignment of four meeting rooms for that motivating example. In a similar application, with say only *one* room available for a given collection of meetings (intervals) with tolerances, a maximum stable set would provide the largest number of meetings from the collection that can be scheduled. In this chapter, we investigate these algorithmic aspects of tolerance graphs.

Narasimhan and Manber (1992) were the first to study the chromatic number, clique and stable set problems for representations of tolerance graphs. Their approach was to first apply known algorithms for cocomparability graphs on the *bounded tolerance subgraph* G_B, induced by the vertices B with bounded tolerance, and then add into the solution the remaining vertices U with unbounded tolerance. It is important to point out that the algorithms in Narasimhan and Manber (1992) find a maximum clique and a maximum stable set, given the tolerance representation, and hence give the chromatic number and the clique cover number since tolerance graphs are perfect. However, this method does *not* supply a coloring for the graph nor a covering by cliques. In Section 9.2, we describe their method for finding cliques, followed by a recent algorithm in Golumbic and Siani (2002) which does give a coloring and has lower time complexity. We give an algorithm for finding a maximum weight stable set in Section 9.3. The complexity analysis presented here takes into account more recent results which improve the efficiency of the algorithms.

9.1 Tolerance and bounded tolerance representations

As we pointed out at the beginning of Chapter 3, the complexity of the recognition problem for the class of tolerance graphs is yet unsolved, and it is not even known how to obtain a tolerance representation when the input graph is known to be a tolerance graph. Recently, Hayward and Shamir (2002) have shown that the recognition of tolerance is in NP.

For these reasons, in this chapter, it must be assumed that a tolerance representation is given as part of the input. Therefore, we make a distinction between coloring a given representation of a tolerance graph, which we are able to do efficiently (Section 9.2), versus coloring tolerance graphs without a representation, which no one is yet able to do efficiently. The same is true for finding a maximum stable set.

Similarly, the recognition problem for bounded tolerance graphs is open and again it is not known how to obtain a bounded tolerance representation when the input graph is known to be a bounded tolerance graph. Despite this difficulty, we can find an optimal coloring and a maximum stable set for bounded tolerance graphs without knowing a representation. We use the fact that bounded tolerance graphs are equivalent to parallelogram graphs (Theorem 2.9) and hence are trapezoid graphs. If G is a bounded tolerance graph (even if we do not know that it is, nor can we test whether it is), then it will nevertheless pass the polynomial time trapezoid graph recognition test of Chapter 8. Therefore, in this case, all the algorithms of Felsner, Müller, and Wernisch (1997) for trapezoid graphs can be applied successfully. This method applies equally to trapezoid graphs which are not bounded tolerance graphs. If G fails to be a trapezoid graph, then it is not a bounded tolerance graph. A similar argument can be made using the fact that bounded tolerance graphs are cocomparability graphs, and applying algorithms for that class.

We return to the more complicated case of tolerance graphs. Determining the chromatic number $\chi(G)$ is computationally equivalent to finding the clique number $\omega(G)$, since tolerance graphs are perfect (Theorem 2.28) and $\chi(G) = \omega(G)$. Thus, we also make a distinction between producing an optimal coloring versus simply finding the chromatic number $\chi(G)$ of a tolerance graph G. The same remarks hold for the clique cover number $\kappa(G)$ and the stability number $\alpha(G)$ which are equal in tolerance graphs.

9.2 Coloring tolerance representations

Let $\langle \mathcal{I}, t \rangle$ be a tolerance representation of $G = (V, E)$, and let $V = B \cup U$ be the partition of the vertices according to those having bounded or unbounded tolerance in this representation. Observe that at most one vertex with unbounded tolerance can participate in any clique, since all the neighbors of such a vertex $u \in U$ must have bounded tolerance, i.e., $\mathcal{N}(u) \subseteq B$. Thus, $\omega(G)$ must be equal to either $\omega(G_B)$ or $\omega(G_B) + 1$, corresponding to the cases where a maximum clique of G is either a maximum clique of G_B or consists of an unbounded $u \in U$ together with a maximum clique of G_B. This can be summarized by the equation,

$$\omega(G) = \max\{\omega(G_B), \omega(G_{\mathcal{N}(u)}) + 1 \mid u \in U\}. \tag{9.1}$$

Note that each of the induced subgraphs used in equation (9.1) is a bounded tolerance graph. Using this fact, Narasimhan and Manber (1992) applied an $O(n^3)$ algorithm for the clique number of a cocomparability graph (Golumbic, 1980), once for each of the $q + 1$ bounded tolerance induced subgraphs in (9.1), to obtain $\omega(G)$, where $n = |V|$ and $q = |U|$. The original complexity stated in Narasimhan and Manber (1992) for the clique problem was, therefore, $O(qn^3)$. However, after Langley (1993) subsequently proved that bounded tolerance graphs are parallelogram graphs (our Theorem 2.9), and Felsner *et al.* (1997) subsequently obtained an $O(n \log n)$ algorithm for producing a maximum clique of a trapezoid graph, the complexity of this method drops to $O(qn \log n)$. We will show that this can be reduced further to $O(qn + n \log n)$.

As pointed out earlier, the method of Narasimhan and Manber (1992) does not provide an optimal coloring of the tolerance graph, but rather the size of such a coloring. In the case where $\omega(G) = \omega(G_B) + 1$, we can color G_B with $\omega(G_B) = \chi(G_B)$ colors, using the $O(n \log n)$ algorithm of Felsner, Müller, and Wernisch (1997), adding one additional color for all the vertices in U, and obtain an optimal coloring. However, when $\omega(G) = \omega(G_B)$ we cannot always extend a coloring of G_B to all of G. Golumbic and Siani (2002) solve this problem by moving some of the vertices from U to B, and reducing some of the tolerances. We now present their coloring algorithm for tolerance graphs.

The algorithm sweeps across the tolerance representation from left to right (*reduction*) acting on each interval I_x having unbounded tolerance by either

(a) lowering its tolerance from t_x down to $|I_x|$ provided that this does not add an edge to the graph, or (b) finding a witness I_z for I_x testifying that the tolerance t_x cannot be reduced to the length of its interval. A second sweep (*coloring*) colors the intervals.

By Lemma 5.18, we may assume that the endpoints of the intervals are distinct, although it is easy to modify the algorithm to drop this assumption. During a sweep across the representation, we maintain the set of *active* intervals, i.e., the intervals whose left endpoint has been scanned but whose right endpoint has not.

Let I_x be an interval with unbounded tolerance $t_x > |I_x|$. An interval I_z is called a *hovering witness* for I_x if $I_x \subset I_z$ and $|I_x| < t_z \le |I_z|$. Thus, $xz \notin E$, since $|I_x \cap I_z| = |I_x| < \min\{t_x, t_z\}$, and z has bounded tolerance, but reducing the tolerance of x from t_x to $|I_x|$ would create a new edge between x and z in the tolerance graph. We say that I_x is *inevitably infinite* in a given tolerance representation if it has at least one hovering witness.

Example 9.1. In the tolerance representation of G on the left side of Figure 9.1, the interval I_d has a hovering witness, namely I_c. Notice that there is no hovering witness for the interval I_b. Therefore, I_d is inevitably infinite but I_b is not.

Algorithm 9.2. Coloring a tolerance representation

> *Input*: A tolerance representation $\langle \mathcal{I}, t \rangle$ for graph $G = (V, E)$.
> *Output*: A minimum coloring of the intervals.
> *Method*: The algorithm is given in Figure 9.2. In the Reduction phase, when an interval I_x of infinite tolerance is encountered, either its tolerance is lowered to the length of the interval (if no new edge would be created) or an arbitrary

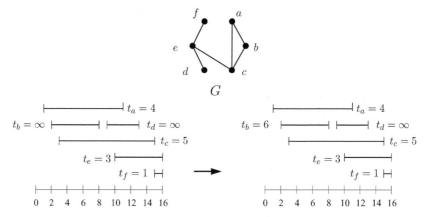

Figure 9.1. A graph G, a tolerance representation of G, and the transformed tolerance representation of G following the reduction phase of Algorithm 9.2.

Reduction:

Sweeping over the endpoints of the intervals from left to right **do**

Let p be the current endpoint, and let I_x be its interval.

if p is a left endpoint **then do**

add I_x to the *active* list

if $t_x > |I_x|$ **and** $\exists I_z \in active$ such that $|I_x| < t_z \le |I_z|$ and $R(x) \le R(z)$

then

$hovering\_witness(I_x) := I_z$

else

$t_x := |I_x|$

end

else remove I_x from the *active* list

end

Coloring:

Color the bounded intervals using the algorithm of Felsner, Muller, and Wernisch (1997).

Assign each inevitably infinite interval the same color as the hovering witness assigned to it.

Figure 9.2. The algorithm for coloring a tolerance representation.

hovering witness is assigned to I_x showing that I_x is inevitably infinite. Following the full sweep of Reduction, the intervals are colored in the Coloring phase. First the (original and newly) bounded intervals are colored, followed by the (now inevitably infinite) unbounded intervals.

Proposition 9.3. *Algorithm 9.2 assigns an optimal coloring to a tolerance graph G, for any given tolerance representation. The complexity of the algorithm is $O(qn + n\log n)$.*

Proof. During the Reduction phase of the algorithm, the unbounded tolerance of a vertex is reduced to the length of its interval if and only if doing so leaves the tolerance graph unchanged. Therefore, the newly reduced representation is a tolerance representation of the same tolerance graph G. Moreover, every interval that remains unbounded has a hovering witness assigned to it.

Let $V = B' \cup U'$ be the partition of the vertices which are bounded and inevitably unbounded in the reduced representation. During the Coloring phase, the algorithm of Felsner, Müller, and Wernisch (1997) for trapezoid graphs gives an optimal coloring of $G_{B'}$ using $\chi(G_{B'})$ colors. We will show that, when extended to the unbounded vertices, the coloring is valid for all of G.

Let $x \in U'$ and let I_z be the hovering witness assigned to I_x, so our algorithm assigns $color(x) = color(z)$. Suppose there is an edge $xy \in E(G)$ where $color(y) = color(x)$. Since the unbounded vertices form a stable set, we have $y \in B'$. On the one hand, y and z have the same color in $G_{B'}$, so $yz \notin E(G)$.

Thus, $|I_y \cap I_z| < \min\{t_y, t_z\} \le t_y$. On the other hand, since $I_x \subseteq I_z$ we have $I_y \cap I_x \subseteq I_y \cap I_z$, from which it follows (x being unbounded and adjacent to y) that $t_y = \min\{t_x, t_y\} \le |I_x \cap I_y| \le |I_y \cap I_z|$, a contradiction. Thus, our coloring of G is valid.

Finally, since no new colors are added to color the unbounded vertices, it follows that our algorithm gives an optimal coloring of G.

The complexity of the Reduction phase is $O(qn)$ since each time one of the q unbounded intervals is encountered, all intervals on the *active* list must be compared with it. The Coloring phase requires $O(n \log n)$ for the bounded part (Felsner, Müller, and Wernisch, 1997), and is completed to the remaining vertices in linear time. □

9.3 Maximum weight stable set of a tolerance representation

In this section, we present an algorithm for finding a maximum weight stable set in a tolerance graph G, given a tolerance representation for G. It is based on the $O(n^2 \log n)$ algorithm by Narasimhan and Manber (1992) for the maximum (cardinality) stable set problem in a tolerance graph. As discussed earlier, an $O(n \log n)$ algorithm for the maximum weight stable set problem for trapezoid graphs, and hence for bounded tolerance graphs, is given in Felsner, Müller, and Wernisch (1997).

Let $\langle \mathcal{I}, t \rangle$ be a tolerance representation for G, and let $w(v) > 0$ be the *weight* of vertex $v \in V(G)$. The weight of a stable set $S \subseteq V(G)$ is the sum $w(S) = \sum_{v \in S} w(v)$. For convenience, we assume all interval endpoints are distinct (Lemma 2.3). We augment the representation by adding two dummy vertices s and t whose intervals I_s and I_t will be disjoint from all the other intervals. Position I_s to the left of all other intervals, and I_t to the right of all other intervals. We assign tolerances and weights as follows: $t_s = |I_s|$, $t_t = |I_t|$ and $w(s) = w(t) = 0$.

What does the set of intervals corresponding to a stable set of G look like in the tolerance representation? It will have some bounded intervals, none of which may contain another, and some unbounded intervals, which may be contained in any other interval but which may not contain any of the bounded members. First we focus our attention on the bounded vertices B.

A transitive orientation of the comparability graph $\overline{G_B}$ can be obtained by directing each edge according to right endpoints (see Exercise 2.3). Formally, let $P = (B \cup \{s, t\}, \prec)$ be the order defined on the bounded vertices B and the two new vertices s and t, where $s \prec x \prec t$ for all $x \in B$, and $x \prec z \Leftrightarrow (xz \notin E(G_B))$ and $(R(x) < R(z))$ for all $x, z \in B$. It is easy to see the following.

Remark 9.4. The stable sets of G_B are in one-to-one correspondence with the chains in P from s to t.

We now define for each comparable pair $x \prec x'$, where $x, x' \in B \cup \{s, t\}$, a set of unbounded vertices $S(x, x')$ which can be "squeezed" into the space between them. Formally,

$$S(x, x') = \{u \in U \mid xu, x'u \notin E(G) \text{ and } R(x) < R(u) < R(x')\}.$$

This definition also includes the special case of $S(s, t) = U$ as well as the sets $S(s, z)$ and $S(z, t)$, for $z \in B$. We show in Lemma 9.6 that $S(x, x')$ is precisely the set of unbounded vertices that could be added to any chain in which x is immediately followed by x'.

Example 9.5. Consider the collection of intervals $I_i = [2i, 2i + 9]$ for $i = 1, \ldots, 10$ with tolerances as follows:

i	1	2	3	4	5	6	7	8	9	10
t_i	5	8	∞	∞	2	6	3	∞	∞	7

Figure 9.3 shows the tolerance graph G, the Hasse diagram P and the sets $S(i, j)$. Notice that no vertices can be added to the chain $s \prec 1 \prec 5 \prec 10 \prec t$, since $S(s, 1) = S(1, 5) = S(5, 10) = S(10, t) = \emptyset$, and thus gives the stable set $\{1, 5, 10\}$. In contrast to this, the shorter chain $s \prec 1 \prec 10 \prec t$ allows $S(1, 10) = \{4, 8\}$ to be inserted, and thus gives the larger stable set $\{1, 4, 8, 10\}$, but even this set is not maximal. However, it is easy to check that $S = \{1, 4, 6, 8, 10\}$ is a maximal stable set in this graph. The bounded vertices in S are $S \cap B = \{1, 6, 10\}$. Note that the unbounded vertices in S are $S \cap U = \{4, 8\}$, which can be obtained by taking the union: $S(s, 1) \cup S(1, 6) \cup S(6, 10) \cup S(10, t) = \{4, 8\}$. The next lemma proves this is true for maximal stable sets in general.

Lemma 9.6. *Let S be a maximal stable set of the tolerance graph G, and let $S \cap B = \{x_1 \prec x_2 \prec \cdots \prec x_k\}$. Assigning $x_0 = s$ and $x_{k+1} = t$, we have $S \cap U = \cup\{S(x_i, x_{i+1}) \mid i = 0, 1, \ldots, k\}$.*

Proof. Suppose S is a maximal stable set of G, i.e., it is not contained in a larger stable set. If $y \in S \cap U$, we have $x_j y \notin E(G)$ for all $0 \leq j \leq k + 1$. Thus, there is some i satisfying $R(x_i) < R(y) < R(x_{i+1})$, and $y \in S(x_i, x_{i+1})$.

Conversely, if $y \in S(x_i, x_{i+1})$ for some i, then $y \in U$ and $R(x_i) < R(y) < R(x_{i+1})$. Furthermore, I_{x_i} and $I_{x_{i+1}}$ are not contained in I_y because $x_i y, x_{i+1} y \notin E(G)$ by definition of $S(x_i, x_{i+1})$, and so $L(x_i) < L(y) < L(x_{i+1})$.

Suppose y is not in S then, since S is a maximal stable set, there must be some $x_j y \in E(G)$ where $j < i$ or $j > i + 1$. In the first case, $R(x_j) < R(x_i) < R(y)$ so $t_{x_j} \leq |I_{x_j} \cap I_y| < |I_{x_j} \cap I_{x_i}| < t_{x_j}$ which is a contradiction. In the second

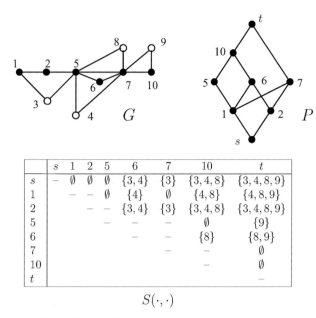

	s	1	2	5	6	7	10	t
s	–	∅	∅	∅	{3, 4}	{3}	{3, 4, 8}	{3, 4, 8, 9}
1		–	–	∅	{4}	∅	{4, 8}	{4, 8, 9}
2			–	–	{3, 4}	{3}	{3, 4, 8}	{3, 4, 8, 9}
5				–	–	–	∅	{9}
6					–	–	{8}	{8, 9}
7						–	–	∅
10							–	∅
t								–

$$S(\cdot, \cdot)$$

Figure 9.3. An illustration for the stable set algorithm.

case, $R(y) < R(x_{i+1}) < R(x_j)$ which implies that $L(x_{i+1}) < L(x_j)$ since $x_{i+1}, x_j \in B$, so $t_{x_j} \le |I_{x_j} \cap I_y| < |I_{x_j} \cap I_{x_{i+1}}| < t_{x_j}$, again a contradiction. $\qquad \square$

Corollary 9.7. *Given a set of positive weights for the vertices of a tolerance graph G, any maximum weight stable set of G consists of a chain of vertices $x_1 \prec x_2 \prec \cdots \prec x_k$ of bounded tolerance together with the union of the sets $\cup \{S(x_i, x_{i+1}) \mid i = 0, 1, \ldots, k\}$, where $x_0 = s$ and $x_{k+1} = t$.*

Proof. The corollary follows from Lemma 9.6 since a maximum weight stable set is a maximal stable set whenever the weights are positive. $\qquad \square$

We are now ready to present the main algorithm of this section.

Algorithm 9.8. Maximum weight stable set of a tolerance representation

Input: A tolerance representation $\langle \mathcal{I}, t \rangle$ for graph $G = (V, E)$, and a weight $w(v) > 0$ for each $v \in V$.

Output: A maximum weight stable set.

Method: Construct the order P using a transitive orientation of the comparability graph $\overline{G_B}$. We use a modification of the standard height calculation technique for any acyclic digraph, which we apply to the order P. With each vertex $z \in B \cup \{s, t\}$, associate a cumulative weight $W(z)$ defined as follows:

$$W(z) = w(z) + \max_{x \prec z}\{W(x) + w(S(x, z))\} \tag{9.2}$$

where the maximum is taken over all $x \in B \cup \{s\}$ less than z (and not just those covered by z.)

Calculating the cumulative weights $W(z)$ is usually implemented bottom up, starting with s and concluding with t, using recursive depth-first search. At each stage, a pointer $p(z)$ is set to record the vertex x which gave the maximum value in equation (9.2). In this way, we can recover the chain in P which achieves $W(t)$, and which leads to a maximum weight stable set in G, a fact that we prove in Theorem 9.10 below.

We demonstrate the algorithm with two different sets of weights.

Example 9.9. (a) Suppose every vertex v in V has weight $w(v) = 1$ and $w(s) = w(t) = 0$. Then a maximum weight stable set is a maximum cardinality stable set. For the graph in Figure 9.3, the algorithm assigns

$$\begin{aligned}
W(s) &= w(s) = 0, \\
W(1) &= W(2) = 1, \\
W(5) &= 2 = w(5) + W(1) + w(\emptyset), \\
W(6) &= 4 = w(6) + W(2) + w(\{3, 4\}), \\
W(7) &= 3 = w(7) + W(2) + w(\{3\}), \\
W(10) &= 5 = w(10) + W(6) + w(\emptyset), \\
W(t) &= 6 = w(t) + W(6) + w(\{8, 9\}).
\end{aligned}$$

This corresponds to the stable set $\{2, 3, 4, 6, 8, 9\}$.

(b) Suppose we are given the weights $w(i)$ below.

i	s	1	2	3	4	5	6	7	8	9	10	t
$w(i)$	0	4	2	1	3	5	2	3	5	6	4	0

The cumulative weights $W(i)$ are obtained for the graph in Figure 9.3 as follows:

$$\begin{aligned}
W(s) &= w(s) = 0, \\
W(1) &= 4, \\
W(2) &= 2, \\
W(5) &= w(5) + W(1) + w(\emptyset) = 5 + 4 = 9, \\
W(6) &= w(6) + W(1) + w(\{4\}) = 2 + 4 + 3 = 9, \\
W(7) &= w(7) + W(1) + w(\emptyset) = 3 + 4 + 0 = 7, \\
W(10) &= w(10) + W(6) + w(\{8\}) = 4 + 9 + 5 = 18,
\end{aligned}$$

and finally, noting that

$$W(s) + w(\{3, 4, 8, 9\}) = 0 + 15 = 15 \text{ corresponding to } \{3, 4, 8, 9\},$$
$$W(1) + w(\{4, 8, 9\}) = 4 + 14 = 18 \text{ corresponding to } \{1, 4, 8, 9\},$$
$$W(2) + w(\{3, 4, 8, 9\}) = 2 + 15 = 17, \text{ corresponding to } \{2, 3, 4, 8, 9\},$$
$$W(5) + w(\{9\}) = 9 + 6 = 15 \text{ corresponding to } \{1, 5, 9\},$$
$$W(6) + w(\{8, 9\}) = 9 + 11 = 20 \text{ corresponding to } \{1, 4, 6, 8, 9\},$$
$$W(7) + w(\emptyset) = 7 + 0 = 7 \text{ corresponding to } \{1, 7\},$$
$$W(10) + w(\emptyset) = 18 + 0 = 18 \text{ corresponding to } \{1, 4, 6, 8, 10\},$$

we take the maximum and obtain $W(t) = w(t) + 20 = 20$ giving the maximum weight stable set $\{1, 4, 6, 8, 9\}$.

Theorem 9.10. *Let* $s = x_0 \prec x_1 \prec x_2 \prec \cdots \prec x_k \prec x_{k+1} = t$ *be the chain produced by Algorithm 9.8. Then* $S = \{x_1, x_2, \ldots, x_k\} \cup \{S(x_i, x_{i+1}) \mid i = 0, 1, \ldots, k\}$ *is a maximum weight stable set of G.*

Proof. By Corollary 9.7, a maximum weight stable set of G corresponds to a maximum weight path from s to t in P where the weight of the path is the sum of the vertex and edge weights. (The weight of an edge (x, y) in P is $w(S(x, y))$.) By induction on the height of an element in P, the equation (9.2) insures that $W(z)$ is the weight of the heaviest path from s to z. Applying this to $W(t)$, the algorithm finds a maximum weight stable set. □

Lastly, we analyze the complexity of Algorithm 9.8. Building the transitive orientation F of G_B to obtain the order P can be done in time $O(|B|^2)$ using the right endpoints of the intervals. Calculating all the sets $S(x, y)$ and their weights $w(S(x, y))$ can easily be accomplished in $O(|B|^2|U|)$ time. This time bound can be improved to $O(|B|^2 \log |U|)$ as stated for the case of equal weights in Narasimhan and Manber (1992) and in general by using techniques borrowed from computational geometry (Moshe Lewenstein, personal communication). Finally, determining each $W(x)$ and maintaining pointers to the maximum child can be done in $O(|B|^2)$ time. Therefore, the overall complexity is $O(n^2 \log(n))$.

9.4 Exercises

Exercise 9.1. Given the tolerance representations in Exercise 1.1(b) and (c), apply the algorithms in this chapter to obtain a minimum coloring, maximum clique, and maximum stable set for each of the tolerance graphs.

Exercise 9.2. Read the paper by Felsner, Müller, and Wernisch (1997) and implement their algorithms for minimum coloring, maximum clique, maximum stable set and minimum clique cover on trapezoid graphs.

Exercise 9.3. Apply the Reduction phase of Algorithm 9.2 to the tolerance representation in Example 9.5 to obtain a bounded tolerance representation for the same graph. (Note the similarity with Proposition 2.29.) Draw the Hasse diagram for P' now that all the vertices have bounded tolerance in the new representation, and find a maximum weight chain in P' for each of the weight functions in Example 9.9 (a) and (b). Do they correspond to the same maximum weight stable sets as in Example 9.9? Why?

Chapter 10

The hierarchy of classes of bounded bitolerance orders

In this chapter we consider the classes of bounded bitolerance orders arranged in a hierarchy in Figure 10.1. We begin by describing the notation and conventions used in Figure 10.1 and justifying the inclusions and equivalences in the hierarchy. In Section 10.3, we restrict attention to bipartite orders. In that setting, the hierarchy collapses and most of the classes are equivalent. Section 10.4 provides the details to show that any example that appears along an edge between two classes provides a separating example between those two classes.

10.1 Introduction

In Section 5.2, we defined subclasses of bounded bitolerance orders by adding restrictions on interval lengths, tolerant points $p(v)$ and $q(v)$, and left and right tolerances. These restrictions are summarized in Table 10.1. The restrictions are listed so that the top entry of each column is the most restrictive, and they are less restrictive as you travel down the column.

Each of the three categories of restrictions is independent. Thus, the restrictions can be combined, by taking one from each column, to give 18 classes of bitolerance orders, some of which turn out to be equivalent. In this chapter we often refer to a class by its abbreviation, for example, (1aii) is the class of unit point-core bitolerance orders, and (3ci) is the class of (bounded) tolerance orders.

Bogart and Trenk (1994) consider nine of the classes (in the order listed below) which arise from allowing one restriction from the first two categories and one restriction from the third: (1ci), (3ai), (1cii), (2ci), (2cii), (3bi), (3bii), (3ci), and (3cii). Fishburn and Trotter (1999) consider some of these classes plus point-core bitolerance orders (3aii) which they call *split interval orders*

146

Table 10.1. *Three categories of restrictions on bounded bitolerance representations.*

interval length	p and q	t_l and t_r
1. unit	a. point-core	i. tolerance
2. proper	b. totally bounded	ii. bitolerance
3. arbitrary	c. arbitrary	

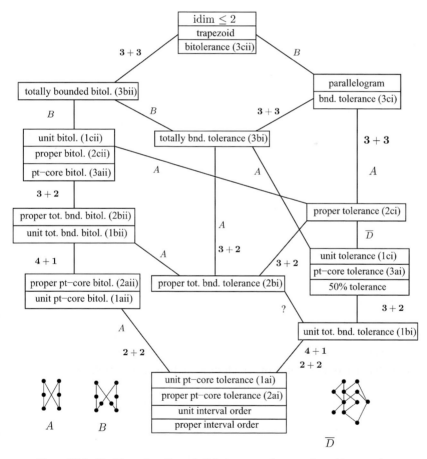

Figure 10.1. The hierarchy of bounded bitolerance orders together with separating examples.

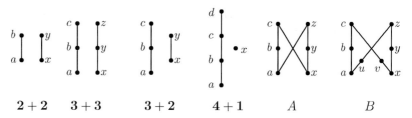

Figure 10.2. Orders which are separating examples in Figure 10.1.

and unit point-core bitolerance orders (1aii), which they call *split semiorders*. All 18 classes are considered in Isaak, Nyman, and Trenk (2001).

The hierarchy of these 18 classes appears in Figure 10.1, and Figure 10.2 shows the Hasse diagrams for the separating examples $2 + 2$, $3 + 2$, $3 + 3$, $4 + 1$, A, and B. The complete cut for each of these separating examples is shown in the hierarchy, that is, for each class in Figure 10.1, one can see exactly which of $2 + 2$, $3 + 2$, $3 + 3$, $4 + 1$, A, and B is a member of that class. The edge between classes (1bi) and (2bi) is the only one for which we have no separating example. The other separating example which appears in 10.1 is $\overline{D}$. In Exercise 10.1, the reader is asked to show that the incomparability graph of $\overline{D}$ is the Dartmouth graph D in Figure 2.9.

Theorem 10.1. *The class hierarchy and separating examples illustrated in Figure 10.1 are correct.*

Proof. First we justify the inclusions between classes. The restrictions in each category of Table 10.1 are listed from most restrictive to least restrictive. For example the "point-core" restriction implies the "totally bounded" restriction, thus we immediately get inclusions of the type (1ai) $\subseteq$ (1bi). Each of the inclusions shown in Figure 10.1 can be explained in this way, and we do not know of any other inclusions.

The classes that appear together in a box in Figure 10.1 are equivalent classes, and several of these results appear earlier in this book. Table 10.2 shows where each of the proofs of these equivalences can be found.

We justify the equivalence of classes in the same box in Figure 10.1 in the next section. We postpone the proofs about separating examples until Section 10.4. □

10.2 Equivalent classes

We begin by proving that all bounded bitolerance orders can be represented so that $q(v) \le p(v)$ for each element v as discussed in Section 5.2.1. The

Table 10.2. *A chart showing where proofs of equivalent*
classes can be found.

Equivalence of classes ...	... proved in
(idim $\leq$ 2)/trapezoid/3cii	Theorem 5.24
parallelogram/3ci	Lemma 5.20
1cii/2cii/3aii	Theorems 5.26 and 6.9
1ci/3ai/50% tolerance	Theorem 6.1
1aii/2aii	Theorem 10.3
1bii/2bii	Theorem 10.3
1ai/2ai/unit interval orders	Theorem 10.4

transformation used in the next proposition has been used by other authors
for different purposes. In Bogart, Fishburn, Isaak, and Langley (1995) and
Langley (1993) it is used to show that the class of unit tolerance orders is equivalent to the class of 50% tolerance orders (our Theorems 2.31 and 6.1), and in
Golumbic, Monma, and Trotter (1984) the authors use this same transformation
to show that tolerance graphs have representations in which all the intervals in
the representation have a common intersection point.

Proposition 10.2. *(Isaak, Nyman, and Trenk, 2001) If $P = (V, \prec)$ is a bounded*
bitolerance order, then P has a representation $\langle \mathcal{I}, p, q \rangle$ in which $q(v) \leq p(v)$
for each $v \in V$.

Proof. Fix a bounded bitolerance representation $\langle \mathcal{I}, p, q \rangle$ of P in which $I_v =$
$[L(v), R(v)]$. For any constant $M \geq 0$, the intervals $I'_v = [L'(v), R'(v)]$ and
tolerant points $p'(v), q'(v)$ defined by $L'(v) = L(v) - M$, $q'(v) = q(v) - M$,
$p'(v) = p(v)$ and $R'(v) = R(v)$ also give a bounded bitolerance representation
of P. By choosing M sufficiently large, we get a bounded bitolerance representation of P in which $q'(v) \leq p'(v)$ for each $v \in V$. $\qquad\square$

Note that the transformation in the proof of Proposition 10.2 increases the
length of every interval by M and thus it preserves the properties of "unit" and
"proper". In addition, it increases both the left and right tolerances by M, thus
it also preserves the "tolerance" property.

There are several results in the literature comparing "unit" classes of interval
and tolerance graphs to the analogous "proper" classes, for example, see Theorems 1.4, 6.9, 12.32 and 13.38. The next theorem gives two additional results
of this type. The latter is noted and used in Fishburn and Reeds (2001), but we
believe that the first explicit proof appears in Isaak, Nyman, and Trenk (2001).

Theorem 10.3. *The classes of unit point-core bitolerance orders* (1aii) *and proper point-core bitolerance orders* (2aii) *are equivalent. The classes of unit totally bounded bitolerance orders* (1bii) *and proper totally bounded bitolerance orders* (2bii) *are equivalent.*

Proof. The inclusions (1aii) $\subseteq$ (2aii) and (1bii) $\subseteq$ (2bii) are immediate, so we need only show the reverse inclusions.

Observe that two bitolerance representations for which the relative order of the interval endpoints and tolerant points is the same represent the same order. We used this idea in Chapter 7 to justify Remark 7.9 ("beads on a wire"). Using this observation, we next show that a proper bitolerance representation can be transformed into a unit bitolerance representation of the same order. Afterwards we note that the transformation preserves the "point-core" and "totally bounded" properties.

We proceed by induction. Assume that any proper bitolerance representation $\langle \mathcal{I}, p, q \rangle$ of an order $P = (V, \prec)$ with $|V| < n$ can be transformed into a unit bitolerance representation of P. Furthermore, assume this can be accomplished so that the relative order of the set of endpoints and tolerant points is unchanged.

Let $P = (V, \prec)$ be a proper bitolerance order with $|V| = n$ and, using Lemma 5.18, fix a proper bitolerance representation $\langle \mathcal{I}, p, q \rangle$ of P in which all endpoints and tolerant points are distinct. Let x be the element with smallest left endpoint. Since the representation is proper, $R(x)$ is also the smallest right endpoint. By induction, fix a unit bitolerance representation $\langle \mathcal{I}', p', q' \rangle$ of $P - x$ in which the points in $\{L'(v), p'(v), q'(v), R'(v) \mid v \in V - x\}$ appear in the same order as the corresponding points in $\{L(v), p(v), q(v), R(v) \mid v \in V - x\}$.

For concreteness, translate and scale the new representation of $P - x$ so that the smallest left endpoint is $L'(y) = 0$ and $|I'_v| = 1$ for all v. Now place $R'(x)$ so that its position with respect to the points in $\{L'(v), p'(v), q'(v), R'(v) \mid v \in V - x\}$ matches the position of $R(x)$ with respect to the corresponding points in $\{L(v), p(v), q(v), R(v) \mid v \in V - x\}$. We know $R'(x)$ will be the smallest right endpoint in $\langle \mathcal{I}', p', q' \rangle$, thus $R'(x) < R'(y) = 1$. Set $L'(x) = R'(x) - 1 < 0$, thus $L'(x)$ will be the smallest left endpoint in $\langle \mathcal{I}', p', q' \rangle$, as desired. Finally, place $p'(x)$ (resp. $q'(x)$) so that its position relative to points in $\{L'(v), p'(v), q'(v), R'(v) \mid v \in V - x\}$ matches the position of $p(x)$ (resp. $q(x)$) with respect to the corresponding points in $\{L(v), p(v), q(v), R(v) \mid v \in V - x\}$.

The new representation is unit. Furthermore, it has the same relative ordering of the interval endpoints and tolerant points as the original, so by our observation above, it represents the same order. If the original representation was point-core ($p(v) = q(v)$ for all v) then since the ordering was maintained, $p'(v) = q'(v)$

for all v, and the new representation is point-core. Likewise, if the original representation was totally bounded ($p(v) \le q(v)$ for all v) then again since the ordering was maintained, $p'(v) \le q'(v)$ for all v, and the new representation is totally bounded. This completes the proof. $\qquad\qquad\qquad\qquad\qquad\qquad\square$

Note that in proving Theorem 10.3 we have provided an alternative proof of the equivalence of (i) and (ii) in Theorem 6.9.

We conclude this section with a proof that the classes in the bottom box of Figure 10.1 are equivalent.

Theorem 10.4. *The following are equivalent statements about an order* P.

(i) P *is a unit interval order.*
(ii) P *is a proper interval order.*
(iii) P *contains neither* $2 + 2$ *nor* $3 + 1$ *as an induced suborder.*
(iv) P *is a unit tolerance order with constant tolerances.*
(v) P *is a unit tolerance order with constant cores.*
(vi) P *is a unit point-core tolerance order* (1ai).
(vii) P *is a proper point-core tolerance order* (2ai).

Proof. The equivalence of (i), (ii) and (iii), is implied by the work of Scott and Suppes (1958), and written explicitly in terms of graphs in Roberts (1969). The proof also appears in Golumbic (1980) and elsewhere. The implication (vi) $\Longrightarrow$ (vii) follows from the definitions of "unit" and "proper" restrictions. The remaining implications appear in Isaak, Nyman, and Trenk (2001) and we present them here.

(i) $\Longrightarrow$ (iv): Using Lemma 1.5, we may fix a unit interval representation $\mathcal{I} = \{I_v \mid v \in V\}$ of $P = (V, \prec)$ in which all endpoints of intervals are distinct and $C = |I_v|$ for all $v \in V$. Let ϵ be the smallest positive difference between distinct endpoints in the representation. It is not hard to show that the intervals $\{I_v \mid v \in V\}$ and the tolerances $t_v = \epsilon/2$ for each $v \in V$ give a unit tolerance representation of P in which all tolerances are constant.

(iv) $\Longrightarrow$ (vi): Let $\langle \mathcal{I}, p, q \rangle$ be a unit tolerance representation of $P = (V, \prec)$ in which all intervals have length C and all tolerances are equal to t. This representation has constant cores since $core(v) = q(v) - p(v) = (R(v) - t) - (L(v) + t) = C - 2t$. Form a new set of intervals $I'_v = [L'(v), R'(v)]$ and tolerances by setting $L'(v) = L(v) - (q(v) - p(v))$, $q'(v) = q(v) - (q(v) - p(v)) = p(v)$, $R'(v) = R(v)$ and $p'(v) = p(v)$ for all $v \in V$. Geometrically, this corresponds to taking the parallelogram representation of P associated with $\langle \mathcal{I}, p, q \rangle$ and sliding all points on the top line to the left by the constant $q(v) - p(v)$ (or to the right if $core(v)$ is negative). The new collection of intervals $\{I'_v\}$ and tolerances give a unit point-core representation of P.

(iv) $\Longleftrightarrow$ (v): In any unit tolerance representation with constant interval length C,

$$core(v) = q(v) - p(v) = (R(v) - t_v) - (L(v) + t_v) = |I_v| - 2t_v = C - 2t_v.$$

Thus, cores are constant if and only if tolerances are constant.

(vii) $\Longrightarrow$ (iii): In a point-core tolerance representation $\langle \mathcal{I}, p, q \rangle$ of $P = (V, \prec)$, the splitting point $f(v)$ is equal to the center point $c(v)$ of interval I_v for all $v \in V$. Thus, $x \prec y$ in P if and only if $R(x) < c(y)$ and $c(x) < L(y)$. A comparability occurs between x and y in P when neither interval contains the other's center, and thus $x || y$ in $P \Longleftrightarrow c(x) \in I_y$ or $c(y) \in I_x$.

In a *proper* tolerance representation, the centers of intervals occur in the same order as the right and left endpoints, i.e., $R(x) < R(y) \Longleftrightarrow L(x) < L(y) \Longleftrightarrow c(x) < c(y)$. With this background, we are now ready to show that the orders $2 + 2$ and $3 + 1$ are *not* proper point-core tolerance orders.

Suppose there were a proper point-core tolerance representation of the order $2 + 2$ with ground set $V = \{x, y, z, w\}$ and whose only comparabilities are $x \prec y$ and $z \prec w$. For $v \in V$, let $I_v = [L(v), R(v)]$ and $c(v)$ be the intervals and center points of the representation. Without loss of generality, we may assume $c(z) \leq c(x)$ and thus $R(z) \leq R(x)$.

Since $x \prec y$ we have $R(z) \leq R(x) < c(y)$, so $c(y) \notin I_z$. But $z \parallel y$, so $c(z) \in I_y$. However, this means $L(y) \leq c(z) \leq c(x) \leq R(x) < c(y) \leq R(y)$, so $c(x) \in I_y$, contradicting $x \prec y$.

Now suppose there were a proper point-core tolerance representation of the order $3 + 1$ with ground set $V = \{x, y, z, w\}$ and whose only comparabilities are $x \prec y \prec z$. Again, let $I_v = [L(v), R(v)]$ and $c(v)$ denote the interval and center point assigned to $v \in V$ in this representation. Since $x \prec y \prec z$, we have $c(x) < L(y)$ and $R(y) < c(z)$. If $L(w) < c(x)$ and $c(z) < R(w)$ then we would have $I_y \subset I_w$, violating the proper restriction. By symmetry, we may assume $R(w) \leq c(z)$.

Since $w \parallel z$ and $c(z) \notin I_w$, we must have $c(w) \in I_z$. Combining this with $x \prec y \prec z$ yields $R(x) < c(y) < L(z) \leq c(w)$, so $c(w) \notin I_x$. But $c(y) < c(w)$ implies $L(y) < L(w)$ so $c(x) < L(y) < L(w)$ and $c(x) \notin I_w$. Together these contradict $w \parallel x$. $\square$

10.3 Bipartite orders

In this section, we restrict attention to orders $P = (V, \prec)$ for which there is no chain $x \prec y \prec z$ for $x, y, z \in V$. Such orders are called *bipartite orders* since

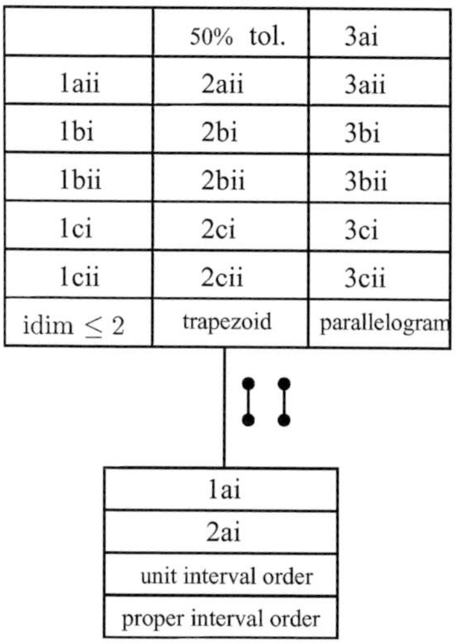

	50% tol.	3ai
1aii	2aii	3aii
1bi	2bi	3bi
1bii	2bii	3bii
1ci	2ci	3ci
1cii	2cii	3cii
idim ≤ 2	trapezoid	parallelogram

1ai
2ai
unit interval order
proper interval order

Figure 10.3. The hierarchy from Figure 10.1 (collapsed) in the case of bipartite bitolerance orders.

their comparability graphs are bipartite. Some authors refer to such orders as has having *height 1*.

In Figure 10.1, all the separating examples shown (except for the two copies of $2 + 2$ at the bottom) are not bipartite orders. Indeed, the following theorem shows that there are no bipartite separating examples because all of these classes (except the four in the bottom box) are equivalent in the bipartite domain. Figure 10.3 illustrates this collapse in the hierarchy of the classes in Figure 10.1 in the case that only bipartite orders are considered.

Theorem 10.5 appears in Isaak, Nyman, and Trenk (2001) and generalizes the results in Bogart and Trenk (1994) and Fishburn and Trotter (1999), where fewer classes were considered. The proof of Theorem 10.5 involves the following condition from Bogart and Trenk (1994) on indexing the maximal and minimal nonisolates in a bipartite order.

A bipartite order P satisfies the *BT-indexing condition* if the minimal nonisolates of P can be indexed $\{x_1, x_2, \ldots, x_m\}$ and the maximal nonisolates can be indexed $\{y_1, y_2, \ldots, y_n\}$ so that whenever $x_i \parallel y_j$ we have either $x_k \parallel y_j$ for all $k : 1 \leq k \leq i$, or $x_i \parallel y_k$ for all $k : 1 \leq k \leq j$. The order in Figure 10.5 satisfies the BT-indexing condition with the indexing given.

Theorem 10.5. *Within the domain of bipartite orders, all the classes in Figure 10.1 are equivalent, except for the four classes in the bottom box.*

3–crown   4–crown

Figure 10.4. The 3-crown and the 4-crown.

Moreover, a bipartite order is a member of these equivalent classes if and only if it satisfies the BT-indexing condition.

Before proving Theorem 10.5, we illustrate its usefulness by proving a result about k-crowns. The 3-crown and the 4-crown are shown in Figure 10.4. The result in Example 10.6 was used in Example 5.15 to show that the complement of an even cycle on six or more vertices is not a bounded bitolerance graph.

Example 10.6. The k-crown P_k is not a bounded bitolerance order for $k \geq 3$.

Proof. Suppose P_k were a bounded bitolerance order for some $k \geq 3$. By Theorem 10.5, we may index the minimal elements $x_1, x_2, \ldots, x_k$ of P_k and the maximal elements $y_1, y_2, \ldots, y_k$ to satisfy the BT-indexing condition.

Since x_1 is comparable to two maximal elements, there exists i ($2 \leq i \leq k$) such that $x_1 \prec y_i$ and similarly, there exists j ($2 \leq j \leq k$) such that $x_j \prec y_1$. In order to satisfy the BT-indexing condition, we have $x_j \prec y_i$. If $x_1 \prec y_1$ then P_k would contain a 2-crown, a contradiction, thus $x_1 \parallel y_1$. Now there is another maximal element y_ℓ ($2 \leq \ell \leq k, \ell \neq i$) such that $x_1 \prec y_\ell$ because x_1 is comparable to two maximal elements. Since x_j is only comparable to two maximal elements (y_1 and y_i), we have $x_j \parallel y_\ell$. But this violates the BT-indexing condition since $x_1 \prec y_\ell$ and $x_j \prec y_1$. □

Proof (of Theorem 10.5). Because of the inclusions in Figure 10.1, it suffices to prove the following two results. (A) If idim(P) ≤ 2 then P satisfies the BT-indexing condition. (B) If P satisfies the BT-indexing condition, then P is contained in each of the classes (1aii) and (1bi).

Proof of (A): Let $P = (V, \prec)$ be a bipartite order with interval dimension at most 2, and let $P_1 = (V, \prec_1)$ and $P_2 = (V, \prec_2)$ be interval orders for which $P = P_1 \cap P_2$. Let $\mathcal{I}_1 = \{[L_1(v), R_1(v)] \mid v \in V\}$ and $\mathcal{I}_2 = \{[L_2(v), R_2(v)] \mid v \in V\}$ be interval representations of P_1 and P_2, respectively, in which all endpoints of intervals are distinct. Index the minimal non-isolated vertices X of P in decreasing order according to their right endpoint in the interval representation of P_1. Thus, $X = \{x_1, x_2, \ldots, x_m\}$ where $i < j$ implies $R_1(x_i) > R_1(x_j)$.

Index the maximal non-isolated vertices Y of P in increasing order according to their left endpoint in the interval representation of P_2. Thus, $Y = \{y_1, y_2, \ldots, y_n\}$ where $i < j$ implies $L_2(y_i) < L_2(y_j)$. (See Exercise 10.2 for an example.)

We show that this indexing satisfies the BT-indexing condition. Suppose $x_i \parallel y_j$ in P. Then either (i) $x_i \nprec_1 y_j$ or (ii) $x_i \nprec_2 y_j$. In case (i) we must have $R_1(x_i) > L_1(y_j)$ and thus for all k in $1 \leq k \leq i$, $R_1(x_k) \geq R_1(x_i) > L_1(y_j)$. So $x_k \nprec_1 y_j$ and hence $x_k \nprec y_j$ for all k in $1 \leq k \leq i$. The elements of Y are maximal in P, so $y_j \nprec x_k$ and therefore $x_k \parallel y_j$ in P for all k in $1 \leq k \leq i$ as desired.

In case (ii) we have $R_2(x_i) > L_2(y_j)$ and thus $R_2(x_i) > L_2(y_j) \geq L_2(y_k)$ for all k in $1 \leq k \leq j$. Hence, $x_i \nprec_2 y_k$ and thus $x_i \nprec y_k$ for all k in $1 \leq k \leq j$. Again $y_k \nprec x_i$ since y_k is maximal in P, therefore $x_i \parallel y_k$ in P for all k in $1 \leq k \leq j$ as desired.

Proof of (B): Let $P = (V, \prec)$ be a bipartite ordered set whose minimal non-isolated elements $X = \{x_1, x_2, \ldots, x_m\}$ and maximal non-isolated elements $Y = \{y_1, y_2, \ldots, y_n\}$ are indexed according to the BT-indexing condition. Let Z be the set of isolated elements in P, thus $V = X \cup Y \cup Z$ is a partition of V. Finally, let $M = \max\{m, n\}$ and fix $C = 8M$.

We will define two different bounded bitolerance representations, one satisfying restrictions (1aii), and one satisfying restrictions (1bi), and show they each represent P. The construction of these representations is illustrated in Figure 10.5. In both representations, all the intervals will have length C. We next define the other common parts of the representations.

- For each $x_i \in X$, let $q(x_i) = -(M + i - 1)$.
- For each $y_j \in Y$, let $p(y_j) = M + j - 1$.
- For each $x_i \in X$, let $L(x_i) = R(x_i) - C$ where

$$R(x_i) = \begin{cases} p(y_1) - 1/2 & \text{if } x_i \prec y_1 \\ p(y_k) - 1/2 & \text{if } x_i \prec y_k \text{ and } x_i \parallel y_l \forall l \\ (1 \leq l < k). \end{cases}$$

- For each $y_j \in Y$, let $R(y_j) = L(y_j) + C$ where

$$L(y_j) = \begin{cases} q(x_1) + 1/2 & \text{if } x_1 \prec y_j \\ q(x_k) + 1/2 & \text{if } x_k \prec y_j \text{ and } x_l \parallel y_j \forall l \\ (1 \leq l < k). \end{cases}$$

- For each isolate $z \in Z$, let $L(z) = -C/2$, $R(z) = C/2$, and $p(z) = q(z) = 0$.

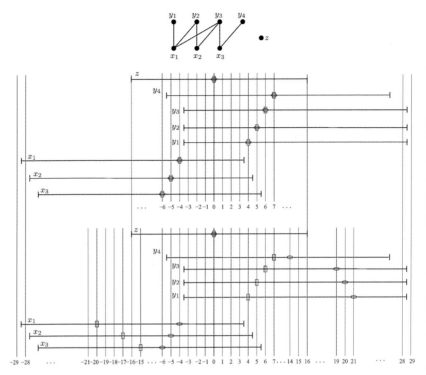

Figure 10.5. An ordered set, a unit point-core bitolerance (1aii) representation of it, and a unit totally bounded tolerance (1bi) representation of it.

It follows from these definitions that $-2M < q(x_i)$, $L(y_j) < 0$ and $0 < p(y_j)$, $R(x_i) < 2M$ for all $x_i \in X$ and $y_j \in Y$. Thus,

$$L(x_i) = R(x_i) - 8M < -6M < q(x_i) < 0 < R(x_i)$$

and

$$R(y_j) = L(y_j) + 8M > 6M > p(y_j) > 0 > L(y_j).$$

It remains to define $p(x_i)$ for each $x_i \in X$ and $q(y_j)$ for each $y_j \in Y$. For the unit point-core bitolerance representation (1aii) we let $p(x_i) = q(x_i)$ for all $x_i \in X$ and $q(y_j) = p(y_j)$ for all $y_j \in Y$. This clearly satisfies the conditions of a point-core bitolerance representation since $L(v) < p(v) = q(v) < R(v)$ for all $v \in V$. Call this *representation* r_1.

For the unit totally bounded tolerance representation (1bi), we instead let $p(x_i) = L(x_i) + (R(x_i) - q(x_i))$ for all $x_i \in X$ and $q(y_j) = R(y_j) - (p(y_j) - L(y_j))$ for all $y_j \in Y$. This clearly satisfies the "tolerance" requirement.

Furthermore, for each $x_i \in X$,

$$p(x_i) = L(x_i) + (R(x_i) - q(x_i)) < -6M + 4M = -2M < q(x_i) < R(x_i)$$

and

$$p(x_i) = L(x_i) + (R(x_i) - q(x_i)) > L(x_i).$$

Similarly, for each $y_j \in Y$,

$$L(y_j) < p(y_j) < q(y_j) < R(y_j).$$

Thus, the representation also satisfies the "totally bounded" requirement and has $p(v), q(v) \in I_v$ for all $v \in V$. Call this *representation r_2*.

Let $Q = (V, \prec')$ be the unit point-core bitolerance order represented by representation r_1. We will show that $Q = P$, thereby showing that P is a member of the class (1aii). Furthermore, in this proof we will use only features of the representation that are common to representation r_2. Thus, the same proof will also show that P is a member of the class (1bi) and complete the proof of the theorem.

We show $v \prec w \iff v \prec' w$ for all $v, w \in V$ using four cases.

Case 1: $v, w \in X$.

Let $v = x_i$ and $w = x_j$. Since X is an antichain in P, we must show $x_i \parallel x_j$ in Q. Note that

$$L(x_j) = R(x_j) - C < (2M - 1) - 8M = -6M - 1 < q(x_i),$$

so $x_i \not\prec' x_j$. By symmetry, $x_j \not\prec' x_i$, thus $x_i \parallel x_j$ in Q.

Case 2: $v, w \in Y$.

The proof is similar to Case 1.

Case 3: $v \in V, w = z \in Z$.

Since z is an isolated element of P, we know $z \parallel v$ in P and must show $z \parallel v$ in Q. We know $p(z) = q(z) = 0 \in I_v$. Thus, $v \not\prec' z$ because $R(v) > 0 = p(z)$ and $z \not\prec' v$ because $q(z) = 0 > L(v)$. Hence $z \parallel v$ in Q.

Case 4: $v = x_i \in X, w = y_j \in Y$.

If $x_i \prec y_j$ then, by the definitions of $R(x_i)$ and $L(y_j)$, we know $R(x_i) < p(y_j)$ and $L(y_j) > q(x_i)$, so $x_i \prec' y_j$.

Otherwise, $x_i \parallel y_j$ in P. By the BT-indexing condition, either (i) $x_i \parallel y_l$ in P for all l in $1 \leq l \leq j$, or (ii) $x_l \parallel y_j$ in P for all l in $1 \leq l \leq i$. In case (i), $R(x_i) > p(y_j)$ by definition of $R(x_i)$, and in case (ii), $L(y_j) < q(x_i)$ by definition of $L(y_j)$. In either case, $x_i \not\prec' y_j$. We know $y_j \not\prec' x_i$ since

$$R(y_j) = L(y_j) + C > (-2M + 1) + 8M = 6M + 1 > R(x_i) \geq p(x_i).$$

Thus, $x_i \parallel y_j$ in Q as desired. This completes the proof of the theorem. $\square$

Trenk (1998) introduced the class of 1-weak orders which contain the bipartite orders. In the domain of 1-weak orders, the classes of bounded bitolerance orders (3cii) and totally bounded bitolerance orders (3bii) are still equivalent (Trenk, 1998).

10.4 Separating examples

In this section, we provide the details that justify the placement of orders $2 + 2$, $3 + 2, 3 + 3, 4 + 1$, A and B as separating examples in Figure 10.1. Whenever one of these orders appears along an edge in Figure 10.1, we prove it is a member of the larger class and not a member of the smaller class. As mentioned earlier, each example defines a cut across the hierarchy. This section is intended for the researcher in need of such examples; other readers may wish to skip this section.

Several of these examples discussed in this section appear in Bogart and Trenk (1994). The order $\overline{D}$ was proven to separate the classes of unit tolerance orders (1ci) and proper tolerance orders (2ci) in Bogart, Fishburn, Isaak, and Langley (1995). The main result of that paper, that the classes of unit tolerance orders and proper tolerance orders are unequal, provides a contrast to the many "unit = proper" results mentioned throughout this book. The edge between (1bi) and (2bi), for which we have no separating example, is another instance of comparing classes of unit and proper tolerance orders.

10.4.1 The orders $2 + 2$ and $3 + 3$

Proposition 10.7. *The orders* $2 + 2$ *and* $3 + 3$ *in Figure 10.2 separate the classes indicated in Figure 10.1.*

Proof. It is easy to check that the order $2 + 2$ is a unit point-core bitolerance order (1aii) and a unit totally bounded tolerance order (1bi) (Exercise 10.4). However, it is not a unit point-core tolerance order (1ai) by Theorem 10.4.

The following bounded tolerance representation of $3 + 3$ uses the labeling in Figure 10.2 and assigns $I_v = [L(v), R(v)]$, $p(v) = R(v)$ and $q(v) = L(v)$ for all $v \in V$: $I_a = [1, 10]$, $I_b = [2, 11]$, $I_c = [3, 12]$, $I_x = [4, 5]$,

$I_y = [6, 7]$, $I_z = [8, 9]$. It then follows that **3 + 3** is a bounded bitolerance order (3cii).

In Bogart and Trenk (1994) it is shown that order **3 + 3** is not a totally bounded bitolerance order (3bii). It then follows that **3 + 3** is not a member of the more restrictive classes (3bi) and (1cii/2cii/3aii). The latter in turn implies that **3 + 3** is not a member of the class (2ci). □

10.4.2 The order 3 + 2

Lemma 10.8. *The order* **3 + 2** *is not a proper totally bounded bitolerance order* (2bii).

Proof. Suppose we had a proper totally bounded bitolerance representation $\langle \mathcal{I}, p, q \rangle$ of the order **3 + 2**, labeled as in Figure 10.2. First we show that the assumption $L(c) \leq q(x)$ leads to a contradiction. If $L(c) \leq q(x)$ then $L(c) \leq q(x) < L(y)$ since $x \prec y$. Using $a \prec b \prec c$ and the fact that the representation is totally bounded yields $R(a) < p(b) \leq q(b) < L(c) < L(y)$. However, this means I_a is completely to the left of I_y, contradicting $a \parallel y$.

Thus, $q(x) < L(c)$. Now $x \parallel c$ so we must have $R(x) \geq p(c)$ and $b \prec c$ so $R(x) \geq p(c) > R(b)$. Since the representation is proper we conclude $L(x) > L(b)$.

Now $a \parallel y$ so either (i) $R(a) \geq p(y)$ or (ii) $L(y) \leq q(a)$. We show that each of these leads to a contradiction. If (i) holds then $R(a) \geq p(y) > R(x)$ since $x \prec y$. Using $L(x) > L(b)$ from above and $a \prec b$ we have $L(a) \leq q(a) < L(b) < L(x)$. Together these imply $I_x \subseteq I_a$, contradicting the fact that the representation is proper. If (ii) holds then $L(y) \leq q(a) < L(b) < L(x) \leq q(x) < L(y)$, a contradiction. □

Proposition 10.9. *The order* **3 + 2** *separates the classes indicated in Figure 10.1.*

Proof. The order **3 + 2**, labeled as in Figure 10.2, is a unit tolerance order (1ci) using the following representation: $I_a = [2, 12]$, $I_b = [3, 13]$, $I_c = [4, 14]$, $I_x = [0, 10]$, $I_y = [6, 16]$, and $t_v = |I_v|$ for $v = a, b, c$ and $t_v = \frac{1}{2}|I_v|$ for $v = x, y$. Therefore, **3 + 2** is also a member of the larger classes (2ci), (1cii/2cii/3aii) and (3bi).

It remains to show that **3 + 2** is not a member of the classes (1bii/2bii), (2bi), and (1bi). Since each of these classes is contained in (2bii), it suffices to show that **3 + 2** is not a member of (2bii), which is done in Lemma 10.8. □

10.4.3 The order $4 + 1$

Lemma 10.10. *The order* $4 + 1$ *is not a proper point-core bitolerance order* (2aii).

Proof. Suppose we had a proper point-core bitolerance representation $\langle \mathcal{I}, f \rangle$ of the order $4 + 1$, labeled as in Figure 10.2. Since $a \prec b \prec c \prec d$ we have the inequalities $f(a) < L(b)$, $R(a) < f(b) < L(c)$, $R(b) < f(c) < L(d)$, $R(c) < f(d)$.

First consider the case in which $L(x) > f(a)$. Since $x \parallel a$ we must have $f(x) < R(a)$. Using the inequalities above and $L(c) < f(c)$, we obtain $f(x) < L(d)$. But $x \parallel d$ so we must have $R(x) > f(d)$. Now we have $R(x) > f(d) > R(c)$ and $L(x) < f(x) < R(a) < L(c)$, which implies that I_c is a proper subset of I_x, a contradiction.

Therefore, we must have $L(x) \leq f(a)$. Using the inequalities above, we have $L(x) < L(b)$, and since the representation is proper we must have $R(x) < R(b)$ which in turn implies $R(x) < L(d) < f(d)$. Since $x \parallel d$ we must have $L(d) < f(x)$ which implies $L(d) < f(x) < R(x) < L(d)$, a contradiction. $\square$

Proposition 10.11. *The order* $4 + 1$ *separates the classes indicated in Figure 10.1.*

Proof. The order $4 + 1$, labeled as in Figure 10.2, is a unit totally bounded tolerance order (1bi) using the following representation: $I_a = [0, 20]$, $I_b = [11, 31]$, $I_c = [22, 42]$, $I_d = [33, 53]$, $I_x = [17, 37]$, and $t_x = 1$ and $t_v = 10$ for $v = a, b, c, d$. Therefore, $4 + 1$ is also a member of the larger class (1bii/2bii).

In Lemma 10.10, we show that $4 + 1$ is not a member of the class (1aii/2aii) and therefore it is not a member of the smaller class (1ai/2ai). $\square$

10.4.4 The order A

Lemma 10.12. *The order* A *in Figure 10.2 is not a proper tolerance order* (2ci).

Proof. Suppose $A = (V, \prec)$ were a proper tolerance order and fix a proper tolerance representation of A in which $v \in V$ is assigned the interval $I_v = [L(v), R(v)]$ and the tolerant points $p(v), q(v) \in I_v$. By symmetry we may assume $L(b) \leq L(y)$, and thus, since the representation is proper, $R(b) \leq R(y)$.

First we show $t_b < t_y$. Since $y \prec z$ we have $R(y) < p(z)$, thus $R(b) \leq R(y) < p(z)$. But $b \parallel z$ so we must have $L(z) \leq q(b)$. In addition, $q(y) < L(z)$ (since $y \prec z$) so $q(y) < q(b)$. Hence, $t_b = R(b) - q(b) < R(y) - q(y) = t_y$ as desired.

Next we show the opposite inequality $t_y < t_b$ must also hold, a contradiction. Since $a \prec b$ we have $q(a) < L(b)$. Combining this with our original assumption $L(b) \le L(y)$ yields $q(a) < L(y)$. But $a \parallel y$ so $p(y) \le R(a)$. Now $R(a) < p(b)$ (since $a \prec b$) so $p(y) < p(b)$. Using this last inequality and our assumption $L(b) \le L(y)$ gives $t_y = p(y) - L(y) < p(b) - L(b) = t_b$, a contradiction. $\square$

Proposition 10.13. *The order A in Figure 10.2 separates the classes indicated in Figure 10.1.*

Proof. Figure 5.2 shows a unit point-core bitolerance representation of order A in which the oval and rectangle in interval I_i mark the location of the splitting point $f(i) = p(i) = q(i)$. Thus A is a member of the class (1aii), and therefore is also a member of the larger class (1bii/2bii).

The order A is a totally bounded tolerance order (3bi) using the representation $I_a = [1, 11]$, $p(a) = q(a) = 6$; $I_b = [7, 17]$, $p(b) = q(b) = 12$; $I_c = [13, 23]$, $p(c) = q(c) = 18$; $I_x = [7, 9]$, $p(x) = q(x) = 8$; $I_y = [9, 15]$, $p(y) = 10$, $q(y) = 14$; and $I_z = [15, 17]$, $p(z) = q(z) = 16$. Thus A is also a member of the larger class (3ci).

In Lemma 10.12 we proved that order A is not a proper tolerance order (2ci). Hence, it is not a member of the smaller classes (2bi), and (2ai). This completes the proof. $\square$

10.4.5 The order B

Lemma 10.14. *The order B in Figure 10.2 is not a point-core bitolerance order* (3aii).

Proof. For a contradiction, suppose B were a point-core bitolerance order. Fix a point-core bitolerance representation of B in which element w is assigned interval $I_w = [L(w), R(w)]$ and splitting point $f(w)$. Recall that $i \prec j$ if and only if $R(i) < f(j)$ and $f(i) < L(j)$. Thus if $i \parallel j$ and one of these inequalities holds, then the other must be reversed.

By symmetry, we may assume that

$$f(b) \le f(y). \tag{10.1}$$

Claim 1: $f(z) < f(c)$.

Since $y \prec z$ we know $f(y) < L(z)$. Combining this with (10.1) yields $f(b) < L(z)$. But $b \parallel z$ so $f(z) \le R(b)$. Now $b \prec c$ so $R(b) < f(c)$, which combined with the previous inequality yields $f(z) < f(c)$ as desired.

Table 10.3. *A totally bounded bitolerance*
representation of order B.

i	$L(i)$	$p(i)$	$q(i)$	$R(i)$
x	1	2	5	14
y	7	15	16	24
z	17	25	27	28
a	3	4	8	11
b	9	12	19	21
c	20	22	26	27
u	10	13	16	23
v	6	15	18	21

Claim 2: $f(x) < f(a)$.

Since $a \prec b$ we know $R(a) < f(b)$ which combined with (10.1) yields $R(a) < f(y)$. However, $a \parallel y$ so we must have $L(y) \leq f(a)$. This last inequality together with $f(x) < L(y)$ (because $x \prec y$) gives $f(x) < f(a)$ as desired.

Now we consider the relative positions of $f(u)$ and $f(v)$. First suppose $f(u) \geq f(v)$. Since $x \prec v$ we have $R(x) < f(v)$ and thus

$$R(x) < f(u). \tag{10.2}$$

But $a \prec u$ so $f(a) < L(u)$ and by Claim 2 we have $f(x) < L(u)$. This last inequality combined with (10.2) imply $x \prec u$, a contradiction.

Otherwise, $f(u) < f(v)$. Since $v \prec c$ we have $f(v) < L(c)$ which gives

$$f(u) < L(c). \tag{10.3}$$

However, $u \prec z$ so $R(u) < f(z)$ and combining this with Claim 1 yields $R(u) < f(c)$. This last inequality together with (10.3) imply $u \prec c$, a contradiction. □

Lemma 10.15. *The order B in Figure 10.2 is not a bounded tolerance order* (3ci).

Proof. Suppose the order B were a bounded tolerance order. Then its incomparability graph $\overline{B}$ (Figure 2.4) would be a bounded tolerance graph by Remark 5.4 and hence a tolerance graph. This contradicts Example 2.19. □

Proposition 10.16. *The order B in Figure 10.2 separates the classes indicated in Figure 10.1.*

Proof. Table 10.3 gives a representation of the order B as a totally bounded bitolerance order (3bii), thus B is also a member of the larger class (3cii).

It remains to show that order B is not a member of the classes (3ci), (3bi) and (1cii/2cii/3aii). In Lemma 10.14, we showed that the order B is not a member of the class of point-core bitolerance orders (3aii/1cii/2cii), and in Lemma 10.15, we showed that the order B is not in the class of bounded tolerance orders (3ci), thus B is not a member of the more restrictive class (3bi). This accounts for all occurrences of B in Figure 10.1. □

10.5 Exercises

Exercise 10.1. Verify that the incomparability graph of the order $\overline{D}$ in Figure 10.1 is the Dartmouth graph D in Figure 2.9.

Exercise 10.2. For the order P and the interval realizer I_1, I_2 given in Figure 8.2, construct the indexing of the vertices of P according to the proof of Theorem 10.5, statement (A). Verify that this indexing satisfies the BT-indexing condition.

Exercise 10.3. Using the BT-indexing of the order P from Exercise 10.2, construct a unit point-core bitolerance representation of P and a totally bounded tolerance representation of P using the construction given in the proof of Theorem 10.5, statement (B).

Exercise 10.4. Construct a unit point-core bitolerance representation and a unit totally bounded tolerance representation of the order $\mathbf{2 + 2}$.

Chapter 11

Tolerance models of paths and subtrees of a tree

11.1 Introduction

We began this book by introducing the class of tolerance graphs, which generalize the intersection graphs of intervals on the line (interval graphs), adding an edge between two vertices in the tolerance graph when the size of the intersection of their intervals exceeds at least one of the tolerances. Subsequently, we studied a further generalization defined by allowing separate right and left tolerances on the intervals (bitolerance graphs).

In this chapter, we present a totally different approach to generalizing tolerance graphs by replacing the real line by a tree and replacing the role of intervals by either paths or other types of subtree. Several classical results are known for classes of intersection graphs of paths and subtrees of a tree, which we review in the next three sections. We then present results on tolerance versions.

11.2 Intersection models

Let $\mathbf{T}$ be a tree and let $\mathcal{T} = \{T_i\}$ be a collection of subtrees (connected subgraphs) of $\mathbf{T}$. We may think of the host tree $\mathbf{T}$ either as a *continuous* model of a tree embedded in the plane, thus generalizing the real line from the one-dimensional case, or as a finite *discrete* model of a tree, namely, a connected graph of vertices and edges having no cycles, thus generalizing the path P_k from the one-dimensional case. Making a distinction between these two models will become important when we measure the size of the intersection of two subtrees. In the continuous model, we will use Euclidean distance along paths in the tree; in the discrete model, we will count the number of common vertices or common edges.

The usual definition of the intersection graph $G = (V, E)$ of a collection $\mathcal{T} = \{T_i \mid i = 1, \ldots, n\}$ of subtrees, which has vertex set $V = \{1, \ldots, n\}$ and edge set $E = \{ij \mid T_i \cap T_j \neq \emptyset\}$, may be interpreted in two ways. On the one hand, if we interpret "intersection" to mean sharing a vertex of the host $\mathbf{T}$ in the discrete case, or a point in the continuous case, we will call G the *vertex intersection* graph. On the other hand, if we interpret "intersection" to mean sharing an edge of $\mathbf{T}$ in the discrete case, or an otherwise measurable segment of $\mathbf{T}$ in the continuous case, we will call G the *edge intersection* graph. As we will see shortly, these two definitions give rise to different classes of graphs.[1]

The following well known characterization, due independently to Buneman (1974), Gavril (1974) and Walter (1972), applies for both the continuous and the discrete models of the tree. A proof can be found in Golumbic (1980).

Theorem 11.1. *A graph is the vertex intersection graph of a set of subtrees of a tree if and only if it is a chordal graph.*

In the next section, we consider the discrete model of the host tree $\mathbf{T}$, and in the subsequent sections the continuous model.

11.3 Discrete models

Let $\mathbf{T}$ be a host tree in the discrete model, and $\mathcal{T}$ a collection of subtrees. As introduced above, we distinguish between the vertex intersection graph and the edge intersection graph. If $\mathbf{T}$ is a star $K_{1,n}$, then each subtree consists of either a substar containing the central node or just a single leaf node. The next result shows that in the restriction to stars, the class of graphs obtained in the case of vertex intersection is precisely the class of split graphs, as observed in McMorris and Shier (1983).

Theorem 11.2. *A graph is the vertex intersection graph of distinct subtrees of a star if and only if it is a split graph.*

Proof. Recall that a graph G is a split graph if its vertices can be partitioned into a stable set $X = \{x_1, \ldots, x_k\}$ and a clique $Y = \{y_1, \ldots, y_\ell\}$. If G is a split graph, consider the star $\mathbf{T}$ formed by a central node u and leaves $\overline{x}_1, \ldots, \overline{x}_k, \overline{y}_1, \ldots, \overline{y}_\ell$. Let the subtree corresponding to $x_i \in X$ be the single leaf $\overline{x}_i$ in $\mathbf{T}$ and let

---

[1] One may regard the expression "vertex intersection" as interchangable with "nonempty intersection", and "edge intersection" as interchangable with "nontrivial intersection".

the subtree corresponding to $y_i \in Y$ be the substar of $\mathbf{T}$ induced by $\{u, \overline{y}_i\} \cup \{\overline{x}_j \mid x_j \in \mathcal{N}(y_i)\}$. Clearly, this is a vertex intersection representation for G. Conversely, if we are given a representation for G as the vertex intersection graph of distinct substars of a star, then those substars containing the central node correspond to a clique in G and the remaining subtrees (the single leaves) correspond to a stable set of G. □

The preceding theorem on subtrees of a star refers to vertex intersection. In contrast to this, Golumbic and Jamison (1985b) observed that all graphs may be obtained in the case of edge intersection, which we state as follows.

Theorem 11.3. *Every graph can be represented as the edge intersection graph of substars of a star.*

Proof. Let $G = (V, E)$ be any graph, and let $E = \{e_1, \ldots, e_m\}$. Consider the star $\mathbf{T}$ formed by a central node u and leaves $\overline{e}_1, \ldots, \overline{e}_m$. Define the substar corresponding to v_i to be the substar T_i of $\mathbf{T}$ induced by $\{u\} \cup \{\overline{e}_\ell \mid v_i \in e_\ell\}$. Clearly, $v_i v_j \in E$ if and only if T_i and T_j share an edge, namely edge $u\overline{e}_k \in \mathbf{T}$ where $e_k = v_i v_j$. □

Two different classes of intersection graphs also arise when considering simple paths of an arbitrary host tree $\mathbf{T}$. A "vertex intersection graph of paths in a tree" is called a *VPT graph*. They are also called *path graphs* in the literature, and Gavril (1978) gave a characterization and a polynomial time recognition algorithm for the class. A graph obtained as the "edge intersection graph of paths in a tree" is called an *EPT graph*. The class of EPT graphs was introduced in Golumbic and Jamison (1985b, 1985a), where they showed the recognition problem to be NP-complete.

Example 11.4. Figure 11.1 shows a tree $\mathbf{T}$, the VPT graph G_{VPT} and the EPT graph G_{EPT} obtained from the ten paths $\{Q_i\} \cup \{P_{i,j}\}$ where Q_i goes from i to i' and $P_{i,j}$ goes from i to j. Note that the graph G_{EPT} contains three chordless 4-cycles.

Neither of the classes VPT and EPT contains the other. A VPT graph must be chordal by Theorem 11.1, whereas an EPT graph may contain chordless cycles of any length (see Exercise 11.1). Thus, $C_k \in$ EPT–VPT, for all $k \geq 4$. It was shown in Golumbic and Jamison (1985a) that the graph G_{VPT} from Figure 11.1 is not an EPT graph, thus $G_{VPT} \in$ VPT–EPT.

The following theorem shows that the classes of VPT and EPT coincide precisely when the host tree has maximum degree 3. We include a further equivalence, namely the graphs which are both chordal and EPT.

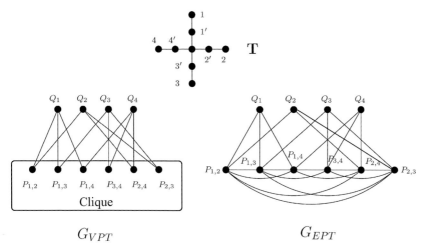

Figure 11.1. The tree and the graphs from Example 11.4.

Theorem 11.5. *The following are equivalent for a graph G:*

(i) $G \in VPT \cap EPT$,
(ii) *G has a vertex intersection representation on a tree of maximum degree 3,*
(iii) *G has an edge intersection representation on a tree of maximum degree 3,*
(iv) $G \in chordal \cap EPT$.

Proof. The equivalence (i) ⇔ (ii) ⇔ (iii) is due to Golumbic and Jamison (1985a). Since VPT graphs are chordal, (i) ⇒ (iv). Syslo (1985) proved the inclusion $chordal \cap EPT \subseteq VPT$ which yields the implication (iv) ⇒ (i). □

The problems of recognition and finding a maximum clique, minimum coloring, maximum stable set and minimum clique cover all have polynomial time complexity for the classes of VPT, chordal, split, interval, permutation and comparability graphs (see Golumbic, 1980). For the class of EPT graphs, maximum clique and maximum stable set have polynomial time algorithms, minimum coloring is NP-complete and the complexity of recognition is an open problem (Golumbic and Jamison, 1985b; 1985a; Tarjan, 1985).

In Jamison and Mulder (2000a; 2000b), the authors have placed these discrete models into a more general setting. An (h, s, p)-subtree representation of $G = (V, E)$ consists of a collection of subtrees $\{T_v\}_{v \in V}$ of a host tree **T** such that (i) the maximum degree of T is at most h, (ii) every subtree T_v has maximum degree at most s, and (iii) $uv \in E \Longleftrightarrow T_u$ and T_v have at least p vertices of the host tree **T** in common. By placing such degree restrictions on the host tree or the representing subtrees, and by considering the p-vertex intersection graphs of these subtrees, Jamison and Mulder have developed a rich theory that includes several of the classes that we have already discussed.

Let $[h, s, p]$ denote the class of all graphs which have an (h, s, p)-subtree representation. We use the term *p-intersection graphs of subtrees of a tree*[2] for the class $[\infty, \infty, p]$, where ∞ indicates no constraint on the degree. Thus, $[\infty, \infty, 1]$ are the chordal graphs by Theorem 11.1, $[\infty, 2, 1]$ are the VPT graphs or path graphs, and $[2, 2, 1]$ are the interval graphs, since a tree with maximum degree 2 is a path. In fact, in the case of chordal graphs, McMorris and Scheinerman (1991) observed that $[\infty, \infty, 1] = [3, 3, 1]$, that is, if G is a chordal graph, then G has a representation as an intersection graph of subtrees of a host tree $\mathbf{T}$ having degree at most 3 (see Exercise 11.3). The class of EPT graphs is equivalent to $[\infty, 2, 2]$ since two intersecting paths which contain two vertices a and b of $\mathbf{T}$ also share all edges of the path between a and b. The class $[3, 3, 3]$ is studied in Jamison and Mulder (2000b). As a further example, Theorem 11.5 can be restated as

$$[\infty, 2, 1] \cap [\infty, 2, 2] = [3, 2, 1] = [3, 2, 2] = [\infty, \infty, 1] \cap [\infty, 2, 2].$$

Remark 11.6. It is easy to show (Exercise 11.4) that for the first two parameters, these classes are monotone:

$$[h, s, p] \subseteq [h', s, p] \text{ and } [h, s, p] \subseteq [h, s', p], \quad (\text{for } h \leq h', s \leq s').$$

For the third parameter p, however, the situation is different. As we have seen above, the classes of VPT graphs $[\infty, 2, 1]$ and EPT graphs $[\infty, 2, 2]$ are not comparable. However, Jamison and Mulder have suggested the following.

Conjecture 11.7. *(Jamison and Mulder, 2000a)* $[h, s, p] \subseteq [h, s, p']$, *for* $2 \leq p \leq p'$.

In Jamison and Mulder (2000a), the authors report that the conjecture holds for $p = 2$ and $p = 3$ and for all $p' \geq p^2 - 4p + 6$.

Containment graphs

Finally, we present some results known about containment graphs. A graph $G = (V, E)$ is a *containment graph* of a collection $\mathcal{F} = \{S_i\}$ of subsets of a set $\mathbf{S}$ if it has vertex set $V = \{1, \dots, n\}$ and edge set $E = \{ij \mid$ either $S_i \subseteq S_j$ or $S_j \subseteq S_i\}$. A graph with such a representation is called a *containment graph*. We recall from Theorem 2.6 that the containment graphs of intervals on the line are equivalent to the permutation graphs and to the

---

[2] Jamison and Mulder have called this class the *tree-tolerance graphs* with constant tolerance p, however, we reserve the usage of tolerance for the continuous model.

Table 11.1. *Graph classes involving trees (in the discrete model).*

Type of interaction	Objects	Host	Graph class
vertex intersection	subtrees	tree	chordal graphs $[\infty, \infty, 1]$
vertex intersection	subtrees	star	split graphs
edge intersection	subtrees	star	all graphs $[\infty, \infty, 2]$
vertex intersection	paths	path	interval graphs $[2, 2, 1]$
vertex intersection	paths	tree	VPT graphs $[\infty, 2, 1]$
edge intersection	paths	tree	EPT graphs $[\infty, 2, 2]$
containment	intervals	line	permutation graphs
containment	paths	tree	? (open question)
containment	subtrees	star	comparability graphs

tolerance graphs where each interval has tolerance equal to its length. Moreover, Golumbic and Scheinerman (1989) observed that every comparability graph can be represented as the containment graph of a collection of subtrees (substars) of a star. This can be shown using a construction similar to the one in the proof of Theorem 11.3. The converse follows immediately from the observation that every containment graph is a comparability graph. Characterizing the containment graphs of paths in a tree is an open question.

Table 11.1 summarizes the graph classes we have discussed in this section.

11.4 Neighborhood subtrees

For the remainder of this chapter, we consider the continuous model of the host tree. Let **T** be a tree embedded in the plane so that it never crosses itself. The embedding is not necessarily a straight line embedding, although any curve in the plane could always be replaced by straight line segments. We regard **T** as a continuum of points, and we define the distance $d(p, q)$ between points p and q in **T** to be the Euclidean distance along the unique path $P(p, q)$ from p to q in **T**. In military terms, this is the infantry distance and not the airforce distance. The *degree* of a point p on the tree is the number of branches emanating from it; an *endpoint* has degree 1 and a *branch point* has degree strictly greater than 2. The *boundary* of a tree consists of all of its endpoints.

Let T be a subtree of **T**, i.e., a continuous subset of points of **T**. The *diameter* of T is the largest distance between any pair of points in T, that is, $\text{diam}(T) = \max\{d(p, q) \mid p, q \in T\}$. Let p and q be a pair of endpoints of T such that $d(p, q) = \text{diam}(T)$ and let c be the midpoint of the path $P(p, q)$. It is an easy

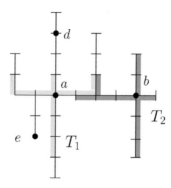

Figure 11.2. A host tree with two subtrees T_1 and T_2.

exercise to show that *every point of T is at a distance at most $\frac{1}{2}$diam(T) from
c, and that c is the common midpoint of all paths of length* diam(T). We call c
the *center* of T. Figure 11.2 shows a host tree with two subtrees T_1 and T_2. The
subtree T_1, shaded in light grey, has center a and diameter 6. The subtree T_2,
shaded in dark grey, has center b and also has diameter 6.

We now introduce the notion of a neighborhood subtree. Let c be an arbitrary
point on the tree **T**, and let $r \geq 0$ be a non-negative real number. Consider the set
of points $T(c, r) = \{q \in \mathbf{T} \mid d(c, q) \leq r\}$ within a distance r from the designated
point c. The subtree $T(c, r)$ is *truncated* if there is an endpoint p of **T** such that
$d(p, c) < r$.

A subtree T of **T** is called a *neighborhood subtree of radius r centered
on c* if $T = T(c, r)$ and diam(T) $= 2r$ where c is the center of T. We define
the *size* of a neighborhood subtree to be $\|T(c, r)\| = 2r$, that is, equal to its
diameter. By convention, a tree consisting of one point has diameter zero, and
the diameter of the empty tree is undefined. For example, in Figure 11.2, T_1 is
not a neighborhood subtree since it does not extend to include points d and e;
however, T_2 is a neighborhood subtree since $T_2 = T(b, 3)$ although two of its
branches are truncated.

It is not hard to show that if the subtree $T(c, r)$ is truncated so much that it fails
to have diameter $2r$, then its parameters can be recalculated such that $T(c, r) =
T(c', r')$ where $r' = \frac{1}{2}$diam($T(c, r)$) and c' is the true center of $T(c, r)$). For
example, in Figure 11.3 with $c = (4, 1)$, the shaded subtree $T = T(c, 3 + \sqrt{2})$
has true center at $c' = (3, 1)$ and $T = T(c', 2 + \sqrt{2})$.

Remark 11.8. We allow neighborhood subtrees $T(c, r)$ to be truncated, but we
assume they have a pair of endpoints at a distance of $2r$.

This differs slightly from Bibelnieks and Dearing (1993) where a stronger assumption is made, namely, that every branch of **T** is (extended to be) sufficiently

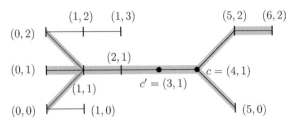

Figure 11.3. Recalculating the true center of a subtree.

long so that whenever we are interested in a subtree $T(c, r)$ it is never cut off prematurely along any branch.

By Remark 11.8, we will assume that whenever the notation $T(c, r)$ is used, the point c is the true center and $2r$ the diameter of the neighborhood subtree $T(c, r)$. The following lemma shows that neighborhood subtrees are preserved by intersection. It will be used repeatedly in this chapter to measure the intersection of neighborhood subtrees.

Lemma 11.9. *(Bibelnieks and Dearing, 1993) The intersection of two neighborhood subtrees is empty or is a neighborhood subtree. Moreover, if $T_1 = T(c_1, r_1)$ and $T_2 = T(c_2, r_2)$, then*

(i) $T_1 \subseteq T_2 \Leftrightarrow \|T_1 \cap T_2\| = 2r_1 \Leftrightarrow r_2 - r_1 \geq d(c_1, c_2)$,
(ii) $T_1 \cap T_2 = \emptyset \Leftrightarrow \|T_1 \cap T_2\|$ *is undefined* $\Leftrightarrow r_1 + r_2 < d(c_1, c_2)$,
(iii) *if $T_1 \nsubseteq T_2$, $T_2 \nsubseteq T_1$ and $T_1 \cap T_2 \neq \emptyset$ then*
$\|T_1 \cap T_2\| = r_1 + r_2 - d(c_1, c_2) < min\{\|T_1\|, \|T_2\|\}.$

Proof. Let $T_0 = T_1 \cap T_2$. If $T_1 \subseteq T_2$ (condition(i)), then clearly T_0 is a neighborhood subtree and the equivalence in (i) is easily verified. In the case of (ii), by convention, $T_0 = \emptyset$ if and only if its diameter (size) is undefined; the equivalence with the inequality is clear. Otherwise, the subtrees overlap (condition (iii)). Since neither contains the other, we have the inequalities

$$d(c_1, c_2) + r_2 > r_1 \quad \text{and} \quad d(c_1, c_2) + r_1 > r_2. \quad (11.1)$$

Thus, $\frac{1}{2}[d(c_1, c_2) + r_2 - r_1]$ and $\frac{1}{2}[d(c_1, c_2) + r_1 - r_2]$ are positive, and we note that they add up to $d(c_1, c_2)$. Therefore, we may define c_0 to be the point on the path $P(c_1, c_2)$ such that

$$d(c_1, c_0) = \tfrac{1}{2}[d(c_1, c_2) + r_1 - r_2]$$
$$d(c_2, c_0) = \tfrac{1}{2}[d(c_1, c_2) + r_2 - r_1]. \quad (11.2)$$

Finally, define

$$r_0 = \tfrac{1}{2}[r_1 + r_2 - d(c_1, c_2)].$$

We will show that $T_0 = T(c_0, r_0)$, i.e., the intersection of T_1 and T_2 is a neighborhood subtree.

Let $p \in T_0 = T_1 \cap T_2$. Since $c_0 \in P(c_1, c_2)$, either $c_0 \in P(c_1, p)$ or $c_0 \in P(c_2, p)$. If $c_0 \in P(c_1, p)$, then

$$d(c_0, p) = d(c_1, p) - d(c_1, c_0) \leq r_1 - \tfrac{1}{2}[d(c_1, c_2) + r_1 - r_2] = r_0$$

so $p \in T(c_0, r_0)$. Similarly, if $c_0 \in P(c_2, p)$, then $p \in T(c_0, r_0)$ which proves $T_0 \subseteq T(c_0, r_0)$.

Now let $p \in T(c_0, r_0)$. We have

$$d(c_0, p) \leq r_0 = \tfrac{1}{2}[r_1 + r_2 - d(c_1, c_2)]$$
$$= \tfrac{1}{2}[r_1 + r_2 - [d(c_1, c_0) + d(c_0, c_2)]].$$

Applying the triangle inequality, then the inequality in the previous line and then (11.2), we obtain

$$d(c_1, p) \leq d(c_1, c_0) + d(c_0, p) \leq \tfrac{1}{2}[r_1 + r_2 + d(c_1, c_0) - d(c_0, c_2)]$$
$$= \tfrac{1}{2}\left[r_1 + r_2 + \left(\tfrac{1}{2}[d(c_1, c_2) + r_1 - r_2]\right)\right.$$
$$\left. - \left(\tfrac{1}{2}[d(c_1, c_2) + r_2 - r_1]\right)\right] = r_1$$

so $p \in T_1$. Similarly, we can show $p \in T_2$, which proves $T(c_0, r_0) \subseteq T_0$. So $T_0 = T(c_0, r_0)$ and is a neighborhood subtree. Thus,

$$\|T_1 \cap T_2\| = \|T_0\| = 2r_0 = r_1 + r_2 - d(c_1, c_2).$$

Finally, by (11.1),

$$d(c_1, c_2) + 2r_2 > r_1 + r_2 \quad \text{and} \quad d(c_1, c_2) + 2r_1 > r_1 + r_2,$$

and we conclude that $r_1 + r_2 - d(c_1, c_2) < \min\{2r_1, 2r_2\} = \min\{\|T_1\|, \|T_2\|\}$. $\qquad\square$

The usual Helly property for trees states that if, in a family $\mathcal{T}$ of subtrees of a tree, each pair of subtrees intersect, then the intersection of all subtrees of $\mathcal{T}$ is nonempty (see Golumbic, 1980). The following corollary may be regarded as the *Helly property for neighborhood subtrees*.

Corollary 11.10. *Let $T_1, \ldots, T_k$ be neighborhood subtrees and let $T = T_1 \cap \cdots \cap T_k$.*

(i) *If $T_i \cap T_j \neq \emptyset$ for all i, j (pairwise intersection), then T is a nonempty neighborhood subtree.*

(ii) *If $\|T_i \cap T_j\| > 0$ for all i, j, then $\|T\| > 0$.*

Proof. (i) By the usual Helly property for trees, T is nonempty. Applying Lemma 11.9 repeatedly, it follows that T is a neighborhood subtree. (ii) The case of nonzero size follows similarly by a more careful argument about measuring the diameters, and is left to the reader. □

A graph $G = (V, E)$ is a *neighborhood subtree intersection graph* if there exists an embedded tree **T** and a mapping of each vertex $v \in V$ to a neighborhood subtree $T(c_v, r_v)$ of **T** such that

$$uv \in E \iff T(c_u, r_u) \cap T(c_v, r_v) \neq \emptyset. \tag{11.3}$$

The neighborhood subtree intersection graphs have yet to be characterized. We will study their properties in Section 11.6.2.

11.5 Neighborhood subtree tolerance (NeST) graphs

We are now ready to add tolerances to the neighborhood subtree model. Lemma 11.9 allows us to measure the size of the intersection of two neighborhood subtrees. This allows us to define neighborhood subtree tolerance graphs in a manner analogous to (interval) tolerance graphs.

A graph $G = (V, E)$ is called a *neighborhood subtree tolerance* (NeST) *graph* if each vertex $v \in V$ can be assigned a neighborhood subtree T_v of some embedded tree **T** and a tolerance $t_v > 0$ such that $xy \in E(G)$ if and only if $T_x \cap T_y \neq \emptyset$ and $\|T_x \cap T_y\| \geq \min\{t_x, t_y\}$. Let $\mathcal{T} = \{T_v\}$ and $t = \{t_v\}$ for $v \in V$. We refer to $\langle \mathcal{T}, t \rangle$ as a *NeST representation* for G.

Figure 11.4 shows a NeST graph G and a representation for it, where a label (c_v, r_v, t_v) at a point c_v on the tree indicates the subtree $T(c_v, r_v)$ assigned to vertex $v \in V(G)$ with tolerance t_v. We omit the shading in this figure to make it more understandable. One can verify easily that this representation gives the graph G. The following simple method allows us to handle larger examples.

For all pairs of centers c_i, c_j do the following:

1. Determine the distance $d(c_i, c_j)$.
2. Calculate $\|T_i \cap T_j\|$ using $d(c_i, c_j)$ and Lemma 11.9.
3. Compare $\|T_i \cap T_j\|$ with $\min\{t_i, t_j\}$ to determine whether there is an edge ij in G.

We illustrate this in Figure 11.5 for the NeST representation of graph G shown in Figure 11.4. The entry ij in the third table in Figure 11.5 is underlined if $\|T_i \cap T_j\| \geq t_i$, thus creating the edge ij in the NeST graph. Notice that the underlining is not symmetric, which could be interpreted as placing a direction

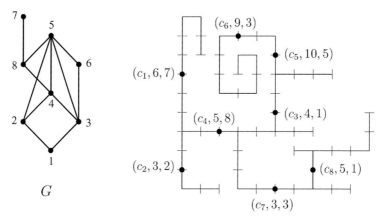

Figure 11.4. A graph G and a NeST representation. The label (c, r, t) at a point c on the tree denotes the subtree $T(c, r)$ having tolerance t.

i	1	2	3	4	5	6	7	8
r_i	6	3	4	5	10	9	3	5
t_i	7	2	1	8	5	6	3	1

	1	2	3	4	5	6	7	8
1	-	5	9	5	12	15	11	14
2	5	-	8	4	11	14	10	13
3	9	8	-	4	3	6	8	11
4	5	4	4	-	7	10	6	9
5	12	11	3	7	-	3	11	14
6	15	14	6	10	3	-	14	17
7	11	10	8	6	11	14	-	3
8	14	13	11	9	14	17	3	-

$$d(c_i, c_j)$$

	1	2	3	4	5	6	7	8
1	-	4	1	6	4	0	0	0
2	4	-	0	4	2	0	0	0
3	1	0	-	5	8	7	0	0
4	6	4	5	-	8	4	2	1
5	4	2	8	8	-	16	2	1
6	0	0	7	4	16	-	0	0
7	0	0	0	2	2	0	-	5
8	0	0	0	1	1	0	5	-

$$\| T_i \cap T_j \|$$

Figure 11.5. Illustrating the verification of a NeST representation for the graph G in Figure 11.4. The underlined entries are those which create an edge in the NeST graph.

(or bidirection) on each edge. This idea is explored for interval tolerance in Chapter 13.

Remark 11.11. As in the case of tolerance graphs, we may assume, without loss of generality, that no two subtrees are equal in a NeST representation; otherwise, we could extend each branch of the host tree by a small $\epsilon > 0$

and, by an argument similar to the one in Lemma 2.3, obtain distinct subtrees. Moreover, the transformation preserves proper containments.

We now present an alternative definition of NeST graphs, due to Hayward, Kearney, and Malton (2002), which we will need in Section 11.8. Consider a NeST representation $\langle T, t \rangle$ of G where $T = \{T_v \mid v \in V\}$ is its set of neighborhood subtrees. The non-neighbors of a vertex v are precisely those vertices u where the intersection between T_v and T_u is too small to exceed either of their tolerances. Among those non-neighbors u of v, we let δ_v be the size of the largest intersection $\|T_v \cap T_u\|$ ($u \notin \mathcal{N}[v]$). In other words, (witnesses have reported that) v is able to endure intersections of size δ_v without there being an edge in the graph G. If v is a *universal* vertex, that is, v is adjacent to all other vertices, we let $\delta_v = 0$.

In Hayward, Kearney, and Malton (2002), the authors refer to this approach as "tolerance-free" representations since the $\{\delta_v\}$, which are used in place of the $\{t_v\}$, are calculated from the subtrees $\{T_v\}$ and the graph G itself. We formalize the alternative definition of NeST graphs in the following proposition.

Proposition 11.12. *(Hayward, Kearney, and Malton, 2002) A graph $G = (V, E)$ is a NeST graph if and only if there is a set $T = \{T_v \mid v \in V\}$ of neighborhood subtrees of an embedded tree T such that $xy \in E(G) \iff \|T_x \cap T_y\| > \min\{\delta_x, \delta_y\}$ where*

$$\delta_v = \begin{cases} 0 & \text{if } v \text{ is universal} \\ \max\{\|T_v \cap T_u\| \mid u \notin \mathcal{N}[v]\} & \text{otherwise.} \end{cases} \tag{11.4}$$

Proof. ($\Longrightarrow$): Let $\langle T, t \rangle$ be a NeST representation of G according to the original definition. We will prove that T satisfies the condition of the Proposition. First observe that $t_v > \|T_v \cap T_u\|$ for all $u \notin \mathcal{N}[v]$, so we have $t_v > \delta_v$ for all $v \in V$.

If $xy \in E$ then either $\|T_x \cap T_y\| \geq t_x > \delta_x$ or $\|T_x \cap T_y\| \geq t_y > \delta_y$, hence

$$\|T_x \cap T_y\| > \min\{\delta_x, \delta_y\}.$$

If $xy \notin E$ then by the definition of δ_v we have $\|T_x \cap T_y\| \leq \delta_x$ and $\|T_x \cap T_y\| \leq \delta_y$.

($\Longleftarrow$): Suppose we are given a set $T = \{T_v \mid v \in V\}$ of neighborhood subtrees which satisfies $xy \in E(G)$ if and only if $\|T_x \cap T_y\| > \min\{\delta_x, \delta_y\}$, where δ_v is defined in (11.4). We now assign a tolerance to each vertex as follows. Let Q be the set of all "important" points on the host tree $\mathbf{T}$, that is, Q contains ($\forall v \in V$) the center c_v, all points at a distance r_v from c_v, and all endpoints of $\mathbf{T}$. Choose $\epsilon = \frac{1}{2}\min\{d(p, q) \mid p, q \in Q, \; p \neq q\}$ and define $t_v = \delta_v + \epsilon$. We will show that $\langle T, t \rangle$ is a NeST representation of G.

Table 11.2. *Subclasses of NeST graphs.*

Subclass	Defining condition
unit NeST	$r_v = r$, $\forall v \in V(G)$ for some constant $r > 0$
proper NeST	no T_i properly contains another T_j
bounded NeST	$t_v \leq 2r_v$, $\forall v \in V(G)$
constant tolerance NeST	$t_v = \tau$, $\forall v \in V(G)$ for some constant $\tau > 0$
NeST containment	$t_v = 2r_v$, $\forall v \in V(G)$

If $xy \in E$, then either $\|T_x \cap T_y\| > \delta_x$ or $\|T_x \cap T_y\| > \delta_y$. Without loss of generality, assume the former holds. Thus, $\|T_x \cap T_y\| - \delta_x > \epsilon$ by the definition of Q and ϵ. Therefore,

$$\|T_x \cap T_y\| > \delta_x + \epsilon = t_x \geq \min\{t_x, t_y\}.$$

If $xy \notin E$, since $\epsilon > 0$, we have $\|T_x \cap T_y\| \leq \delta_x < t_x$ and $\|T_x \cap T_y\| \leq \delta_y < t_y$, which completes the proof. □

Proposition 11.12 allows us to speak about representations without mentioning the tolerances explicitly. Thus, on one hand, given the subtrees $\mathcal{T}$ and the tolerances t (the original definition), we can construct the graph G, yet, on the other hand, given the subtrees $\mathcal{T}$ and the graph G (the alternative definition), we can construct suitable tolerances t. The recognition problem, which is still open, is given just the graph G, can we can construct the collection of subtrees $\mathcal{T}$?

We abuse terminology slightly and also call the collection $\mathcal{T} = \{T_v\}$ a NeST representation for G when it satisfies the alternative definition of Proposition 11.12.

11.6 Subclasses of NeST graphs

In the same manner as for (interval) tolerance graphs, we can define the subclasses of unit, proper, bounded, constant, and containment NeST graphs by placing restrictions on the tolerance representation. A graph $G = (V, E)$ is said to be in the NeST subclass listed in Table 11.2 if it has a NeST representation satisfying the associated condition.

We first show the equivalence of the classes of unit, proper and bounded NeST graphs. Next, we survey what is known for constant tolerance NeST graphs, which are equivalent to the neighborhood subtree intersection graphs.

Lastly, we study NeST containment graphs which are shown to be equivalent to the neighborhood subtree containment graphs.

11.6.1 Unit, proper and bounded NeST graphs

Bibelnieks and Dearing (1993) raised the question of whether the class of NeST graphs equals the class of bounded NeST graphs. This question is still open. However, unlike the situation for tolerance graphs where unit tolerance, proper tolerance and bounded tolerance are distinct classes, here the three classes unit NeST, proper NeST and bounded NeST are equivalent, as we show in the next theorem. The equivalence of (ii) and (iii) is proved in Bibelnieks and Dearing (1993) and the equivalence of (i) and (ii) is proved in Hayward, Kearney, and Malton (2002).

Theorem 11.13. *The following are equivalent for a graph G.*

(i) *G is a unit NeST graph.*
(ii) *G is a proper NeST graph.*
(iii) *G is a bounded NeST graph.*

Proof. (i)$\Longrightarrow$(ii): Unit NeST representations are easily seen to be proper.

(ii)$\Longrightarrow$(iii): Let $\langle \mathcal{T}, t \rangle$ be a proper NeST representation for G on host tree $\mathbf{T}$, where $T_v = T(c_v, r_v)$ has tolerance t_v for all $v \in V(G)$. Let us call vertices $x, y \in V(G)$ *clones* of the NeST representation when $T_x = T_y$ and $t_x = t_y$. We may assume that $\langle \mathcal{T}, t \rangle$ has no clones, for otherwise delete all but one from each set of clones, change the NeST representation into a bounded one as described below, and reintroduce the clones in the new representation.

If all tolerances are bounded, then we are done. Otherwise, let $x \in V(G)$ have tolerance $t_x = \infty$.[3] We will show that replacing t_x with the diameter $2r_x$ creates a new NeST representation of G which is proper and has fewer vertices of unbounded tolerance. Repeating this argument for each vertex having unbounded tolerance will establish (iii).

Let $y \in V(G)$ be any other vertex. If $T_x \cap T_y = \emptyset$, then $xy \notin E(G)$ regardless of changing tolerances. Therefore, we assume $T_x \cap T_y \neq \emptyset$, so that $\|T_x \cap T_y\|$ is defined.

If $xy \in E(G)$, then

$$\|T_x \cap T_y\| \geq \min\{t_x, t_y\} = t_y \geq \min\{2r_x, t_y\},$$

as required.

---

[3] As usual, for any unbounded tolerance $t_x > \|T_x\|$, we may assume $t_x = \infty$.

If $xy \notin E(G)$, then $\|T_x \cap T_y\| < \min\{t_x, t_y\} = t_y$. But since the representation is proper and has no clones, we may apply part (iii) of Lemma 11.9 to obtain $\|T_x \cap T_y\| < \|T_x\| = 2r_x$. Therefore,

$$\|T_x \cap T_y\| < \min\{2r_x, t_y\},$$

as required.

(iii)$\Longrightarrow$(i): Let $\langle \mathcal{T}, t \rangle$ be a bounded NeST representation for G on a host tree $\mathbf{T}$, where $T_v = T(c_v, r_v)$ has tolerance t_v for all $v \in V(G)$. Choose $r > \max\{r_v \mid v \in V(G)\}$. Enlarge $\mathbf{T}$ by adding, for each $v \in V(G)$, a new branch $\mathbf{L}_v$ of length $2r$ attached at the point c_v. Let $\mathbf{T}'$ be the resulting host tree. (Be careful not to let the new branches $\mathbf{L}_v$ intersect any other points of $\mathbf{T}$ or of each other.) Finally, construct a new NeST representation $\langle \mathcal{T}', t' \rangle$ on $\mathbf{T}'$ as follows: $\forall v \in V(G)$, let $T_v' = T(c_v', r_v')$ where

c_v' is located on $\mathbf{L}_v$ such that $d(c_v', c_v) = r - r_v$,

$r_v' = r$, and

$t_v' = t_v$.

Note, by the construction, that

$$T_x' \cap T_y' = \emptyset \iff T_x \cap T_y = \emptyset. \tag{11.5}$$

Claim I: $\langle \mathcal{T}', t' \rangle$ **is a unit representation.** Since $r_v' = r$, it suffices to show that each neighborhood subtree T_v' indeed has diameter $2r$. Since the original subtree T_v has diameter $2r_v$, there is a point p in $\mathbf{T}$ such that $d(p, c_v) = r_v$. Now let $q \in \mathbf{L}_v$ be the point on $\mathbf{L}_v$ where $d(q, c_v') = r$. It follows that in $\mathbf{T}'$, the path from p to q satisfies $P(p, q) = P(p, c_v) \cup P(c_v, c_v') \cup P(c_v', q)$, so $d(p, q) = r_v + (r - r_v) + r = 2r$, which proves Claim I.

Claim II: $\langle \mathcal{T}', t' \rangle$ **is a NeST representation for G.** We must show that for all $x, y \in V(G)$, $xy \in E(G) \Leftrightarrow \|T_x' \cap T_y'\| \geq \min\{t_x', t_y'\}$. We first note that since $\langle \mathcal{T}', t' \rangle$ is a unit representation, it is also proper. Thus, part (iii) of Lemma 11.9 can be applied again to obtain: if $T_x' \cap T_y' \neq \emptyset$ then

$$\|T_x' \cap T_y'\| = r_x' + r_y' - d(c_x', c_y') = 2r - d(c_x', c_y').$$

But

$$d(c_x', c_y') = d(c_x', c_x) + d(c_x, c_y) + d(c_y, c_y')$$
$$= (r - r_x) + d(c_x, c_y) + (r - r_y).$$

Thus,

$$\|T_x' \cap T_y'\| = 2r - [2r - r_x - r_y + d(c_x, c_y)] = r_x + r_y - d(c_x, c_y). \tag{11.6}$$

We complete the proof of Claim II by considering three cases.

Case 1: One of T_x or T_y contains the other: Without loss of generality, assume $T_x \subseteq T_y$. In this case, $xy \in E(G)$ since the representation is bounded, that is, $\|T_x \cap T_y\| = \|T_x\| = 2r_x \geq t_x \geq \min\{t_x, t_y\}$. We must show that $\|T'_x \cap T'_y\| \geq \min\{t'_x, t'_y\}$.

Choose endpoints p and q of T_x where $d(p, q) = 2r_x = \operatorname{diam}(T_x)$. At least one of the paths $P(p, c_x)$ or $P(q, c_x)$ has only the point c_x in common with the path $P(c_x, c_y)$, so assume it is $P(p, c_x)$. Now, $p \in T_y$ which implies that

$$r_y \geq d(p, c_y) = d(p, c_x) + d(c_x, c_y) = r_x + d(c_x, c_y).$$

Combining this with equation (11.6), we obtain

$$\|T'_x \cap T'_y\| = r_x + [r_y - d(c_x, c_y)] \geq 2r_x \geq t_x = t'_x \geq \min\{t'_x, t'_y\}$$

as required.

Case 2: T_x and T_y do not intersect. In this case we apply (11.5) so $xy \notin E(G)$ in both representations.

Case 3: T_x and T_y overlap with no containment. In this case, we apply Lemma 11.9 to obtain $\|T_x \cap T_y\| = r_x + r_y - d(c_x, c_y)$. So, by (11.6), $\|T'_x \cap T'_y\| = \|T_x \cap T_y\|$, and

$$xy \in E(G) \Leftrightarrow \|T_x \cap T_y\| \geq \min\{t_x, t_y\} \Leftrightarrow \|T'_x \cap T'_y\| \geq \min\{t'_x, t'_y\}$$

as required. This completes the proof of Claim II. $\square$

11.6.2 Constant tolerance NeST graphs

Recall from Theorem 2.5 that the class of tolerance graphs with constant tolerance is equivalent to the class of interval graphs. The analogous result for NeST graphs appears in the following proposition.

Proposition 11.14. *(Hayward and Kearney, 1993) The following are equivalent for a graph G.*

(i) *G has a NeST representation with constant tolerances.*
(ii) *G is a unit neighborhood subtree intersection graph.*
(iii) *G is a neighborhood subtree intersection graph.*

Proof. It is easy to modify the proof of Theorem 2.5 to show (i) $\Longleftrightarrow$ (ii).

(ii) $\Longrightarrow$ (iii): This follows directly from the definitions.

(iii) $\Longrightarrow$ (ii): Let G be a neighborhood subtree intersection graph. Then we can choose a small $\epsilon > 0$ so that G has a representation as a NeST graph with constant and bounded tolerance ($t_v = \epsilon$ for all v). Applying the same construction as in the proof of (iii) $\implies$ (i) of Theorem 11.13 gives a unit NeST representation of G with tolerances unchanged. If one ignores the tolerances, this representation is a neighborhood subtree intersection representation for G.

$\square$

Bibelnieks and Dearing (1993) show that constant tolerance NeST graphs are strongly chordal, and they give a separating example between these two classes. Their proof relies on a characterization of strongly chordal graphs as those whose neighborhood matrix is totally balanced (Farber, 1983) and the result in Tamir (1983) that the neighborhood matrix of a neighborhood subtree intersection graph is totally balanced. We present an alternative proof due to Uri Peled (2002).

For $k \geq 3$, a $(k, 2)$-*configuration* in a graph $G = (V, E)$ is a subgraph consisting of distinct vertices $x_1, \ldots, x_k$ and $y_1, \ldots, y_k$ such that $y_i x_j \in E$ if and only if $j = i$ or $j = i + 1$, where $k + 1$ is understood as 1. The edges $x_i x_j$ and $y_i y_j$, for $i \neq j$, may or may not be present. Clearly, a k-sun (see Figure 1.17) is the special case of a $(k, 2)$-configuration where the $\{x_i\}$ form a stable set and the $\{y_i\}$ form a clique.

Theorem 11.15. *A neighborhood subtree intersection graph G does not contain a $(k, 2)$-configuration.*

Proof. By Proposition 11.14, we may assume that $G = (V, E)$ has a neighborhood subtree intersection representation on an embedded host tree $\mathbf{T}$, in which all the radii r_v are equal to 1. For convenience, we denote the center c_v of T_v by $\dot{v}$, for all $v \in V(G)$. Thus,

$$uv \in E \iff d(\dot{u}, \dot{v}) \leq 1. \tag{11.7}$$

Claim I: For distinct vertices, u, v, z, w, if $uz, vz \in E, wz \notin E$, then $\dot{w} \notin P(\dot{u}, \dot{v})$.

Proof of Claim I. Suppose $\dot{w}$ lies on the path $P(\dot{u}, \dot{v})$, then by the properties of a tree, either $\dot{w} \in P(\dot{u}, \dot{z})$ or $\dot{w} \in P(\dot{v}, \dot{z})$. Without loss of generality, assume the former. Then, $1 < d(\dot{w}, \dot{z}) \leq d(\dot{u}, \dot{z}) \leq 1$, a contradiction, which proves Claim I.

We are now ready to prove the theorem. Suppose that, G has a $(k, 2)$-configuration, where

$$d(y_i, x_j) \leq 1 \iff j = i \text{ or } j = i + 1. \tag{11.8}$$

For all $j \neq i, i + 1$, applying Claim I with $u = x_i, v = x_{i+1}, w = x_j, z = y_i$, we have $\dot{x}_j \notin P(\dot{x}_i, \dot{x}_{i+1})$. Let T^* be the embedded subtree of $\mathbf{T}$ consisting of all the $\dot{x}_j$ and all the paths joining them.

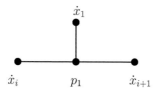

Figure 11.6. The subtree T^{**}.

Claim II: Each $\dot{x}_j$ is an endpoint of T^*.

Proof of Claim II. Suppose some $\dot{x}_j$ is an internal point of T^*. Deleting $\dot{x}_j$ from T^* would split T^* into at least two subtrees each containing a member of $\{\dot{x}_1, \ldots, \dot{x}_k\}$. Hence, there is some i such that $\dot{x}_i$ and $\dot{x}_{i+1}$ occur in different subtrees. But this implies that $\dot{x}_j$ lies on the path $P(\dot{x}_i, \dot{x}_{i+1})$, a contradiction, which proves Claim II.

For each j, let p_j be the branch point closest to $\dot{x}_j$ in T^*. By re-indexing, we may assume that

$$d(\dot{x}_1, p_1) = \min_j d(\dot{x}_j, p_j). \tag{11.9}$$

By removing p_1, we would obtain at least three subtrees, one of which contains $\dot{x}_1$ alone, and the others each contain at least one of $\{\dot{x}_2, \ldots, \dot{x}_k\}$. Hence, there must be an index i such that $\dot{x}_1$ and the consecutive terms $\dot{x}_i, \dot{x}_{i+1}$ belong to different subtrees.

Since $1 < i < k$, from (11.8) we obtain

$$d(\dot{y}_i, \dot{x}_i), d(\dot{y}_i, \dot{x}_{i+1}) \leq 1 < d(\dot{y}_i, \dot{x}_1). \tag{11.10}$$

Since $\dot{x}_1, \dot{x}_i, \dot{x}_{i+1}$ are in different subtrees, the paths between them form a subtree T^{**} as illustrated in Figure 11.6.

Thus $p_i \in P(\dot{x}_i, p_1)$ and $p_{i+1} \in P(\dot{x}_{i+1}, p_1)$, and (11.9) yields

$$d(\dot{x}_1, p_1) \leq d(\dot{x}_i, p_1), d(\dot{x}_{i+1}, p_1). \tag{11.11}$$

Let q be the point of T^{**} nearest to $\dot{y}_i$ (possibly $\dot{y}_i$ itself). If $q \in P(\dot{x}_i, p_1)$, then by (11.11) we have

$$d(\dot{y}_i, \dot{x}_{i+1}) = d(\dot{y}_i, q) + d(q, p_1) + d(p_1, \dot{x}_{i+1})$$
$$\geq d(\dot{y}_i, q) + d(q, p_1) + d(p_1, \dot{x}_1) = d(\dot{y}_i, \dot{x}_1),$$

contrary to (11.10). Similarly $q \in P(\dot{x}_{i+1}, p_1)$ leads to a contradiction. Finally, if $q \in P(\dot{x}_1, p_1)$, then by (11.10) we have

$$d(\dot{y}_i, q) + d(q, p_1) + d(p_1, \dot{x}_i) = d(\dot{y}_i, \dot{x}_i) \leq 1 < d(\dot{y}_i, \dot{x}_1)$$
$$= d(\dot{y}_i, q) + d(q, \dot{x}_1)$$

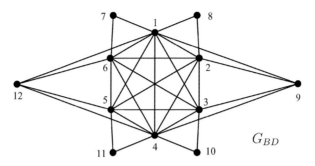

Figure 11.7. A strongly chordal graph G_{BD} which is not a constant NeST graph, but is a bounded NeST graph.

and therefore

$$d(q, p_1) + d(p_1, \dot{x}_i) < d(q, \dot{x}_1).$$

But $d(p_1, \dot{x}_1) \geq d(q, \dot{x}_1)$ and $d(q, p_1) + d(p_1, \dot{x}_i) \geq d(p_1, \dot{x}_i)$, hence $d(p_1, \dot{x}_1) > d(p_1, \dot{x}_i)$, contradicting (11.11). □

Corollary 11.16. *(Bibelnieks and Dearing, 1993) All constant tolerance NeST graphs are strongly chordal.*

Proof. By Proposition 11.14, constant tolerance NeST graphs are equivalent to the neighborhood subtree intersection graphs, which are chordal and sun-free, by Theorems 1.1 and 11.15. Thus, they are strongly chordal by Theorem 1.16. □

The converse of Corollary 11.16 is false, and a separating example from Bibelnieks and Dearing (1993) is given in Figure 11.7. This example is a bounded NeST graph, and a representation is given in Bibelnieks and Dearing (1993). A characterization of the neighborhood subtree intersection graphs as a subclass of the strongly chordal graphs is not known.

11.6.3 NeST containment graphs

In Theorem 2.6, we showed that the interval containment graphs are precisely the (interval) tolerance graphs which have a representation with tolerances equal to interval length, and which are also equivalent to the permutation graphs. Neighborhood subtree containment graphs may be characterized similarly, namely, as those having a NeST representation in which each tolerance is equal to the diameter of the assigned subtree, as proved in the next proposition. By Remark 11.11, we may assume distinct subtrees.

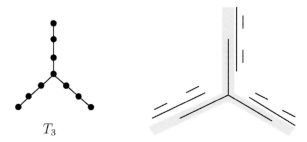

Figure 11.8. The graph T_3 and a NeST containment representation for it.

Proposition 11.17. *Let* $\mathcal{T} = \{T_v\}_{v \in V}$ *be a collection of distinct neighborhood subtrees. The following are equivalent.*

(i) $G = (V, E)$ *is the containment graph of* $\mathcal{T}$.

(ii) $G = (V, E)$ *is the NeST graph of the representation* $\langle \mathcal{T}, \{2r_v\} \rangle$ *where* $2r_v = \|T_v\|$.

Proof. We prove the following equivalence, where the left side defines an edge xy in the containment graph of $\mathcal{T}$ and the right side defines an edge xy in the NeST graph of $\langle \mathcal{T}, \{2r_v\} \rangle$,

$$T_x \subset T_y \text{ or } T_y \subset T_x \iff \|T_x \cap T_y\| \geq \min\{\|T_x\|, \|T_y\|\}.$$

If $T_x \cap T_y = \emptyset$, then $\|T_x \cap T_y\|$ is undefined, and the equivalence holds (both sides are false).

If $T_x \cap T_y \neq \emptyset$, then by Lemma 11.9(i) and (iii), we have

$$T_x \not\subset T_y \text{ and } T_y \not\subset T_x \iff \|T_x \cap T_y\| < \min\{\|T_x\|, \|T_y\|\}. \qquad \square$$

Recall from Theorem 3.7 that the tree T_3 in Figure 11.8 is not a tolerance graph. However, it is a neighborhood subtree containment graph, as shown by the representation in Figure 11.8 from Bibelnieks and Dearing (1993). It is not yet known which trees are NeST containment graphs nor which comparability graphs fail to be NeST containment. Figure 11.9 shows separation examples for the classes of interval containment, NeST containment and subtree containment.

11.7 The hierarchy of NeST graphs

Recall that a graph is weakly chordal if it contains no chordless cycle of length $k \geq 5$ nor the complement of a chordless cycle of length $k \geq 5$. NeST graphs are weakly chordal (Bibelnieks and Dearing, 1993), hence they are perfect graphs.

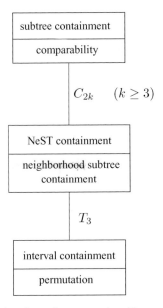

Figure 11.9. A hierarchy of containment graphs with examples separating them.

The proof uses methods similar to those outlined at the beginning of Section 2.5 for tolerance graphs, using the characterization of weakly chordal graphs by two-pairs. We state this result here.

Theorem 11.18. *(Bibelnieks and Dearing, 1993) Every NeST graph is weakly chordal.*

The graph class inclusion in this theorem is proper. An infinite family of weakly chordal graphs, called *m-stars*, which are not NeST graphs, is given by Hayward, Kearney, and Malton (2002) and presented here. The first three members of this family are illustrated in Figure 11.10. Note that the black vertices induce a complete bipartite graph. This is the essential feature of the family.

An m-star G_{4m} ($m \geq 2$) has vertex set $V = S \cup C$ where $S = \{s_0, \ldots, s_{2m-1}\}$ is a stable set, $C = \{c_0, \ldots, c_{2m-1}\}$ induces a complete bipartite graph with $c_i c_j \in E \Leftrightarrow i$ and j have opposite parity, and for $0 \leq k \leq 2m - 1$, we have s_k is adjacent to c_k and $c_{k+1} \pmod{2m}$.

Proposition 11.19. *(Hayward, Kearney, and Malton, 2002) The m-star G_{4m} is weakly chordal but is not a NeST graph.*

The reader can easily show that G_{4m} is weakly chordal (Exercise 11.10). The fact that it is not NeST relies on another characterization of NeST graphs

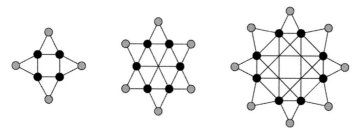

Figure 11.10. Weakly chordal graphs which are not NeST graphs.

involving representations as phylogenetic trees, and can be found in Hayward, Kearney, and Malton (2002).

Figure 11.11 shows a hierarchy of classes of NeST graphs along with related classes. This hierarchy is not complete, and we list the open questions to complete it in Remark 11.21. It is not yet known whether the classes of strongly chordal and chordal graphs are incomparable with bounded NeST and NeST. It is also not yet known if there exists a separating example between bounded NeST and NeST.

Theorem 11.20. *The class hierarchy edges and the separating examples illustrated in Figure 11.11 are correct.*

Proof. The equivalence of the classes unit, proper and bounded NeST was proven in Theorem 11.13, and the equivalence of the classes of neighborhood subtree intersection graphs, constant tolerance NeST graphs and unit neighborhood subtree intersection graphs was proven in Proposition 11.14.

The inclusions and separating examples for *interval* $\subset$ *unit tolerance* $\subset$ *proper tolerance* $\subset$ *bounded tolerance* $\subset$ *tolerance* are repeated from Figure 2.8, and the inclusion *NeST* $\subset$ *weakly chordal* is given in Theorem 11.18 with the separating example given by Proposition 11.19.

The inclusion *constant tolerance NeST* $\subset$ *strongly chordal* is given in Corollary 11.16 with the separating example G_{BD} found in Figure 11.7 due to Bibelnieks and Dearing (1993), who also give a bounded NeST representation for this graph. In Exercise 11.8, the reader will show that every tree is a constant tolerance NeST graph. Thus, the tree T_3 is contained in all of the NeST families, however, T_3 is not a tolerance graph as shown in Theorem 3.7. The remaining inclusions shown in the figure: *tolerance* $\subset$ *NeST, bounded tolerance* $\subset$ *bounded NeST* $\subseteq$ *NeST, interval* $\subset$ *constant tolerance NeST* $\subset$ *bounded NeST*, and *strongly chordal* $\subset$ *chordal* $\subset$ *weakly chordal*, follow from the definitions. The placement of the remaining separating examples is easily verified. □

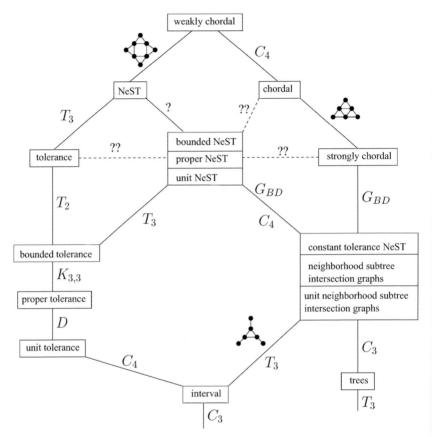

Figure 11.11. The hierarchy of NeST graphs. The symbol "?" indicates that we do not know whether the inclusion has a separating example or not. The symbol "??" indicates that we do not know whether the classes are incomparable or not.

Remark 11.21. The following open questions remain in the NeST graph hierarchy.

(i) Is there a separating example between the classes of bounded NeST and NeST graphs, or are they equivalent?

(ii) Is the class of tolerance graphs incomparable with the class of bounded NeST graphs, or is the former contained in the latter?

(iii) What is the relationship between the classes of chordal and strongly chordal graphs compared to the NeST and bounded NeST graphs?

Question 11.22. Looking ahead, Theorem 11.25 shows that the class of threshold tolerance (TT) graphs is contained in the class of bounded NeST graphs and that the class of threshold graphs is precisely the intersection of TT graphs and constant tolerance NeST graphs. How does the class of TT graphs relate to the other classes in the hierarchy in Figure 11.11?

11.8 A connection with threshold and threshold tolerance graphs

Definition 11.23. A graph $G = (V, E)$ is threshold tolerance (TT) if its vertices can be assigned positive weights $\{w_v \mid v \in V\}$ and positive tolerances $\{t_v \mid v \in V\}$ such that

$$xy \in E(G) \Leftrightarrow w_x + w_y \geq \min\{t_x, t_y\}. \tag{11.12}$$

Figure 11.12 shows a threshold tolerance graph G and an assignment of weights and tolerances $[w_v, t_v]$ is given next to each vertex v.

Threshold tolerance graphs were introduced in Monma, Reed, and Trotter (1988). The special case in which all the t_v are equal is equivalent to one of the well known characterizations of the class of threshold graphs, as we saw in Theorem 1.17.

The complements of threshold tolerance graphs are called *coTT graphs,* which we will encounter in Chapter 12 as being equivalent to the sum-tolerance chain graphs. Monma, Reed, and Trotter (1988) proved the following characterization of coTT graphs.

Lemma 11.24. *A graph* $\overline{G} = (V, \overline{E})$ *is a coTT graph if and only if positive numbers* a_v *and* b_v *can be assigned to each vertex* $v \in V$ *such that*

$$xy \in E(\overline{G}) \Leftrightarrow a_x \leq b_y \text{ and } a_y \leq b_x.$$

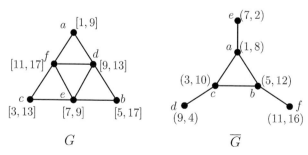

Figure 11.12. (a) A threshold tolerance graph G with an assignment $[w_v, t_v]$ for each $v \in V(G)$. (b) A coTT graph $\overline{G}$ with an assignment (a_v, b_v) for each $v \in V(G)$.

Proof. First note that, without loss of generality, we may require the inequality in (11.12) to be strict, by adding a very small $\epsilon > 0$ to each w_v if needed. Second, we may also assume that $t_v > w_v$ for all $v \in V$ by adding a very big number B to each w_v and $2B$ to each t_v if needed. With these assumptions, the lemma can easily be verified by setting $a_v = w_v$ and $b_v = t_v - w_v > 0$ since

$$a_x \leq b_y \text{ and } a_y \leq b_x \quad \Leftrightarrow \quad w_x + w_y \leq \min\{t_x, t_y\}. \qquad \Box$$

The graph $\overline{G}$ in Figure 11.12 is coTT, and an assignment of positive numbers (a_v, b_v) is given next to each vertex v. This figure illustrates the transformation between $[w_v, t_v]$ and (a_v, b_v) used in the proof of Lemma 11.24.

We leave it as an exercise to show that the tree T_2 is neither TT nor coTT (Exercise 11.11).

The main result in this section characterizes the threshold tolerance graphs (TT graphs) as a special type of NeST graph, specifically, those having a NeST representation in which all pairs of centers of the neighborhood subtrees are equidistant. Formally, a NeST representation has *equidistant centers* if $d(c_x, c_y) = d$ for some constant d and all $x \neq y$.

In the next theorem, the equivalence (i) $\Leftrightarrow$ (v) is due to M. Saks as cited in Monma, Reed, and Trotter (1988), and the other equivalences are due to Hayward, Kearney, and Malton (2002). An *embedded star* is a tree with at most one branch point.

Theorem 11.25. *The following are equivalent conditions for a graph G.*

(i)   *G is a threshold tolerance graph.*
(ii) *G has a NeST representation with equidistant centers.*
(iii) *G has a NeST representation with equidistant centers which is proper and whose host* **T** *is an embedded star, with one subtree center per branch.*
(iv) *There exists an assignment of positive numbers $\{r_v \mid v \in V(G)\}$ such that*

$$xy \in E(G) \Leftrightarrow \begin{cases} r_z < r_x \ \forall z \notin \mathcal{N}[y] \\ or \\ r_u < r_y \ \forall u \notin \mathcal{N}[x]. \end{cases}$$

(v)   *There exists a total ordering $\prec$ of the vertices $V(G)$ such that*

$$xy \in E(G) \Leftrightarrow \begin{cases} z \prec x \ \forall z \notin \mathcal{N}[y] \\ or \\ u \prec y \ \forall u \notin \mathcal{N}[x]. \end{cases}$$

Proof. (i) $\Longrightarrow$(iii): Suppose $G = (V, E)$ is a threshold tolerance graph and we have an assignment of weights and tolerances such that $xy \in E \Leftrightarrow w_x + w_y \geq \min\{t_x, t_y\}$. Choose some $m > \max\{w_v \mid v \in V\}$, and let **T** be a star embedded in the plane with $n = |V|$ line segments each of length $3m$. Place the center c_v of each neighborhood subtree T_v at a distance of m from the central node of the star, one per line segment, so the centers will be distinct. Define $r_v = m + w_v$ and $T_v = T(c_v, r_v)$. Clearly, by our construction, $\|T_x \cap T_y\| = w_x + w_y$. So, $xy \in E \Leftrightarrow \|T_x \cap T_y\| \geq \min\{t_x, t_y\}$. Therefore, $\langle \{T_v\}, \{t_v\} \rangle$ is a NeST representation for G on the star **T** with all subtree centers equidistant. Finally, we note that this representation is proper since $m < r_v < 2m$ for all v.

(iii) $\Longrightarrow$(ii): Trivial.

(ii) $\Longrightarrow$(iv): Suppose we have a NeST representation $\mathcal{T}$ for G, where $T_v = T(c_v, r_v)$ with δ_v defined as in Proposition 11.12, and such that all pairs of neighborhood subtree centers are equidistant.[4] We will prove (iv) using the radii r_v for all $v \in V(G)$.

Since the centers are equidistant, we have, for all distinct u, v, w,

$$\|T_u \cap T_v\| < \|T_u \cap T_w\| \Rightarrow r_v < r_w.$$

If $xy \in E(G)$ then either $\delta_x < \|T_x \cap T_y\|$ or $\delta_y < \|T_x \cap T_y\|$. In the first case,

$$\|T_x \cap T_u\| \leq \delta_x < \|T_x \cap T_y\| \quad \forall u \notin \mathcal{N}[x]$$

which implies that $r_u < r_y$, $\forall u \notin \mathcal{N}[x]$. Similarly, in the second case, $r_z < r_x$, $\forall z \notin \mathcal{N}[y]$. This proves the forward direction of (iv).

If $xy \notin E(G)$ then $x \notin \mathcal{N}[y]$ and $y \notin \mathcal{N}[x]$. So the first condition of (iv), $r_z < r_x \forall z \notin \mathcal{N}[y]$, fails with $z = x$ and the second condition fails with $u = y$.

(iv) $\Longrightarrow$(v): Choose any total ordering $\prec$ of $V(G)$ such that $v \prec w \Rightarrow r_v \leq r_w$. Thus, $r_w < r_v \Rightarrow w \prec v$.

If $xy \in E(G)$ then either $r_z < r_x \forall z \notin \mathcal{N}[y]$ or $r_u < r_y \forall u \notin \mathcal{N}[x]$. The former implies $z \prec x \forall z \notin \mathcal{N}[y]$ and the latter implies $u \prec y \forall u \notin \mathcal{N}[x]$.

Conversely, if $xy \notin E(G)$ then $x \notin \mathcal{N}[y]$ and $y \notin \mathcal{N}[x]$. So the first condition $z \prec x \forall z \notin \mathcal{N}[y]$ fails with $z = x$ and the second condition fails with $u = y$.

(v) $\Longrightarrow$(i): Let $\prec$ be a total ordering of the vertices satisfying the conditions of (v). We will construct a coTT assignment for $\overline{G}$ using Lemma 11.24, thus showing that G is a threshold tolerance graph. Let a_v equal the position of v in

---

[4] In the transformation between the two definitions of NeST in Proposition 11.12, the host tree and the neighborhood subtrees remained unchanged. Therefore, the property of equidistant centers would also remain unchanged.

the order $\prec$ and defining $b_v = \frac{1}{2}$ if v is a universal vertex and $b_v = \max\{a_u \mid u \notin \mathcal{N}[v]\}$ otherwise. We first note that one can use two equivalent ways to express the meaning that *all non-neighbors of v occur strictly below w in the ordering*, this is,

$$u \prec w \ \forall u \notin \mathcal{N}[v] \Leftrightarrow b_v < a_w. \tag{11.13}$$

Now applying (v), if $xy \in E(G)$ (that is, $xy \notin E(\overline{G})$), then either $z \prec x \ \forall z \notin \mathcal{N}[y]$ or $u \prec y \ \forall u \notin \mathcal{N}[x]$. In the first case, setting $w = x$, $v = y$ and $u = z$ in (11.13) we have $b_y < a_x$; in the second case, setting $w = y$ and $v = x$ gives $b_x < a_y$.

Conversely, if $xy \notin E(G)$, that is, $xy \in E(\overline{G})$, then both conditions of (v) fail, implying that $a_x \leq b_y$ (using (11.13) with $w = x$, $v = y$ and $u = z$) and $a_y \leq b_x$ (using $w = y$ and $v = x$). Therefore, $xy \in E(\overline{G})$ if and only if $a_x \leq b_y$ and $a_y \leq b_x$. By Lemma 11.24, $\overline{G}$ is coTT and G is threshold tolerance. □

We leave it as an exercise to show that the graph $\overline{G}$ in Figure 11.12 is not TT and hence G is not coTT.

Theorem 11.26. *(Hayward, Kearney, and Malton, 2002) A graph G is a threshold graph if and only if G has a NeST representation in which both (a) all pairs of centers are equidistant and (b) all tolerances are equal.*

Proof. ($\Longrightarrow$): Recall that threshold graphs can be viewed as the special case of threshold tolerance where all the t_v must be equal. If G is a threshold graph, then the construction that we used to prove the implication (i) $\Longrightarrow$ (iii) in Theorem 11.25 can also be applied here with the added assumption that all tolerances t_v are equal.

($\Longleftarrow$): Consider a NeST representation for G with neighborhood subtrees $\{T(c_v, r_v)\}$ and tolerances $\{t_v\}$ in which all pairs of centers are equidistant and all tolerances are equal, say $t_v = \tau \ \forall v \in V(G)$. Using the hereditary property that all induced subgraphs also satisfy (a) and (b) together with the characterization of threshold graphs in Theorem 1.17, it is sufficient to show that G has either a universal or an isolated vertex.

Let $y \in V(G)$ satisfy $r_y = \max\{r_v \mid v \in V(G)\}$, and let x be any other vertex. If x is isolated, then we are done. Otherwise, let z be a neighbor of x, so $\|T_x \cap T_z\| \geq \tau$. However, since $r_y \geq r_z$ we must have $\|T_x \cap T_y\| \geq \|T_x \cap T_z\| \geq \tau$, so $xy \in E(G)$. Thus, y is universal. □

11.9 Exercises

Exercise 11.1. Give a representation for the chordless cycle C_k as the edge intersection graph of paths in a tree (EPT).

Exercise 11.2. Give an EPT representation for the 4-sun S_4 on a host tree with maximum degree 3.

Exercise 11.3. Prove that if G is a chordal graph, then G has a representation as an intersection graph of subtrees of a tree **T** having degree at most 3, that is, $[\infty, \infty, 1] = [3, 3, 1]$.

Exercise 11.4. Prove the monotone inclusions between the classes in Remark 11.6.

Exercise 11.5. *The Pie Lemma* (Golumbic and Jamison, 1985b) An n-pie in an EPT representation $\mathcal{T} = \{T_v\}_{v \in V(G)}$ of G on a host tree **T** consists of n distinct edges $a_i b$ ($i = 1, \ldots, n$) of **T** all incident on a common vertex b, with n distinct $T_i \in \mathcal{T}$ such that $T_i \cap \{a_j b \mid j = 1, \ldots, n\} = \{a_i b, a_{i+1} b\}$, arithmetic modulo n. Prove that the only way to represent a chordless cycle C_n in an EPT graph is with an n-pie in the host tree. That is, if G contains a chordless n-cycle, then any EPT representation of G must contain an n-pie.

Exercise 11.6. (Jamison and Mulder, 2000a) Extend the Pie Lemma to any $(h, s, 2)$-subtree representation, and show that if $G \in [h, s, 2]$, then G contains no chordless cycle of length strictly greater than h.

Exercise 11.7. For the tree in Figure 11.3 let $c = (4, 1)$, $c' = (3, 1)$, $c'' = (1, 1)$ and $T_1 = T(c, \sqrt{2})$, $T_2 = T(c', 2 + 2\sqrt{2})$, $T_3 = T(c'', 3)$. Find the center and the radius of each of the neighborhood subtrees $T_1 \cap T_2$, $T_1 \cap T_3$, $T_2 \cap T_3$ and $T_1 \cap T_2 \cap T_3$.

Exercise 11.8. Prove that every tree is a neighborhood subtree intersection graph, and hence a constant tolerance NeST graph.

Exercise 11.9. Prove condition (ii) of Corollary 11.10.

Exercise 11.10. Prove that the m-star G_{4m} is a weakly chordal graph.

Exercise 11.11. Show that the tree T_2 (shown in Figure 2.2) is neither TT nor coTT.

Exercise 11.12. Show that the graph $\overline{G}$ in Figure 11.12 is not TT and hence G is not coTT.

Exercise 11.13. Given a unit neighborhood subtree intersection representation for G, prove the following using the notation in the proof of Theorem 11.18. For distinct vertices u, v, w, z, if $uz, vz \in E$ and $uw, vw \notin E$, then $\dot{w} \notin P(\dot{u}, \dot{v})$.

Chapter 12
ϕ-tolerance graphs

We have already seen several generalizations of tolerance and bounded tolerance graphs in this book. In defining bounded bitolerance graphs (Chapter 5), we allowed the assignment of different tolerances to the right and left sides of the intervals. In defining NeST tolerance graphs (Chapter 11), we replaced the real line by a tree and the intervals were replaced by neighborhood subtrees. The class of bounded bitolerance graphs properly contains the bounded tolerance graphs, and class of NeST tolerance graphs properly contains the class of tolerance graphs.

We have also presented a number of restrictions of tolerance graphs such as the subclasses of unit tolerance graphs (Chapter 2), probe graphs (Chapter 4), and threshold tolerance graphs (Section 11.8). All of these, like interval graphs and permutation graphs, are properly contained in the class of tolerance graphs.

In all of our tolerance representations so far, an edge is added to the graphs when the size of the intersection of two intervals is large enough to "bother" one of them. In the case of tolerance, $ij \in E \Leftrightarrow |I_i \cap I_j| \geq \min\{t_i, t_j\}$. In this chapter, we turn our attention to variations of this condition where the operation "min" is replaced by another binary function ϕ, for example "max" or "sum".

12.1 Introduction

Let ϕ be a symmetric binary function, positive valued on positive arguments. A graph $G = (V, E)$ is a ϕ-*tolerance graph* if there is an interval representation $\{I_i\}$ with positive tolerances $\{t_i\}$ satisfying

$$ij \in E \Leftrightarrow \left|I_i \cap I_j\right| \geq \phi(t_i, t_j).$$

The resulting ϕ-tolerance graphs may be quite different from our usual tolerance graphs (i.e., min-tolerance graphs). For example, tolerance graphs are weakly chordal (Theorem 2.17) whereas, as we will see in Corollary 12.12, max-tolerance and sum-tolerance graphs are not weakly chordal and may, in fact, contain chordless cycles of all sizes. This remains true for a large family of functions denoted by Φ, which we call Archimedean functions, and we discuss many results about this family in Section 12.3. Since the families of max-tolerance and sum-tolerance graphs may contain odd length chordless cycles, they are not perfect graphs.

Rather little is known about max-tolerance and sum-tolerance graphs, although we will develop some additional properties for them in this chapter. We begin with some observations which are analogous to our early results on min-tolerance, justifying each after its statement.

Remark 12.1. All max-tolerance and sum-tolerance graphs have bounded representations.

Any vertex with tolerance strictly greater than the length of its interval must be an isolated vertex in a max-tolerance or sum-tolerance graph. For any such vertex, we may reassign to it a different interval, disjoint from all others, and an arbitrary bounded tolerance. Indeed, this remark holds for any ϕ-tolerance graph where $\phi(x, y) \geq x, \forall x$. It also applies to unit and proper representations.

Remark 12.2. For any function ϕ, the ϕ-tolerance graphs with constant tolerances ($t_i = c, \forall i$) yield precisely the class of interval graphs.

Just as was the case for min-tolerance (Theorem 2.5), the result follows because $\phi(t_i, t_j) = \phi(c, c)$ is also a constant.

Remark 12.3. For representations where all tolerances equal the length of their intervals ($t_i = |I_i|, \forall i$), max-tolerance graphs consist of a union of disjoint cliques and sum tolerance graphs are stable sets.

In the former case, identical intervals (duplicates) are the only adjacencies, and in the latter case there are no edges.

In Section 12.2. we study the *tolerance chain graphs* (min, max and sum) which are defined to be those that have a representation consisting of a nested set of intervals (i.e., a set of intervals ordered by inclusion). After introducing Archimedean functions in Section 12.3, we then turn our attention to results where ϕ is a polynomial function, which are discussed in Sections 12.4 and 12.5. Then, in Section 12.6 we prove there is a universal function ϕ^* so that every graph is a ϕ^*-tolerance graph. Finally, in Section 12.7 we present results on unit and proper ϕ-tolerance graphs. Sections 12.3, 12.5, and 12.6 are based on Golumbic, Jamison, and Trenk (2002).

12.2 φ-tolerance chain graphs

In this section, we consider the special case of φ-tolerance graphs in which the representation consists of a nested family of intervals (i.e., a set of intervals ordered by inclusion). This class was introduced in Jacobson, McMorris, and Mulder (1991). We define the class of *φ–tolerance chain graphs* to be the φ–tolerance graphs of such a nested family. We concentrate here on the cases where φ equals "min", "max", and "sum". Further generalizations of φ-tolerance chain graphs are investigated in Golumbic and Jamison (2003).

The classes of min-tolerance chain, max-tolerance chain and sum-tolerance chain are shown in the hierarchy in Figure 12.1 as a series of inclusions, along

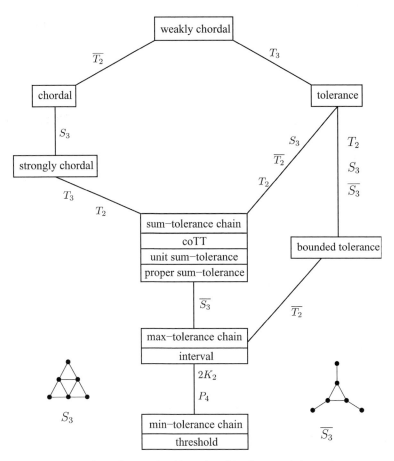

Figure 12.1. The complete hierarchy of φ-tolerance chain graphs.

with their relationships to other familiar classes of graphs. We begin by proving
the equivalences of classes appearing in the same box.

In proving results about ϕ-tolerance chain graphs, it is useful to have a *normalized* representation, which we now define. Any nested family of intervals
$\{I_i\}$ can be *normalized* by replacing each interval $I_i = [L(i), R(i)]$ by the interval $I_i' = [0, r_i]$ where $r_i = R(i) - L(i) = |I_i|$ and observing that this preserves
the lengths of the intervals and their nesting. Thus, without loss of generality,
we may choose our nested family to be of the form $\mathcal{N} = \{ [0, r_i] \mid i = 1, \ldots, n \}$
with $0 < r_1 \leq r_2 \leq \cdots \leq r_n$ which we call a *normalized* representation.

We next present three equivalence theorems. Recall that a vertex is called
universal if it is adjacent to all other vertices, and is called *isolated* if it is
adjacent to no other vertex. We need the following characterization of threshold
graphs from Chvátal and Hammer (1977) which appears in our Theorem 1.17.
A graph $G = (V, E)$ is a threshold graph if and only if for each subset $X \subseteq V$
there exists a vertex $x \in X$ which is either universal or isolated in the induced
subgraph G_X.

Theorem 12.4. *(Jacobson, McMorris, and Mulder, 1991) A graph is a min-tolerance chain graph if and only if it is a threshold graph.*

Proof. Let $\langle \mathcal{N}, t \rangle$ be a normalized min-tolerance chain graph representation
of $G = (V, E)$, where interval $I_i = [0, r_i]$ has tolerance t_i and is assigned to
vertex $v_i \in V$ and with $0 < r_1 \leq r_2 \leq \cdots \leq r_n$. Since being a min-tolerance
chain graph and being a threshold graph are hereditary properties, it suffices to
show that G has a universal vertex or an isolated vertex.

If $t_1 \leq r_1$ (bounded tolerance), then v_1 is a universal vertex since $|I_1 \cap I_j| =$
$|I_1| = r_1 \geq t_1 \geq \min\{t_1, t_j\}$ for all j. If $t_1 > r_1$ (unbounded tolerance) and v_1
is not an isolated vertex, then v_1 has a neighbor v_k which implies that $t_k \leq r_1$.
Therefore, v_k is a universal vertex since $|I_k \cap I_j| = \min\{r_k, r_j\} \geq r_1 \geq t_k \geq$
$\min\{t_k, t_j\}$ for all j. This proves that G is a threshold graph.

Conversely, suppose that G is a threshold graph and that x is either
an isolated or universal vertex. By induction, assume we have a normalized min-tolerance chain graph representation $\langle \mathcal{N}', t \rangle$ of $G - x$, and let $m = \min\{t_i, r_i \mid i = 1, \ldots, n$ and $v_i \neq x\}$. To obtain a min-tolerance representation
for G, we add the interval $I_x = [0, m/2]$ and assign tolerance $t_x = m$ if x is an
isolated vertex or $t_x = m/2$ if x is a universal vertex. The details of verifying
this representation are straightforward and are left to the reader. $\square$

Theorem 12.5. *(Jacobson, McMorris, and Mulder, 1991) A graph is a max-tolerance chain graph if and only if it is an interval graph.*

Proof. Suppose that $G = (V, E)$ is an interval graph with interval representation $\mathcal{J} = \{J_i\}_{v_i \in V}$ satisfying $v_i v_j \in E \Leftrightarrow J_i \cap J_j \neq \emptyset$. Let $J_i = [\ell_i, r_i]$ and let us assume, by shifting and reordering the intervals if necessary, that $\ell_i > 0$ for all i and $0 < r_1 \leq r_2 \leq \cdots \leq r_n$. In this case, the new set of intervals $I_i = [0, r_i]$ form a nested chain. We assign tolerance ℓ_i to vertex $v_i \in V$ and easily verify that for all $i < j$,

$$J_i \cap J_j \neq \emptyset \iff r_i \geq \ell_j \iff |I_i \cap I_j| = r_i \geq \max\{\ell_i, \ell_j\}.$$

Thus, G is max-tolerance chain graph.

Conversely, assume we are given a normalized max-tolerance chain graph representation of $G = (V, E)$, where vertex $v_i \in V$ is assigned interval $I_i = [0, r_i]$ and tolerance t_i. We may assume that $t_i \leq r_i$ (since this is a max-tolerance representation and any vertex v_i with unbounded tolerance would be isolated so its interval $[0, r_i]$ and tolerance t_i could be replaced by $[0, r_i']$ where $t_i' = r_i' > \max\{r_j \mid 1 \leq j \leq n\}$). Thus v_i is adjacent to v_j for $i < j$ if and only if $|I_i \cap I_j| = r_i \geq \max\{t_i, t_j\}$ which occurs precisely when $r_i \geq t_j$ since $r_i \geq t_i$. It is easy to verify that the set of intervals

$$\mathcal{J} = \{J_k \mid J_k = [t_k, r_k], k = 1, \ldots, n\}$$

is an interval representation for G, since

$$J_i \cap J_j \neq \emptyset \ (\forall i < j) \iff t_j \leq r_i.$$

Therefore, G is an interval graph. □

The class of threshold tolerance graphs and their complements, the coTT graphs, were introduced by Monma, Reed, and Trotter (1988) who also give a polynomial time recognition algorithm for the class. Recall from Lemma 11.24 that a graph $G = (V, E)$ is a *coTT* graph if positive numbers a_v and b_v can be assigned to each vertex $v \in V$ such that $xy \in E \Leftrightarrow a_x \leq b_y$ and $a_y \leq b_x$. In Theorem 11.25, we characterized threshold tolerance graphs. We present here the observation that coTT graphs are equivalent to the sum-tolerance chain graphs.

Theorem 12.6. *(Jacobson, McMorris, and Mulder, 1991) A graph is a sum-tolerance chain graph if and only if it is a coTT graph.*

Proof. By simple arithmetic, we observe that

$$\min\{r_i, r_j\} \geq t_i + t_j \text{ if and only if } t_i \leq r_j - t_j \text{ and } t_j \leq r_i - t_i.$$
$$\tag{12.1}$$

Let $\langle \mathcal{N}, t \rangle$ be a normalized sum-tolerance chain graph representation of $G = (V, E)$ where interval $I_i = [0, r_i]$ has tolerance t_i ($\leq r_i$) satisfying

$$v_i v_j \in E \Leftrightarrow |I_i \cap I_j| = \min\{r_i, r_j\} \geq t_i + t_j.$$

We may ensure that $t_i < r_i$ for all i by adding a large number B to each t_i and adding $2B$ to each r_i if needed. Using (12.1), the assignment $a_i = t_i$ and $b_i = r_i - t_i$ satisfies

$$t_i \leq r_j - t_j \quad \text{and} \quad t_j \leq r_i - t_i$$

and thus G is a coTT graph.

The steps may be reversed to prove the converse.   □

This, together with Theorem 12.32, completes the proofs of the equivalences between classes in the same box in the hierarchy in Figure 12.1. Next we prove results to justify containments between classes.

Theorem 12.7. *(Monma, Reed, and Trotter, 1988)*

(i) *Every interval graph is a coTT graph.*

(ii) *Every coTT graph is a (min-)tolerance graph.*

Proof. (i) Let $G = (V, E)$ be an interval graph with interval representation $\mathcal{J} = \{J_v = [\ell_v, r_v]\}_{v \in V}$. By defining $a_v = \ell_v$ and $b_v = r_v$, we obtain a coTT assignment, since $xy \in E \Leftrightarrow J_x \cap J_y \neq \emptyset \Leftrightarrow \ell_x \leq r_y$ and $\ell_y \leq r_x$.

(ii) Let $G = (V, E)$ be a coTT graph with coTT assignment $\{a_v, b_v \mid v \in V\}$. Define the interval $I_v = [a_v, a_v + b_v]$ and tolerance $\tau_v = a_v$. We show that $\langle \{I_v\}, \tau \rangle$ is a min-tolerance representation for G.

This can be seen as follows. Assume, without loss of generality, that $a_x \leq a_y$. If $a_x \leq b_y$ and $a_y \leq b_x$, then $I_x \cap I_y \neq \emptyset$ so $|I_x \cap I_y| = \min\{b_y, a_x + b_x - a_y\}$. But $t_x = a_x \leq b_y$ and $t_x = a_x \leq a_x + (b_x - a_y)$, so $|I_x \cap I_y| \geq t_x \geq \min\{t_x, t_y\}$.

Conversely, if $|I_x \cap I_y| \geq t_x$, then both $b_y \geq t_x = a_x$ and $a_x + b_x - a_y \geq t_x$ so $b_x \geq a_y$.   □

Recall from Section 1.7.2 that a vertex x is called *simple* if for every pair of neighbors y and z of x, either $\mathcal{N}[y] \subseteq \mathcal{N}[z]$ or $\mathcal{N}[z] \subseteq \mathcal{N}[y]$. An ordering of the vertices $[v_1, \ldots, v_n]$ is called a *simple elimination ordering* of G if for all i, v_i is a simple vertex in the remaining induced subgraph $G_{\{v_i, \ldots, v_n\}}$ after $v_1, \ldots, v_{i-1}$ have been eliminated. We use the characterization from Theorem 1.16 which states that a graph $G = (V, E)$ is strongly chordal if and only if G has a simple elimination ordering.

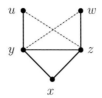

Figure 12.2. A graph used in the proof of Theorem 12.8 where dotted lines indicate nonedges.

Theorem 12.8. *(Monma, Reed, and Trotter, 1988) Every sum-tolerance chain graph (i.e., coTT graph) is a strongly chordal graph.*

Proof. Consider a normalized sum-tolerance chain graph representation of $G = (V, E)$, where v_i is assigned the interval $I_i = [0, r_i]$ and tolerance t_i. Since being a sum-tolerance chain graph is a hereditary property, it suffices to show that G has a simple vertex; the theorem then follows by induction.

Claim: Any vertex x whose tolerance t_x is largest in the representation is a simple vertex.

Let y and z be neighbors of x. We first show that $yz \in E$. Since

$$t_y + t_z \leq t_x + t_z \leq |I_x \cap I_z| \leq r_z \quad \text{and}$$
$$t_y + t_z \leq t_y + t_x \leq |I_y \cap I_x| \leq r_y$$

we obtain

$$t_y + t_z \leq \min\{r_y, r_z\} = |I_y \cap I_z|$$

implying that $yz \in E$. Suppose, to obtain a contradiction, there are vertices $u \in \mathcal{N}[y] - \mathcal{N}[z]$ and $w \in \mathcal{N}[z] - \mathcal{N}[y]$ as illustrated in Figure 12.2 (dotted lines indicate nonedges). On one hand, we have

$$r_y \geq \min\{r_y, r_x\} = |I_y \cap I_x| \geq t_y + t_x \geq \boxed{t_y + t_w}$$
$$> |I_y \cap I_w| = \min\{r_y, r_w\} = r_w \geq \min\{r_z, r_w\} \geq \boxed{t_z + t_w}$$

so $t_y > t_z$.

By symmetry, we could apply the same inequalities, interchanging y with z and w with u, to obtain $t_z > t_y$, a contradiction. This proves the claim and completes the proof of the theorem.   □

Now we are ready to present the main result of this section.

Theorem 12.9. *The class hierarchy and separating examples illustrated in Figure 12.1 are correct. Moreover, the hierarchy is complete.*

Table 12.1. *Containments in the hierarchy of Figure 12.1.*

Relationship between classes ...	... proved in
threshold $\subseteq$ interval	Theorem 1.18
interval $\subseteq$ coTT	Theorem 12.7(i)
interval $\subseteq$ bounded tolerance	Theorem 2.5
coTT $\subseteq$ tolerance	Theorem 12.7(ii)
bounded tolerance $\subseteq$ tolerance	definition
coTT $\subseteq$ strongly chordal	Theorem 12.8
strongly chordal $\subseteq$ chordal $\subseteq$ weakly chordal	definition
tolerance $\subseteq$ weakly chordal	Theorem 2.17

Proof. As noted earlier, the equivalences of classes in the same box in Figure 12.1 are justified in Theorems 12.4, 12.5, 12.6, and 12.32. Table 12.1 shows where the proofs of the containments between classes can be found. Next we prove that the examples shown along edges of the diagram are separating examples.

   The graph $\overline{T_2}$. The graphs T_2 and $\overline{T_2}$ are shown in Figure 2.2. From Figure 2.8, $\overline{T_2}$ is a bounded tolerance graph and hence it is a tolerance graph and is weakly chordal. Since $\overline{T_2}$ has a chordless 4-cycle, it is not chordal, hence it is neither an interval graph nor coTT.

   The graph T_3. The graph T_3 separates tolerance from weakly chordal by Figure 2.8, hence it is not coTT. It is strongly chordal by Theorem 1.16.

   The graph T_2. Again by Theorem 1.16, the graph T_2 is strongly chordal and from Figure 2.8 it is tolerance but not bounded tolerance. Exercise 11.11 states that T_2 is not coTT.

   The graph S_3. The graph S_3 is clearly a chordal graph, but it is not strongly chordal by Theorem 1.16. Figure 4.1 and Theorem 4.3 together demonstrate that S_3 is a tolerance graph, but it is not a cocomparability graph (Figure 3.1) and hence not a bounded tolerance graph. Exercise 11.12 states that S_3 is not coTT.

   The graph $\overline{S_3}$. Figure 11.12 shows that $\overline{S_3}$ is coTT, however, it is not a cocomparability graph and hence not a bounded tolerance graph.

   The graphs $2K_2$ and P_4. It is easy to see that these graphs are interval graphs. By Theorem 1.17, they are not threshold graphs.

   Finally, the hierarchy is complete since together T_3, $\overline{T_2}$ and $\overline{S_3}$ cut across the hierarchy in a manner which shows all the incomparabilities. $\square$

12.3 Archimedean ϕ-tolerance graphs

We call a function $\phi : \mathbf{R}^+ \times \mathbf{R}^+ \to \mathbf{R}^+ \cup \{0\}$ *Archimedean*[1] if for all $c > 0$,

$$\lim_{x \to \infty} \phi(x, c) = \infty \quad \text{and} \quad \lim_{x \to \infty} \phi(c, x) = \infty,$$

that is, for any M there exists a number s such that $\phi(x, c)$, $\phi(c, x) > M$ for all $x \geq s$. Without loss of generality, it will be useful to extend this property to any finite set of numbers $c_1, \ldots, c_k$ choosing s such that $\phi(x, c_i)$, $\phi(c_i, x) > M$ for all c_i and all $x \geq s$.

The functions sum, max, and product are obvious examples of Archimedean functions, whereas min is not Archimedean.

Let Φ denote the set of all *positive* valued *symmetric* Archimedean functions, which we call simply the *Archimedean tolerance functions*. We will see later that for every graph G there exists an Archimedean function ϕ (which depends on G) such that G is a ϕ-tolerance graph. When this happens for all Archimedean functions, we will call the graph an Archimedean ϕ-tolerance graph.

Definition 12.10. A graph G is an *Archimedean ϕ–tolerance graph* (or more simply, an *Archimedean graph*) if G is a ϕ-tolerance graph for all functions $\phi \in \Phi$.

Jacobson, McMorris, and Scheinerman (1991) showed that every tree is an Archimedean ϕ-tolerance graph. (In contrast, trees which are min-tolerance graphs were characterized in Theorem 3.7.) In this section, we follow Golumbic, Jamison, and Trenk (2002) in generalizing this result to a much larger class of graphs which includes all chordless suns and cacti, and the complete bipartite graphs $K_{2,k}$. In general, almost all graphs fail to be Archimedean.

The *chordless k-sun* $S_{k,k}$ ($k \geq 3$) has $2k$ vertices and consists of a chordless cycle C_k with vertices circularly ordered $v_1, \ldots, v_k$ and an independent set $u_1, \ldots, u_k$ such that u_i is adjacent to v_i and v_{i+1} (arithmetic modulo k). Figure 12.3 shows the first three chordless suns.

---

[1] Our choice of the name *Archimedean* function is motivated by the Axiom of Archimedes: given two lengths, a and b, there is always a multiple of the smaller that is greater than the larger. Although the axiom is due to Eudoxus who lived before Archimedes, H. W. Turnbull states that it probably bears the name of Archimedes "because of its application on a grand scale, when he showed that the amount of sand in the world was finite", in Archimedes' famous work *The Sand Reckoner*, "important for its influence on arithmeticians in the nineteenth century". See *The Great Mathematicians*, by H. W. Turnbull, in *The World of Mathematics*, James R. Newman, ed., Simon and Schuster, New York, 1956, vol. 1, pp. 98–99, 106, and *The Sand Reckoner*, by Archimedes, (ibid, pp. 420–429).

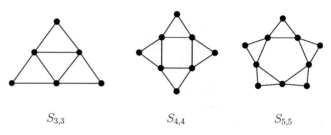

$$S_{3,3} \qquad\qquad S_{4,4} \qquad\qquad S_{5,5}$$

Figure 12.3. The chordless suns $S_{3,3}$, $S_{4,4}$, $S_{5,5}$.

We begin with the following result from Golumbic, Jamison, and Trenk (2002).

Lemma 12.11. *The chordless sun $S_{k,k}$ is an Archimedean φ-tolerance graph.*

Proof. Let $\phi \in \Phi$ be any Archimedean function. We first construct a representation of the path P_{k-1} with vertices ordered $v_1, \ldots, v_{k-1}$ and then close the cycle by adding v_k adjacent only to v_1 and v_{k-1}. Let $m = \phi(1, 1)$ and choose s such that $\phi(s, 1) > 3m$. Finally, let $m' = \phi(s, 1)$, so $m' > 3m$.

Consider the representation of the path P_{k-1} with intervals

$$I_1 = [-m', m]$$

$$I_i = [(2i - 4)m, (2i - 1)m] \text{ (for } i = 2, \ldots, k - 2)$$

and

$$I_{k-1} = [(2k - 6)m, (2k - 5)m + m']$$

and with tolerances $t_{v_i} = 1$ (for $i = 1, \ldots, k - 1$). Since $|I_i \cap I_{i+1}| = m = \phi(1, 1)$ for $i = 1, \ldots, k - 2$, and all other pairs of intervals are disjoint, this is a φ-tolerance representation of P_{k-1}. (Figure 12.4 shows the construction for the case $k = 6$.) Now add interval

$$I_k = [-m', (2k - 5)m + m']$$

and let $t_{v_k} = s$. This adds only the two edges connecting v_k with v_1 and v_{k-1} since $|I_1 \cap I_k| = |I_{k-1} \cap I_k| = m + m' > \phi(s, 1)$ and $|I_i \cap I_k| = 3m < \phi(s, 1) = \phi(t_{v_k}, t_{v_i})$ for $i = 2, \ldots, k - 2$. Thus, we have a φ-tolerance representation for C_k.

We now extend our construction to a φ-tolerance representation of $S_{k,k}$ by adding an interval J_i and a tolerance t_{u_i} associated with each vertex u_i. For each $i = 1, \ldots, k - 2$, we define

$$J_i = I_i \cap I_{i+1} = [(2i - 2)m, (2i - 1)m]$$

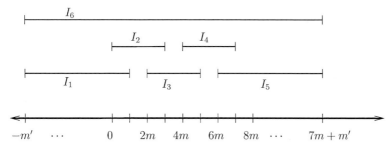

Figure 12.4. The construction of a ϕ-tolerance representation for the cycle C_6.

and $t_{u_i} = 1$ and define

$$J_{k-1} = [(2k-5)m, (2k-5)m + m']$$

and $t_{u_{k-1}} = 1$. It is straightforward to verify that u_i is adjacent only to v_i and v_{i+1} (for $i = 1, \ldots, k-1$). To complete the construction, we must add J_k.

Choose a number s' such that $\min\{\phi(s', s), \phi(s', 1)\} > m + m'$ and let $m'' = \max\{\phi(s', s), \phi(s', 1)\}$. We extend both I_1 and I_k to the left by changing their left endpoints from $-m'$ to $-m' - m''$, and we call these intervals I_1' and I_k'. This does not change any adjacencies. Now define

$$J_k = I_k' = [-m' - m'', (2k-5)m + m']$$

and $t_{u_k} = s'$. First, we note that u_k is adjacent to v_1 and v_k, since $|J_k \cap I_k'| > |J_k \cap I_1'| > m'' \geq \{\phi(s', s), \phi(s', 1)\}$. Second, we see that u_k is not adjacent to $v_2, \ldots, v_{k-1}$ nor $u_1, \ldots, u_{k-1}$ since for all $i = 2, \ldots, k-1$ and $j = 1, \ldots, k-1$, we have $|J_k \cap I_i|, |J_k \cap J_j| \leq \max\{3m, m'\} \leq m + m' < \phi(s', 1)$. $\qquad \square$

Since the functions max and sum are Archimedean functions and C_k is an induced subgraph of $S_{k,k}$, we obtain the following.

Corollary 12.12. *The chordless cycle C_k is a max-tolerance and a sum-tolerance graph.*

We next introduce two new concepts – strong umbrella vertices and (ordinary) umbrella vertices which will allow us to construct additional families of Archimedean graphs.

Let v be a vertex of an Archimedean graph G. We call v a *strong umbrella vertex* if for all $\phi \in \Phi$, G has a ϕ-tolerance representation in which (*i*) the interval $I_v = [L(v), R(v)]$ contains every other interval and (*ii*) the neighbors of v can be partitioned into at most two cliques A and B such that (a) as we travel from $L(v)$ to the right we see all the left endpoints from A before seeing a non-neighbor of v (i.e., the left endpoints of A form a cluster at the left), and (b) as we travel from $R(v)$ to the left we see all the right endpoints from B before seeing a non-neighbor of v (the right endpoints of B form cluster at the right). The term "umbrella" comes from condition (i) and we call $\mathcal{N}(v) = A \cup B$ the *left-right clique partition of* v. We state conditions (a) and (b) of (ii) formally as follows: for every non-neighbor $x \notin \mathcal{N}[v]$ we have

$$L(a) \leq L(x) \ \forall a \in A \quad \text{and} \quad R(x) \leq R(b) \ \forall b \in B.$$

For example, using the representation for $S_{k,k}$ constructed in the proof of Lemma 12.11, we will show that both vertices v_k and u_k are strong umbrella vertices. Condition (i) is satisfied because $I'_k = J_k$ contains all the intervals, and condition (ii) is satisfied using the partitions $\mathcal{N}(v_k) = \{v_1, u_k\} \cup \{v_{k-1}, u_{k-1}\}$ and $\mathcal{N}(u_k) = \{v_1, v_k\} \cup \emptyset$ since $L(v_k) = L(v_1) = L(u_k) = -m' - m''$, $R(v_k) = R(v_{k-1}) = R(u_k) = R(u_{k-1}) = (2k-5)m + m'$. Therefore, by the symmetry of the graph, we have proven the following.

Lemma 12.13. *Every vertex of the chordless sun $S_{k,k}$ is a strong umbrella vertex.*

All the induced subgraphs of the chordless sun have this property too, which is a consequence of this simple but useful observation.

Remark 12.14. The following are hereditary properties, i.e., inherited by all induced subgraphs:

1. a graph being an Archimedean graph,
2. a vertex being a strong umbrella vertex.

It is easy to see that a complete graph K_n is also Archimedean and every vertex is a strong umbrella vertex, by taking $I_v = [0, \phi(1, 1)]$ and $t_v = 1$ for all $v \in V(K_n)$.

One method for constructing larger and more interesting Archimedean graphs is by "gluing together" several disjoint Archimedean graphs at umbrella vertices. The ability to do so, as we will see, is provided by a somewhat weaker condition which we state formally as follows.

We call v an *(ordinary) umbrella vertex* of G if for all $\phi \in \Phi$, there exists a constant $s(v, G, \phi) > 0$ such that $\forall s' \geq s(v, G, \phi)$, G has a ϕ-tolerance representation in which (i) the interval I_v contains every other interval and (ii) $t_v = s'$.

Remark 12.15. Every strong umbrella vertex is an (ordinary) umbrella vertex.

Proof. Let $\phi \in \Phi$ be any Archimedean tolerance function, and let v be a strong umbrella vertex of G. We take a ϕ-tolerance representation $\langle \mathcal{I}, t \rangle$ for G in which the interval I_v assigned to v contains all other intervals in $\mathcal{I}$ and where $\mathcal{N}(v) = A \cup B$ is its left-right clique partition satisfying (a) and (b) from the definition of strong umbrella vertex. Let $M = |I_v|$. By the definition of an Archimedean function, we can choose $s > 0$ such that for all $s' \geq s$ and all $y \in V(G)$ we have $\phi(s', t_y) > M$. Finally, define $s(v, G, \phi) = s$.

We now will adjust the representation $\langle \mathcal{I}, t \rangle$ by stretching the left and right sides as we did in Lemma 12.11, to obtain an equivalent representation $\langle \mathcal{I}', t' \rangle$ which satisfies the requirement of an ordinary umbrella vertex as follows.

Let $s' \geq s$ and let $m = \max\{\phi(s', t_y) \mid y \in V(G)\}$. Next, replace the left endpoints $L(v)$ and $L(a)$ for all $a \in A$ by $L(v) - m$ and the right endpoints $R(v)$ and $R(b)$ for all $b \in B$ by $L(v) + m$. All other intervals remain unchanged. Finally, redefine the tolerance $t'_v = s'$, and leave all other tolerances unchanged. Since only the intervals of the neighbors of v were extended, we observe that $vz \in E(G) \Rightarrow |I'_v \cap I'_z| = |I'_z| > m \geq \phi(s', t_z) = \phi(t'_v, t'_z)$ and $vx \notin E(G) \Rightarrow |I'_v \cap I'_x| = |I_x| \leq M < \phi(s', t_x) = \phi(t'_v, t'_x)$. Thus, the neighborhood of v remains unchanged under the transformation. Moreover, it is easy to verify that no other edges were added or deleted since the neighbors of v were already partitioned into a left clique and a right clique, so lengthening their intersection does not change adjacencies. Therefore, we have constructed the desired representation, and the conclusion follows. □

We now present the "gluing lemma".

Lemma 12.16. *Suppose $H_1, \ldots, H_\ell$ are disjoint Archimedean graphs and u_i is an umbrella vertex of H_i for each i. Then the graph G, obtained by coalescing the vertices $u_1, \ldots, u_\ell$ into one vertex v, is also an Archimedean graph with umbrella vertex v.*

Proof. Let ϕ be any Archimedean function, and let $s(u_i, H_i, \phi)$ be as in the definition of umbrella vertex. Define $s' = \max_i\{s(u_i, H_i, \phi)\}$. We take disjoint ϕ-tolerance representations $\langle \mathcal{I}, t \rangle$ for each H_i in which the interval I_{u_i} assigned to u_i in $\mathcal{I}_i$ contains all other intervals in $\mathcal{I}_i$ and with $t_{u_i} = s'$.

Since the tolerances of $u_1, \ldots, u_\ell$ in their respective representations are all equal to s', in this case, simply extend the disjoint intervals I_{u_i} together into one long interval I_v which contains the disjoint representations $\mathcal{I}_i$ and assign the tolerance $t_v = s'$ to be that common value $s' = t_{u_i}$ $\forall i$. Since the size of all pairs of interval intersections and all tolerances remain unchanged, we obtain the desired ϕ-tolerance representation for G satisfying the claim. Finally, v is an umbrella vertex with $s(v, G, \phi) = s'$. □

Recall that a vertex v of a graph is called a *cutpoint* if removing v from the graph increases the number of connected components. A *block* of a graph is a maximal induced subgraph which has no cutpoint. In a tree, the cutpoints are the internal nodes and each edge is a block. It is well known in graph theory that the edges of a graph are partitioned by its set of blocks. A graph is called a *cactus* if it is connected and each of its blocks is either a single edge or a chordless cycle C_k ($k \geq 3$). Clearly, every tree is a cactus.

Theorem 12.17. *(Golumbic, Jamison, and Trenk, 2002) Every cactus is an Archimedean φ-tolerance graph and every vertex is an umbrella vertex.*

Proof. Let G be a cactus and let $\phi \in \Phi$. We will construct a φ-tolerance representation for G recursively, proving the following by induction on the number of blocks.

Claim: For an arbitrary vertex v of a cactus G, there is a φ-tolerance representation for G in which the interval I_v assigned to v contains all other intervals, and v is an umbrella vertex.

The claim is trivial for a single edge and it is true for any cycle C_k ($k \geq 3$) by our construction in the proof of Lemma 12.11. Thus, the claim holds for a cactus with only one block. We will prove the claim for G, assuming that it holds for all cacti having fewer blocks than G. Let v be an arbitrary vertex of G.

Case 1: *v is not a cutpoint of G.* Since v is not a cutpoint, it is a member of only one block, which we denote $G_B = (B, E_B)$.

Case 1a: *G_B is a single edge.* In this case, v must be a leaf, i.e., it is adjacent to only one vertex u, and u is a cutpoint of G. By the induction hypothesis on the claim, there is a φ-tolerance representation $\langle \mathcal{I}, t \rangle$ for $G - v$ in which the interval $I_u = [L(u), R(u)]$ assigned to u contains all other intervals (because u is an umbrella vertex of $G - v$). Let $m = |I_u|$ which is the length of the longest interval in this representation, and let M be selected so that $M \geq m$. Now choose s to be large enough so that $\phi(s, t_w) > M$ for all $w \in V(G - v)$. Next, let $m' = \phi(s, t_u)$ and extend the interval I_u replacing it with the interval $I'_u = [L(u), R(u) + m']$ which leaves the φ-tolerance representation for $G - v$ essentially unchanged. Finally, we add a new interval $I_v = [L(u) - 1, R(u) + m' + 1]$ and assign the tolerance $t_v = s$. This completes the φ-tolerance representation of G, since $|I_v \cap I'_u| = |I'_u| = m + m' > \phi(s, t_u) = \phi(t_v, t_u)$ and $|I_v \cap I_w| = |I_w| < m \leq M < \phi(s, t_w) = \phi(t_v, t_w)$ for $w \neq u$. Finally, we observe that in this construction, v is a strong umbrella vertex (using the partition $A = \emptyset$ and $B = \{u\}$).

Case 1b: *G_B is a cycle.* Let this cycle be of length k and index its vertices $v_1, v_2, \ldots, v_{k-1}, v_k$ where $v = v_k$. Note that in this case, v's only neighbors in G are v_1 and v_{k-1}. For each of the vertices $v_1, \ldots, v_{k-1}$ consider the connected component G_i containing v_i obtained after (temporarily) erasing the two edges $v_{i-1}v_i$ and v_iv_{i+1} on the cycle. Notice that G_i is an isolated vertex if v_i is not a cutpoint, and that $V(G_1) \cup \cdots \cup V(G_{k-1}) \cup \{v\}$ is a partition of the vertex set $V(G)$. By induction, we may take disjoint ϕ-tolerance representations $\langle \mathcal{I}_i,$ $t\rangle$ for G_i ($i = 1, \ldots, k-1$) in which the interval I_{v_i} assigned to v_i contains all other intervals in $\mathcal{I}_i$ (because v_i is an umbrella vertex of G_i). Since the representations are disjoint, we may assume that they are ordered $\mathcal{I}_1, \cdots, \mathcal{I}_{k-1}$ from left to right. Furthermore, we may assume they are spaced far enough apart so that we can extend the intervals $I_{v_1}, \ldots, I_{v_{k-1}}$ to the right and to the left within the gap so that I_{v_i} overlaps $I_{v_{i+1}}$ by at least $\phi(t_{v_i}, t_{v_{i+1}})$ for $i = 1, \ldots, k-2$, thus causing the vertices $v_1, \ldots, v_{k-1}$ to form a chordless path. Notice the similarity between our construction here and that of C_k, in the proof of Lemma 12.11.

Let m be the length of the longest interval in all of these representations, and as before select a value M so that $M \geq m$. Choose s be large enough so that $\phi(s, t_w) > M$ for all $w \in V(G - v)$. Next, let $m' = \max\{\phi(s, t_{v_1}), \phi(s, t_{v_{k-1}})\}$ and extend the intervals I_{v_1} and $I_{v_{k-1}}$ replacing them with the intervals $I'_{v_1} = [L(v_1) - m', R(v_1)]$ and $I'_{v_{k-1}} = [L(v_{k-1}), R(v_{k-1}) + m']$, respectively. This leaves the represented graph unchanged. Finally, we add a new interval $I_v = [L(v_1) - m' - 1, R(v_{k-1}) + m' + 1]$ and assign the tolerance $t_v = s$ which completes the ϕ-tolerance representation of G, since $|I_v \cap I'_{v_1}| = |I'_{v_1}| > m' \geq \phi(s, t_{v_1})$, $|I_v \cap I'_{v_{k-1}}| = |I'_{v_{k-1}}| > m' \geq \phi(s, t_{v_{k-1}})$ and for $w \neq v_1, v_{k-1}$, $|I_v \cap I_w| = |I_w| \leq m \leq M < \phi(s, t_w)$. Finally, as in case 1a, we observe that in this construction, v is a strong umbrella vertex (using the partition $A = \{v_1\}$ and $B = \{v_{k-1}\}$).

Case 2: *v is a cutpoint of G.* Let $G_1, \ldots, G_k$ be the connected components of $G - v$, and let H_i be the connected subgraph induced by $V(G_i) \cup v$. Thus, the sets $E(H_1), \ldots, E(H_k)$ partition the edge set of G. Since v is not a cutpoint of H_i (for all i), each H_i falls into one of the previous cases 1a or 1b. Therefore, as noted at the end of cases 1a and 1b, v is a strong umbrella vertex in H_i (for all i). We may now apply Lemma 12.16 and Remark 12.15 to prove the claim in this final case. $\square$

Corollary 12.18. *(Jacobson, McMorris, and Scheinerman, 1991) All trees are ϕ-tolerance graphs for every function $\phi \in \Phi$.*

Although the gluing lemma allows one to construct many other Archimedean graphs, we do not know, for example, whether graphs all of whose blocks are chordless k-suns (or simply 3-suns) are Archimedean. However, the construction in our cactus theorem does not appear to generalize to such graphs.

Jacobson, McMorris and Scheinerman (1991) have also shown that almost all graphs are non-Archimedean. Nevertheless, no explicit example is currently known, although unpublished work of Jacobson and Lehel (personal communication) suggests that small bipartite graphs (possibly even $K_{3,3}$) may have no max-tolerance representation, which would make them non-Archimedean, since the max function is Archimedean. Note, however, that the complete graphs $K_{2,k}$ are Archimedean, which we will now show.

Proposition 12.19. *(Golumbic, Jamison, and Trenk, 2002) The complete bipartite graph $K_{2,k}$ is an Archimedean ϕ-tolerance graph.*

Proof. Let $\phi \in \Phi$ be any Archimedean tolerance function, and let $V(K_{2,k}) = \{a, b\} \cup \{v_1, \ldots, v_k\}$ be the bipartition of the vertices into independent sets. We will construct a ϕ-tolerance representation for $K_{2,k}$.

Let $t_1 = 1$, and choose $s_1 > \phi(1, t_1) = \phi(1, 1)$. By induction, having $t_1, \ldots, t_{i-1}$ and $s_1, \ldots, s_{i-1}$, we choose t_i so that

$$\min_{j \le i-1}\{\phi(t_j, t_i)\} > 2s_{i-1}$$

and then choose s_i so that

$$s_i > \max_{j \le i-1}\{\phi(t_j, t_i)\} > 2s_{i-1}.$$

Finally, for $1 \le i \le k$, we define the interval $I_i = [-s_i, s_i]$ assigned to v_i with tolerance t_i, and define $I_a = [-s_k, 0]$ and $I_b = [0, s_k]$ with tolerances $t_a = t_b = 1$.

We now verify that our construction is a ϕ-tolerance representation for $K_{2,k}$. For all i, we have

$$|I_a \cap I_i| = s_i > \max_{j \le i-1}\{\phi(t_j, t_i)\} \ge \phi(t_1, t_i) = \phi(1, t_i) = \phi(t_a, t_i)$$

so a is adjacent to each v_i. Similarly, b is adjacent to each v_i. Next, since $|I_a \cap I_b| = 0$, there is no edge between a and b. Finally, since $s_1 < s_2 < \cdots < s_k$, we have $I_\ell \subset I_i$ for all $\ell < i$. Therefore,

$$|I_\ell \cap I_i| = |I_\ell| = 2s_\ell \le 2s_{i-1} < \min_{j \le i-1}\{\phi(t_j, t_i)\} \le \phi(t_\ell, t_i)$$

so v_ℓ is not adjacent to v_i. $\square$

Remark 12.20. By the symmetry of $K_{2,k}$ we note that each of the vertices $v_1, \ldots, v_k$ is a strong umbrella vertex since in the construction for $K_{2,k}$ the vertex v_k could be chosen arbitrarily.

12.4 Polynomial functions ϕ

In this section we consider ϕ-tolerance graphs when ϕ is a polynomial function. We present the result of Jacobson, McMorris, and Scheinerman (1991) that for any polynomial ϕ on two variables, almost all graphs fail to be ϕ-tolerance graphs. This result also applies to the functions min and max.

We begin by stating a theorem due to Warren (1968) in a version used in Jacobson, McMorris, and Scheinerman (1991). A *sign pattern* for polynomials $p_1, \ldots, p_r$ in s variables is an r-tuple of pluses, minuses and zeroes of the form $\langle \text{sign}[p_1(\mathbf{x})], \ldots, \text{sign}[p_r(\mathbf{x})] \rangle$ where $\mathbf{x} \in \mathbb{R}^s$. For example, the sign pattern for $p_1(x_1, x_2) = x_1 x_2$, $p_2(x_1, x_2) = x_2^5 - 75x_1^2$, $p_3(x_1, x_2) = 4x_1^3 - 281x_2 + 62$ evaluated at $\mathbf{x} = (5, 2)$ is $\langle +, -, 0 \rangle$. A trivial upper bound on the number of sign patterns is 3^r. The following stronger bound holds as r becomes large, and depends on the degree of the polynomials. Note that in the expression below, e is the base of the natural logarithm.

Warren's Theorem. Let $p_1, \ldots, p_r$ be polynomials of degree d in s variables. If $r \geq s$, then the number of different sign patterns for these polynomials is at most $\lceil 8edr/s \rceil^s$.

Using Warren's Theorem, we are now able to prove the main results of this section.

Theorem 12.21. *(Jacobson, McMorris, and Scheinerman, 1991) If ϕ can be expressed as a polynomial in two variables, then there are at most $(cn)^{3n}$ ϕ-tolerance graphs with n vertices, where c is a constant that depends on ϕ.*

Proof. Let us assume that ϕ can be expressed as a polynomial in two variables of degree d. Every ϕ-tolerance representation $\langle \{[a_i, b_i]\}, \{t_i\} \rangle$ for a graph on n vertices may be regarded as a $3n$-tuple $\mathbf{x}$ of real numbers on the variables $\{a_i\}, \{b_i\}, \{t_i\}$, where $i = 1, \ldots, n$. Moreover, the edges of the ϕ-tolerance graph G, for this representation, can be constructed just by knowing the sign pattern of the following set of $5n^2$ polynomials (for all ordered pairs $1 \leq i, j \leq n$) evaluated at $\mathbf{x}$, that is, without knowing the values of $\mathbf{x}$:

$$f_{i,j}(\mathbf{x}) = a_j - a_i,$$
$$g_{i,j}(\mathbf{x}) = b_j - b_i,$$
$$h_{i,j}(\mathbf{x}) = b_j - a_i,$$
$$p_{i,j}(\mathbf{x}) = \phi(t_j, t_i) - b_i + a_j,$$
$$q_{i,j}(\mathbf{x}) = \phi(t_j, t_i) - b_j + a_j.$$

This can easily be seen since the signs of these polynomials give all the information needed to determine the order of the interval endpoints and whether $|[a_i, b_i] \cap [a_j, b_j]| \geq \phi(t_i, t_j)$ for all i and j. Thus, we have shown that if the representation of G_1 and the representation of G_2 give the same sign pattern, then $G_1 = G_2$. Therefore, the number of sign patterns is an upper bound on the number of φ-tolerance graphs on n vertices.

Using Warren's Theorem, since the number of polynomials is $r = 5n^2 \geq 3n = s$, we conclude that the number of sign patterns, and hence the number of φ-tolerance graphs on n vertices, is at most $\lceil 8ed(5n^2)/3n \rceil^{3n} = (cn)^{3n}$, for some constant c that depends on the degree of ϕ. □

Corollary 12.22. *For any polynomial* ϕ, *almost all split graphs are not* φ-*tolerance graphs.*

Proof. The number of split graphs on n vertices exceeds $2^{\frac{n^2}{4}}$. (Take a clique with $\lceil n/2 \rceil$ vertices and an independent set with $\lfloor n/2 \rfloor$ vertices and arbitrary edges between the two sets.) The result follows since $\lim_{n \to \infty}[2^{\frac{n^2}{4}}/(cn)^{3n}] = \infty$.

□

Remark 12.23. Theorem 12.21 and Corollary 12.22 also hold when $\phi = \min$ or $\phi = \max$ since the two polynomials $p_{i,j}(\mathbf{x})$ and $q_{i,j}(\mathbf{x})$ may be replaced by the four polynomials $p'_{i,j}(\mathbf{x}) = t_j - b_i + a_j$, $p''_{i,j}(\mathbf{x}) = t_i - b_i + a_j$, $q'_{i,j}(\mathbf{x}) = t_j - b_j + a_j$, $q''_{i,j}(\mathbf{x}) = t_i - b_j + a_j$ to obtain the edges of the graph.

12.5 Every graph can be represented by an Archimedean polynomial

What *are* the φ-tolerance graphs? Are φ-tolerance graphs frequent or rare occurrences, and in what way might this depend upon the function ϕ? We saw in Section 12.4 that for polynomial functions almost all graphs fail to be φ-tolerance graphs. In contrast to this, we show in this section that every graph G can be represented as a ϕ_G-tolerance chain graph for some Archimedean polynomial ϕ_G, and in the next section that there is a universal function ϕ^* such that all graphs are ϕ^*-tolerance graphs.

Recall that a function $f : \mathbf{R}^+ \times \mathbf{R}^+ \to \mathbf{R}^+ \cup \{0\}$ is called *Archimedean* if for all $c > 0$,

$$\lim_{x \to \infty} f(x, c) = \infty \quad \text{and} \quad \lim_{x \to \infty} f(c, x) = \infty.$$

Strictly speaking, we required our tolerance functions to be symmetric and positive valued, and the tolerance functions ϕ that we construct in this section will have these properties. But Archimedean functions, in general, do not require this assumption. We note the following.

Remark 12.24. The class of Archimedean functions is closed under sums, products and positive scalar multiplication.

As is customary, a polynomial consists of a sum of a finite number of terms, each of which is a product of a scalar and a finite number of variables raised to positive integer powers. In this section and the next, we will prove two results showing that Archimedean polynomials ϕ can represent all graphs as ϕ-tolerance chain graphs. In particular, we first prove that, for every graph G, there exists an Archimedean polynomial ϕ_G (which depends on G) such that G is a ϕ_G-tolerance chain graph. Then we show the existence of an Archimedean "infinite polynomial" ϕ^* such that every graph G is a ϕ^*-tolerance chain graph.

We begin with a lemma and a construction.

Lemma 12.25. *Let λ be any positive function defined on a finite symmetric set $S \subset \mathbf{R}^+ \times \mathbf{R}^+$ of pairs of positive real numbers. There exists an Archimedean polynomial ϕ_λ that agrees with λ on all points of S and is positive valued outside of S. Furthermore, if λ is a symmetric function then so is ϕ_λ.*

Proof. Let λ be a positive function defined on S. We will construct a polynomial $\phi_\lambda \in \Phi$ defined on all of $\mathbf{R}^+ \times \mathbf{R}^+$ such that

$$\phi_\lambda(x, y) = \lambda(x, y) \quad \forall (x, y) \in S$$

and

$$\phi_\lambda(x, y) > 0 \quad \text{if} \quad x > 0, y > 0 \quad \text{and} \quad (x, y) \notin S.$$

Define, for any pair $(a, b) \in \mathbf{R} \times \mathbf{R}$, the function

$$f_{a,b}(x, y) = (x - a)^2 + (y - b)^2$$

that is, $f_{a,b}$ is the Euclidean distance function between the points (a, b) and (x, y). Note that $f_{a,b}$ is Archimedean and that $f_{a,b} \geq 0$ for all x and y with equality if and only if $x = a$ and $y = b$. Now for any finite set P of pairs of real numbers, we define

$$F_P(x, y) = \prod_{(a,b) \in P} f_{a,b}(x, y).$$

As a product of Archimedean functions which are polynomials, F_P is also Archimedean and a polynomial. Moreover, F_P is 0 at all pairs in P, and positive everywhere else.

Now consider our arbitrary positive function $\lambda : S \to \mathbf{R}^+$, and define

$$\phi_\lambda(x, y) = \sum_{p=(a,b)\in S} \lambda(p) \frac{F_{S\setminus\{p\}}(x, y)}{F_{S\setminus\{p\}}(a, b)}.$$

Notice that each summand in this sum is just the Archimedean polynomial $F_{S\setminus\{p\}}$ multiplied by the positive (constant) scalar:

$$\frac{\lambda(a, b)}{F_{S\setminus\{p\}}(a, b)}$$

Hence ϕ_λ is an Archimedean polynomial. Since for $(x, y) \notin S$ all summands are positive, we observe that ϕ_λ is positive outside of S. Moreover, recalling that $F_{S\setminus\{p\}}(x, y)$ is zero for $(x, y) \in S\setminus\{p\}$ and positive otherwise, it follows that the only nonzero summand occurs at $p = (a, b)$ and has value $\lambda(p)$. Thus, ϕ_λ takes on precisely the values given by λ on the set S.

Finally, if S is a symmetric set and λ is a symmetric function on S, i.e., $(a, b) \in S \Leftrightarrow (b, a) \in S$ and $\lambda(a, b) = \lambda(b, a)$, then ϕ_λ is symmetric because every term for (a, b) is balanced by a corresponding term for (b, a). This proves the lemma.   □

We are now ready to give the main result of this section.

Theorem 12.26. *For every graph G there exists an Archimedean polynomial ϕ_G such that G is a ϕ_G-tolerance chain graph.*

Proof. Let G be any finite graph and index the vertices $v_1, v_2, \ldots, v_n$. We will assign intervals and tolerances to the members of $V(G)$ and define an Archimedean polynomial ϕ_G so that G is a ϕ_G-tolerance chain graph.

For each v_k, assign $I_k = [0, k]$ and $t_k = k$. Let $S = \{(a, b) \mid 1 \le a, b \le n, \ a, b \in \mathbf{Z}\}$, and define

$$\lambda(i, j) = \begin{cases} \min(i, j) & \text{if} \quad v_i v_j \in E(G) \\ \max(i, j) & \text{if} \quad v_i v_j \notin E(G) \\ i & \text{if} \qquad i = j. \end{cases}$$

Now let ϕ_λ be the Archimedean polynomial constructed in the proof of Lemma 12.25 and recall that ϕ_λ agrees with λ on the set S.

Since the intervals form a nested chain, we have $|I_i \cap I_j| = \min(i, j)$. Thus, if $v_i v_j \in E(G)$, then

$$\phi_\lambda(t_i, t_j) = \phi_\lambda(i, j) = \lambda(i, j) = \min(i, j) = |I_i \cap I_j|$$

and if $v_i v_j \notin E(G)$ for distinct i, j, then

$$\phi_\lambda(t_i, t_j) = \phi_\lambda(i, j) = \lambda(i, j) = \max(i, j) > \min(i, j) = |I_i \cap I_j|.$$

Therefore, we have a ϕ_λ-tolerance chain representation for G. So defining $\phi_G = \phi_\lambda$ yields the desired result. $\qquad\qquad\qquad\qquad\qquad\square$

12.6 Construction of a universal Archimedean tolerance function

We now use similar ideas to prove the existence of a universal Archimedean tolerance function. We recall here that the "universal" representing function ϕ^* that we are about to construct cannot be a (finite) polynomial due to Theorem 12.21.

Theorem 12.27. *There exists an Archimedean function ϕ^* such that every graph G is a ϕ^*-tolerance chain graph.*

Proof. Make a list $G_1, G_2, \ldots$ consisting of one copy of each finite graph up to isomorphism. Let H be the disjoint union of all these graphs, that is, H is the countably infinite graph where $V(H) = \bigcup V(G_i)$ and $E(H) = \bigcup E(G_i)$. Note that there are no edges between copies of the different finite graphs. Thus, H contains every finite graph as an induced subgraph, and every vertex of H has finite degree.

We now show that H can be realized as a ϕ^*-tolerance chain graph for an Archimedean function ϕ^*. In fact, we will prove a more general result which we state here as Lemma 12.28. From this we may conclude that every finite graph G is a ϕ^*-tolerance chain graph for this "universal" Archimedean function ϕ^*, which will thus prove the theorem.

Lemma 12.28. *For every countable graph H with all vertices of finite degree, there exists an Archimedean function ϕ_H such that H is a ϕ_H-tolerance chain graph.*

Proof of the lemma. Let $v_1, v_2, \ldots$ be an indexing of the countably many vertices of H. Define $\lambda : \mathbf{Z}^+ \times \mathbf{Z}^+ \to \mathbf{Z}^+$ as before by

$$\lambda(i, j) = \begin{cases} \min(i, j) & \text{if} \quad v_i v_j \in E(H) \\ \max(i, j) & \text{if} \quad v_i v_j \notin E(H) \\ i & \text{if} \quad i = j \end{cases}$$

and assign $I_k = [0, k]$ and $t_k = k$, for each v_k. The same argument as before shows this represents the countably infinite graph H. Let us note that λ satisfies a discrete version of the Archimedean property. Indeed, for any index k, the vertex v_k is adjacent by hypothesis to only finitely many other vertices. Hence, there is an $N_k > k$ such that if $j > N_k$, then $v_k v_j \notin E$. Thus for $j > N_k > k$, we get

$$\lambda(k, j) = \max(k, j) = j,$$

so

$$\lim_{j \to \infty} \lambda(k, j) = \infty.$$

The task now is to find an Archimedean function ϕ that extends λ from a discrete function on lattice points to a continuous function on the whole first quadrant. We will do this by linear interpolation over a suitable triangulation of the first quadrant. In order to define the values of $\phi(s, t)$ when $0 < s < 1$, we need some boundary values when $s = 0$. Hence, we extend the above definition of λ and set $\lambda(0, k) = \lambda(k, 0) = k$ and take $N_0 = 1$.

We provide some intuition for the remainder of the proof. Imagine the first quadrant as divided into unit squares with integral corners. Insert the positive diagonal into each of these squares to obtain a triangulation. Now, imagine a pin stuck into each lattice point (i, j) of the first quadrant whose height is $\lambda(i, j)$. We will stretch a canvas over these pin tops to produce a continuous surface which is the graph of the extension function ϕ. The canvas will stretch into a flat triangular roof over each triangle in our triangulation of the (closed) first quadrant. As we pass to infinity along any line parallel to the y-axis, we will pass through triangles whose corner pins also go to infinity in height by the discrete Archimedean property verified above. Thus, the function values of ϕ will also go to infinity, establishing the Archimedean property for ϕ.

Formally, we extend λ to a continuous ϕ by linear interpolation over the triangles of our triangulation. When x lies in a triangle with corners p, q, r, there are *unique* scalars $\alpha, \beta, \gamma \geq 0$ with $\alpha + \beta + \gamma = 1$ such that $x = \alpha p + \beta q + \gamma r$. The α, β, γ are the *barycentric coordinates* of x in the triangle pqr. Let $\phi(x) = \alpha \lambda(p) + \beta \lambda(q) + \gamma \lambda(r)$. We must show that ϕ is well-defined at boundary points (which lie in two or more triangles). Note that the three edges qr, pr, and pq of the triangle pqr are defined by $\alpha = 0, \beta = 0$, and $\gamma = 0$, respectively. Thus, a point on an edge between two triangles will have nonzero barycentric coordinates only for those vertices which the two triangles share, and there they will agree. Therefore, the ϕ is well-defined.

Now for any positive constant c, let $i \leq c \leq i + 1$, and let $M = \max(N_i, N_{i+1})$. If $t > M + 1$, the point (c, t) lies in a square with integral corners $(i, j), (i, j + 1), (i + 1, j), (i + 1, j + 1)$ with $j > N_i$ and $j + 1 > N_{i+1}$. Thus, λ equals the function max on the four corners of the square and hence extends by linear interpolation to just max over the whole square. Thus, $\phi(c, t) = t$ when $t > M$, so the Archimedean property of ϕ follows. This proves the lemma, which concludes the proof of the theorem. $\square$

12.7 Unit and proper representations

The "unit vs. proper" question has been a running theme throughout this book. For min-tolerance graphs, the classes are different (see Figure 2.8). For max-tolerance graphs, the question is still open. For sum-tolerance, the classes are the same, as we will see below. We begin, however, with a surprising result about chain representations.

One may be inclined to think that chain representations must be very different from proper representations. After all, in the former, the intervals are nested and ordered by inclusion, whereas, in the latter, the intervals must be "inclusion-free". But this intuition is incorrect, as we see from the next result which holds for arbitrary tolerance functions ϕ.

Theorem 12.29. *(Jacobson and McMorris, 1991) Every ϕ-tolerance chain graph is a proper ϕ-tolerance graph.*

Proof. Let G be a ϕ-tolerance chain graph with vertices $v_1, v_2, \ldots, v_n$. If G is a complete graph, then a proper ϕ-tolerance representation for G is obtained by assigning to vertex $v_i \in V(G)$ the interval $I_i = [-\frac{1}{i}, c + i]$ and $t_i = 1$ where $c = \phi(1, 1)$. Otherwise, assume that G is not complete and consider a normalized ϕ-tolerance chain graph representation for G where interval $I_i = [0, r_i]$ and tolerance t_i are assigned to vertex v_i such that $r_1 \leq r_2 \leq \cdots \leq r_n$. Let $\epsilon = \min\{\phi(t_i, t_j) - |I_i \cap I_j| \mid v_i v_j \notin E(G)\}$. Clearly, ϵ is well-defined and positive, since there are non-edges. Now, for all $1 \leq j \leq n$, define $I'_j = [\frac{-(n-j)\epsilon}{2n}, r_j + \frac{j\epsilon}{2n}]$. Note that $|I'_j| = r_j + \epsilon/2$, for all j. One can show that

$$|I'_i \cap I'_j| \geq \phi(t_i, t_j) \iff |I_i \cap I_j| \geq \phi(t_i, t_j)$$

from which it follows that $\langle \{I'_i\}, \{t_i\} \rangle$ is a proper ϕ-tolerance representation of G. $\square$

Theorems 12.4 and 12.29 give an alternative proof that every threshold graph is a proper (min-)tolerance graph, although the stronger result that *threshold graphs are unit probe graphs* was shown in Proposition 4.7. Theorem 12.5, together with Theorem 12.29, implies that *every interval graph is a proper max-tolerance graph*. It is not yet known if this can be strengthened to unit max-tolerance. The converse statement does not hold, C_4 being a separating example. In fact, the following result is due to Doug West (personal communication, May 2000).

Example 12.30. The chordless cycle C_k is a unit max-tolerance graph.

Proof. For $1 \le i \le k$, define $I_i = [i, i + k]$. Assign tolerances $t_1 = t_k = 1$ and, for $2 \le j \le k - 1$, $t_j = k - 1$. It is a simple exercise to verify that this is a unit max-tolerance representation for C_k. □

In unpublished notes from 1990, Jacobson and Lehel made the following additional observation.

Proposition 12.31. *All trees are unit max-tolerance graphs.*

Proof. The proof is by induction on the number of vertices. The result certainly holds for trees with one or two vertices. Let $\langle \{I_x\}, \{t_x\} \rangle_{x \in V(T) - \{v\}}$ be a unit max-tolerance representation for $T - \{v\}$ where v is a leaf of the tree T such that $|V(T)| \ge 3$ and $uv \in E(T)$. We further assume in the induction hypothesis that the interval endpoints are distinct and that the representation is bounded, with $|I_x| = 1$ and $0 < t_x \le 1$ for all $x \in V(T) - \{v\}$. We will show how to add v to the representation.

Let $\epsilon > 0$ be the smallest difference between the interval endpoints; hence, $0 \le |I_x \cap I_y| \le 1 - \epsilon$ for all $x \ne y$. Since every $x \in V(T) - \{v\}$ must have some neighbor y in the tree $T - \{v\}$, it follows that,

$$t_x \le \max\{t_x, t_y\} \le \left| I_x \cap I_y \right| \le 1 - \epsilon$$

so $t_x \le 1 - \epsilon$. Now extend the unit max-tolerance representation to all of T by defining $I_v = [L(u) + \frac{1}{4}\epsilon, R(u) + \frac{1}{4}\epsilon]$ and $t_v = 1 - \frac{1}{4}\epsilon$. For u, we have $|I_u \cap I_v| = 1 - \frac{1}{4}\epsilon = \max\{t_u, t_v\}$ so $uv \in E(T)$. For $w \ne u$, we have $|I_w \cap I_v| \le 1 - \frac{3}{4}\epsilon < t_v$ so $wv \notin E(T)$. Finally, note that the assumptions of the induction are maintained. □

We now turn our attention to sum-tolerance graphs. The following result settles the unit vs. proper question for sum-tolerance graphs, and strengthens the statement in Theorem 12.29 in the case of sum-tolerance. The equivalences (i)⇔(ii)⇔(iii) are due to Jacobson and McMorris (1991) and (iii)⇔(iv) was already given in Theorem 12.6.

Theorem 12.32. *The following statements are equivalent.*

  (i) *G is a unit sum-tolerance graph.*
 (ii) *G is a proper sum-tolerance graph.*
(iii) *G is a sum-tolerance chain graph.*
(iv) *G is a coTT graph.*

Proof. The implication (i) $\implies$ (ii) follows immediately, and (iii) $\implies$ (ii) follows from Theorem 12.29.

(ii) $\implies$ (i): Let $\langle\{I_i\}, \{t_i\}\rangle (i = 1, \ldots, n)$ be a proper sum-tolerance representation for the graph G, where $I_i = [L(i), R(i)]$ and $L(i) < L(j) \iff R(i) < R(j) \iff i < j$. It is sufficient to show that G has a representation using intervals of equal length. Let us assume that the first k ($k \geq 1$) intervals have the same length, and proceed by induction. If $k = n$, then we are done. Otherwise, we will show how to construct a sum-tolerance representation $\langle\{I_i^*\}, \{t_i^*\}\rangle$ for G where the first $k + 1$ intervals have the same length.

If $|I_k| > |I_{k+1}|$, let $\delta = |I_k| - |I_{k+1}|$. For $j \leq k$, define $I_j^* = I_j$ and $t_j^* = t_j$. For $j > k$, define $I_j^* = [L(j) - \frac{\delta}{2}, R(j) + \frac{\delta}{2}]$ and $t_j^* = t_j + \frac{\delta}{2}$. It is straightforward to verify that this is a sum-tolerance representation for G. Similarly, if $|I_k| < |I_{k+1}|$, let $\delta = |I_{k+1}| - |I_k|$, and for $j > k$, define $I_j^* = I_j$ and $t_j^* = t_j$, and for $j \leq k$, define $I_j^* = [L(j) - \frac{\delta}{2}, R(j) + \frac{\delta}{2}]$ and $t_j^* = t_j + \frac{\delta}{2}$.

(ii) $\implies$ (iii): Again, let $\langle\{I_i\}, \{t_i\}\rangle (i = 1, \ldots, n)$ be a proper sum tolerance represention for the graph G, where $I_i = [L(i), R(i)]$ and $L(i) < L(j) \iff R(i) < R(j) \iff i < j$. We may assume that $L(1) > 0$, by adding a large constant to each interval endpoint if necessary. A sum-tolerance chain representation is obtained by defining $I_i' = [0, L(i) + R(i)]$ and $t_i' = t_i + L(i)$, since, for $i < j$,

$$|I_i' \cap I_j'| \geq t_i' + t_j' \iff L(i) + R(i) \geq t_i + L(i) + t_j + L(j)$$
$$\iff R(i) - L(j) \geq t_i + t_j \iff |I_i \cap I_j| \geq t_i + t_j.$$

$\square$

Question 12.33. We know from the fact that max-tolerance and sum-tolerance graphs may contain chordless cycles that they are not perfect graphs. Do they also contain complements of cycles? Are bounded tolerance graphs also max-tolerance? What about the status of the complements of trees?

12.8 Exercises

Exercise 12.1. Prove that every ϕ-tolerance chain graph has a normal representation.

Exercise 12.2. (*Jacobson, McMorris, and Mulder, 1991*) Let H be the graph obtained from $\overline{S_3}$ by adding a pendant edge onto each of the three vertices of degree 1. Show that H is not a coTT graph.

Exercise 12.3. (*Jacobson, McMorris, and Mulder, 1991*) Prove that a tree T is a coTT graph if and only if it contains no induced T_3.

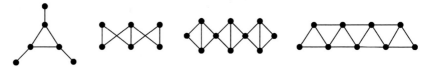

Figure 12.5. Graphs for Exercise 12.4.

Exercise 12.4. Show that each of the graphs in Figure 12.5 is an Archimedean φ-tolerance graph.

Exercise 12.5. Construct an example of a split graph which is not a φ-tolerance graph for any polynomial. (Hard?)

Exercise 12.6. For the graphs C_4, T_3 and $K_{3,3}$, derive the Archimedean polynomials ϕ_{C_4}, ϕ_{T_3} and $\phi_{K_{3,3}}$ from Theorem 12.26 which makes them φ-tolerance chain graphs with respect to their respective functions φ.

Exercise 12.7. Show that if a tolerance function φ satisfies $\phi(x, y) \geq x$, $\forall x$, then φ is Archimedean.

Exercise 12.8. Let φ be a tolerance function satisfying $\phi(cx, cy) = c\phi(x, y)$ for all $c > 0$. Show that a graph G has a unit φ-tolerance representation if and only if, for all $c > 0$, G has a φ-tolerance representation in which all intervals are of equal length c.

Exercise 12.9. Prove the same result as in Exercise 12.8 for the function $\phi(x, y) = x^3 + y^3$. Question for thought: are there functions φ for which a graph may have a representation where all intervals have the same length c for some c, but not for all c? What about the function $\phi(x, y) = x^3 + x + x^2 y^2 + y + y^3$ or $\phi(x, y) = 3^{xy}$?

Chapter 13

Directed tolerance graphs

In Chapter 1, we introduced bounded tolerance representations like the one in Figure 13.1 where the shading along the interval assigned to v indicates the tolerance t_v. We have seen that this representation can be interpreted as a representation of the bounded tolerance graph G or as a representation of the bounded tolerance order P, both shown in Figure 13.1. In this chapter, we consider a third interpretation as a representation of a directed graph $\vec{G}$, whose underlying simple graph is G. The direction on each edge keeps track of which tolerance is met or exceeded when an edge appears in G, that is, there will be a directed arc (x, y) if and only if $|I_x \cap I_y| \geq t_y$.

A *directed graph* (or *digraph*) $\vec{G} = (V, A)$ has vertex set V and arc set A (sometimes written $A(\vec{G})$). If there is an arc from x to y in $\vec{G}$ we write $(x, y) \in A$. An arc $(x, x) \in A$ is called a *loop*. Throughout this chapter we depict a loop at a vertex by drawing a circle around that vertex. If both $(x, y) \in A$ and $(y, x) \in A$ for distinct x, y, we say there is a *double arc* between x and y and write $x \rightleftharpoons y$. If $(x, y) \in A$ and $(y, x) \notin A$, we write $x \rightarrow y$ and say there is a *single arc* between x and y, pointing from x to y. A digraph $\vec{G}$ is called *symmetric* if all arcs (other than loops) are double arcs, that is, $(x, y) \in A(\vec{G})$ whenever $(y, x) \in A(\vec{G})$. The *reversal* of digraph $\vec{G}$ is the digraph $\vec{G}^r$ on the same vertex set so that $(x, y) \in A(\vec{G}) \iff (y, x) \in A(\vec{G}^r)$. The set of *out-neighbors* or *successors* of u is $\mathcal{N}^+(u) = \{v \mid (u, v) \in A(\vec{G})\}$. The *underlying (simple) graph* of $\vec{G} = (V, A)$ is the graph $G = (V, E)$ where $xy \in E$ if and only if x and y are distinct and either $(x, y) \in A$ or $(y, x) \in A$, or both.

In a directed tolerance graph, $(x, y) \in A(\vec{G}) \iff |I_x \cap I_y| \geq t_y$. *Note that in a bounded representation, there is a loop at every vertex.* We can think of the presence of arc (x, y) as telling us that "x bothers y". Thus, whenever we have a tolerance representation of graph G and digraph $\vec{G}$, then an edge between x and y in G can mean one of three things: (i) $x \rightarrow y$ (so x bothers y but y does not bother x), (ii) $y \rightarrow x$ (so y bothers x but x does not bother y) or (iii) $x \rightleftharpoons y$

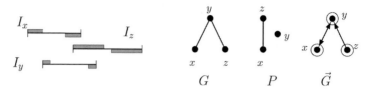

Figure 13.1. A bounded tolerance representation of the graph G, the order P, and the digraph $\vec{G}$.

(so x and y bother each other). Whereas an edge between x and y in G only indicates that there is a conflict, the arc(s) between x and y in $\vec{G}$ tell us who is responsible for the conflict.

The central family we will study are the bounded bitolerance digraphs which are introduced in Section 13.2 and characterized in Section 13.4. We introduce them as a restricted class of the family of *interval 2-point digraphs* which is equivalent to the class of digraphs of Ferrers dimension at most 2. We study these in Section 13.1. Digraph analogues can be defined for any of the classes of bounded bitolerance graphs and orders discussed in Chapter 10. We present the digraph hierarchy in Section 13.5. Digraphs whose underlying graphs are trees or cycles are discussed in Sections 13.6 and 13.7 and forbidden subgraph characterizations for unit, proper and bounded bitolerance digraphs are given for those cases. Finally, in Section 13.8, we revisit the unit vs. proper question in the case of bounded bitolerance digraphs.

13.1 Ferrers dimension 2

Definition 13.1. A digraph $\vec{G} = (V, A)$ is an *interval 2-point digraph* if each vertex $v \in V$ can be assigned an interval $I_v = [L(v), R(v)]$ and two additional real numbers $p(v), q(v)$, so that $A = \{(x, y) \mid L(x) \le q(y) \text{ and } R(x) \ge p(y)\}$. We call the representation $\{[L(v), R(v)], p(v), q(v) \mid v \in V(\vec{G})\}$ an *interval two-point representation* of $\vec{G}$.

We will assume that the points $\{L(v), R(v), p(v), q(v) \mid v \in V\}$ in an interval 2-point representation are distinct, for if not we may make $L(v), p(v)$ slightly smaller and $q(v), R(v)$ slightly larger without altering the digraph represented.

The notation in this definition is suggestive of our usual notation for the representation of bounded bitolerance orders, and we will see this connection more clearly in Section 13.2. Note that here we do *not* require $p(v)$ and $q(v)$ to lie inside I_v, however, if $p(v), q(v) \in I_v$ then there will be a loop at vertex v. In fact, there will be a loop at vertex v whenever $p(v) \le R(v)$ and $q(v) \ge$

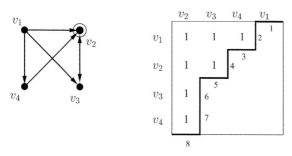

Figure 13.2. An example to illustrate the proof of Proposition 13.2.

$L(v)$. First we define Ferrers digraphs and show that the class of interval 2-point digraphs is equivalent to the class of digraphs of *Ferrers dimension at most 2*.

A directed graph $\vec{G} = (V, A)$ is a *Ferrers digraph* if its successor sets are linearly ordered by inclusion, that is, the vertices can be labeled $V = \{v_1, v_2, \ldots, v_n\}$ so that $\mathcal{N}^+(v_1) \subseteq \mathcal{N}^+(v_2) \subseteq \cdots \subseteq \mathcal{N}^+(v_n)$. The term "Ferrers" comes from the following equivalent definition: $\vec{G}$ is a Ferrers digraph if the rows and columns of the adjacency matrix can be independently permuted so that the 1s form a Ferrers Diagram, that is, the 1s in each row appear consecutively at the beginning of the row and the 1s in each column appear consecutively at the top of the column (see Figure 13.2).

There are many equivalent definitions of Ferrers digraphs due to Cogis (1982), and half a chapter of Mahadev and Peled (1995) is devoted to the subject. If $\vec{G}$ is a symmetric Ferrers digraph, then its underlying simple graph G is a threshold graph (Exercise 13.1). We find the following characterization most useful to us. Our proof is based on Sen, Sanyal, and West (1995).

Proposition 13.2. *A directed graph $\vec{G} = (V, A)$ is a Ferrers digraph if and only if there exist functions $f_1, f_2 : V \to \mathbf{R}$ so that $(x, y) \in A \iff f_1(x) \le f_2(y)$.*

Proof. ($\Longleftarrow$): Suppose there exist functions $f_1, f_2 : V \to \mathbf{R}$ so that $(x, y) \in A$ if and only if $f_1(x) \le f_2(y)$. Order the vertices $V = \{v_1, v_2, \ldots, v_n\}$ in decreasing order of their f_1 values, thus $f_1(v_1) \ge f_1(v_2) \ge \cdots \ge f_1(v_n)$. We show $\mathcal{N}^+(v_i) \subseteq \mathcal{N}^+(v_j)$ for $i < j$. If $y \in \mathcal{N}^+(v_i)$ then $f_1(v_i) \le f_2(y)$. So for all $j > i$ we have $f_1(v_j) \le f_1(v_i) \le f_2(y)$ and thus $y \in \mathcal{N}^+(v_j)$ as desired.

($\Longrightarrow$): Consider the adjacency matrix of $\vec{G}$ with the vertices ordered so that the 1s in each row are consecutive and appear at the beginning of the row and the 1s in each column are consecutive and appear at the beginning of the column. Draw a vertical line segment immediately to the right of the last 1 in each row, and draw a horizontal line segment immediately below the last 1 in each column.

These $2n$ segments form a staircase structure that extends from the top right corner of the matrix to the bottom left corner (see Figure 13.2). Label the segments $1, 2, \ldots, 2n$ starting at the top right corner. Now set $f_1(u)$ equal to the label on the vertical segment in u's row and $f_2(v)$ equal to the label on the horizontal segment in v's column. Then $(u, v) \in A(\vec{G})$ precisely when $f_1(u) \leq f_2(v)$. $\square$

The *Ferrers dimension* of a digraph $\vec{G} = (V, A)$, denoted by $\text{Fdim}(\vec{G})$, is the minimum number of Ferrers digraphs on vertex set V whose intersection gives $\vec{G}$. In particular, a graph $\vec{G} = (V, A)$ has Ferrers dimension at most 2 if there exist Ferrers digraphs $\vec{G}_1 = (V, A_1)$ and $\vec{G}_2 = (V, A_2)$ so that $A = A_1 \cap A_2$. By Proposition 13.2, $\vec{G} = (V, A)$ has Fdim ≤ 2 if and only if there exist functions $f_1, f_2, g_1, g_2 : V \to \mathbf{R}$ so that $(x, y) \in A$ if and only if $f_1(x) \leq f_2(y)$ and $g_1(x) \leq g_2(y)$. This last characterization will be most useful to us and we refer to the four functions f_1, f_2, g_1, g_2 as providing an Fdim ≤ 2 *representation* of digraph $\vec{G}$.

If f_1, f_2, g_1, g_2 is an Fdim ≤ 2 representation of $\vec{G}$, then so is $f_1, f_2, g_1 - c, g_2 - c$, for any constant c. Thus, we may subtract a sufficiently large constant from g_1 and g_2 so that $-f_1(v) > g_1(v)$, or equivalently $f_1(v) < -g_1(v)$ for all $v \in V(\vec{G})$. We call such a representation a *normal* Fdim ≤ 2 *representation*.

Proposition 13.3. *A digraph $\vec{G} = (V, A)$ is an interval 2-point digraph if and only if $\vec{G}$ is a digraph of Ferrers dimension at most 2.*

Proof. Given a digraph $\vec{G}$ with an interval 2-point representation $\{[L(v), R(v)], p(v), q(v) \mid v \in V\}$, define the functions f_1, f_2, g_1, g_2 as follows:

$$f_1(v) = L(v), \quad f_2(v) = q(v), \quad g_1(v) = -R(v), \quad \text{and} \quad g_2(v) = -p(v).$$
$$(13.1)$$

Then $L(x) \leq q(y)$ if and only if $f_1(x) \leq f_2(y)$, and $R(x) \geq p(y)$ if and only if $-g_1(x) \geq -g_2(y)$, or equivalently, $g_1(x) \leq g_2(y)$ and thus $\vec{G}$ has Fdim ≤ 2.

Conversely, if $\vec{G}$ has Fdim ≤ 2, we may choose a normal Fdim ≤ 2 representation of $\vec{G}$. The assignments in (13.1) give an interval 2-point representation of $\vec{G}$. $\square$

13.2 Bounded bitolerance digraphs

Bounded bitolerance digraphs were introduced in Bogart and Trenk (1995) and (2000) and studied further in Shull and Trenk (1997a), (1997b), and (2001). In this section we define bounded bitolerance digraphs as a special case of interval 2-point digraphs, following Shull and Trenk (2001). We prove that the definition here agrees with the one in Bogart and Trenk (2000) and discuss the connection with bounded bitolerance graphs and orders.

Definition 13.4. A directed graph $\vec{G} = (V, A)$ is a *bounded bitolerance digraph* if each vertex $v \in V$ can be assigned an interval $I_v = [L(v), R(v)]$ and two additional real numbers $p(v), q(v) \in I_v$ with $p(v) \neq L(v)$ and $q(v) \neq R(v)$, so that $A = \{(x, y) \mid L(x) \leq q(y)$ and $R(x) \geq p(y)\}$. The representation $\langle \mathcal{I}, p, q \rangle$ is a *bounded bitolerance representation* of $\vec{G}$.

Observe that bounded bitolerance digraphs have a loop at every vertex since $L(x) \leq q(x)$ and $R(x) \geq p(x)$ for all x. Bounded bitolerance representations are interval 2-point representations with the added restriction that the tolerant points $p(v), q(v)$ lie inside the interval I_v. The conditions $p(v) \neq L(v)$ and $q(v) \neq R(v)$ are added to maintain a parallel with tolerance graphs in requiring tolerances to be strictly positive. Their omission would not change the class.

The original definition of bounded bitolerance digraphs in Bogart and Trenk (1995) and (2000) is the following.

Definition 13.5. A directed graph $\vec{G} = (V, A)$ is a bounded bitolerance digraph if each vertex $v \in V$ can be assigned an interval $I_v = [L(v), R(v)]$ and two points $p(v), q(v) \in I_v$ so that $(x, y) \in A$ if and only if (i) $I_x \cap I_y \not\subseteq [L(y), p(y))$ and (ii) $I_x \cap I_y \not\subseteq (q(y), R(y)]$.

Definition 13.5 has the advantage that it is easy to see the connection between bounded bitolerance digraphs and their undirected counterparts. We can think of the intervals $[L(y), p(y))$ and $(q(y), R(y)]$ as being the portions of I_y in which intersection with another interval is tolerated, and an arc (x, y) occurs when the intersection of I_x and I_y is not confined to these regions. However, the Definition 13.5 can be cumbersome and we next show it is equivalent to Definition 13.4.

Proposition 13.6. *The definitions of bounded bitolerance digraphs given in Definitions 13.4 and 13.5 are equivalent.*

Proof. First note that (i) of Definition 13.5 implies (a) $R(x) \geq p(y)$ and (ii) implies (b) $L(x) \leq q(y)$, so if (x, y) is an arc according to Definition 13.5, then it is also an arc according to Definition 13.4. Conversely, (a) and (b) imply $I_x \cap I_y \neq \emptyset$, and this together with (a) implies (i) and together with (b) implies (ii). $\square$

Since a bounded bitolerance representation can be interpreted as representing a graph or a digraph, the next remark follows from the definitions of bounded bitolerance graphs and bounded bitolerance digraphs.

Remark 13.7. (1) If $\vec{G}$ is a bounded bitolerance digraph, then its simple underlying graph G is a bounded bitolerance graph. (2) If G is a bounded bitolerance graph, then there exists a bounded bitolerance digraph $\vec{G}$ whose underlying graph is G.

As mentioned at the beginning of this chapter, the intervals and tolerances shown in Figure 13.1 give a bounded bitolerance representation of the graph G, the order P, and the digraph $\vec{G}$ shown. The simple graph G underlying $\vec{G}$ is the incomparability graph of P, and the next proposition shows this result holds in general.

Proposition 13.8. *If $\langle \mathcal{I}, p, q \rangle$ is a representation of bounded bitolerance digraph $\vec{G}$ and also of bounded bitolerance order P, then the simple graph G underlying $\vec{G}$ is the incomparability graph of P.*

Proof. By definition, $xy \in E(\overline{G})$ if and only if neither (x, y) nor (y, x) is an arc of $\vec{G}$, that is, if and only if (i) $L(x) > q(y)$ or $R(x) < p(y)$, and (ii) $L(y) > q(x)$ or $R(y) < p(x)$. The first inequality from (i) is incompatible with the first inequality from (ii) because of the assumption that $p(v), q(v) \in I_v$ for each v, and similarly for the second inequalities in (i) and (ii). Thus, $xy \in E(\overline{G})$ if and only if (a) $L(x) > q(y)$ and $R(y) < p(x)$ or (b) $L(y) > q(x)$ and $R(x) < p(y)$. The former of these is equivalent to $y \prec x$ in P and the latter is equivalent to $x \prec y$ in P. Thus, $xy \in E(\overline{G})$ if and only if x and y are comparable in P, and therefore G is the incomparability graph of P. □

13.3 Recognition of bounded bitolerance digraphs

The connection between bounded bitolerance digraphs and digraphs of Ferrers dimension at most 2 gives rise to a polynomial time recognition algorithm for the former class based on one for the latter class. We begin with a lemma which is analogous to Proposition 10.2.

Lemma 13.9. *If $\vec{G} = (V, A)$ is an interval 2-point digraph, then $\vec{G}$ has a representation $\langle \mathcal{I}, p, q \rangle$ in which $q(v) \leq p(v)$ for each $v \in V$.*

Proof. The proof is analogous to the proof of Proposition 10.2 and is left as an exercise. □

The following result was observed by Randy Shull (personal communication).

Theorem 13.10. *If a digraph $\vec{G}$ has a loop at each vertex, then the following are equivalent:*

(i) $\vec{G}$ is a bounded bitolerance digraph,

(ii) $\vec{G}$ has Ferrers dimension at most 2.

Proof. (i) $\Longrightarrow$ (ii): This direction follows from Proposition 13.3 and the observation that by definition, bounded bitolerance digraphs are a special case of interval 2-point digraphs.

(ii) $\Longrightarrow$ (i): Using Proposition 13.3 and Lemma 13.9, we may fix an interval 2-point representation $\langle \mathcal{I}, p, q \rangle$ of $\vec{G} = (V, A)$ for which $q(v) \leq p(v)$ for all $v \in V$. By hypothesis, there is a loop at every vertex of $\vec{G}$, hence by definition of an interval 2-point digraph, we have $L(v) \leq q(v) \leq p(v) \leq R(v)$ for each $v \in V$ and the representation is indeed a bounded bitolerance representation of $\vec{G}$. $\square$

Two papers, Cogis (1979) and Doignon, Duchamp, and Falmagne (1984) give polynomial time algorithms for recognizing digraphs of Ferrers dimension at most 2. The algorithms are complex and we omit them here. These algorithms, together with Theorem 13.10, immediately give a polynomial time algorithm for recognizing bounded bitolerance digraphs: if $\vec{G}$ does not have a loop at each vertex, then it is not a bounded bitolerance digraph, if it does, then apply Theorem 13.10. We record this as a remark.

Remark 13.11. The class of bounded bitolerance digraphs can be recognized in polynomial time.

13.4 Characterizations of bounded bitolerance digraphs

Recall from Theorem 5.24 that the classes of bounded bitolerance orders and orders of interval dimension at most 2 (trapezoid orders) are equivalent. In other words, P is a bounded bitolerance order if and only if there exist two interval orders P_1 and P_2 so that $P = P_1 \cap P_2$. Theorem 13.15 is a directed graph analog of this. We begin with preliminary definitions and notation.

Recall that the notation $I_x \ll I_y$ means that the interval I_x is completely to the left of the interval I_y. If $P = (V, \prec)$ is an ordered set, let $\widetilde{P} = (V, A)$ be the directed graph with $A = \{(x, y) \mid x \prec y \text{ or } x \parallel y \text{ in } P\}$. Thus, the arc (x, y) is present in $\widetilde{P}$ precisely when x is *less than or incomparable to* y in P. Note that the underlying simple graph of $\widetilde{P}$ is always the complete graph. We intersect two digraphs (that have the same vertex set) by intersecting their arc sets. In particular, if $P_1 = (V, \prec_1)$ and $P_2 = (V, \prec_2)$ then $\widetilde{P_1} \cap \widetilde{P_2} = (V, A)$ where $A = \{(x, y) \mid x \prec_i y \text{ or } x \parallel_i y \text{ for } i = 1, 2\}$.

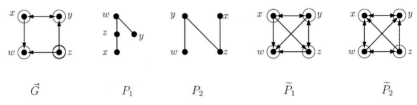

Figure 13.3. A digraph $\vec{G}$ and two interval orders P_1 and P_2 for which $\vec{G} = \tilde{P}_1 \cap \tilde{P}_2$.

Example 13.12. Figure 13.3 shows a digraph $\vec{G}$ together with two ordered sets, P_1 and P_2, and the digraphs $\tilde{P}_1$ and $\tilde{P}_2$. Note that the simple graphs underlying $\tilde{P}_1$ and $\tilde{P}_2$ are complete and that $\vec{G} = \tilde{P}_1 \cap \tilde{P}_2$. The general result is presented in Theorem 13.15.

The following remark follows directly from the definition of $\tilde{P}_1 \cap \tilde{P}_2$.

Remark 13.13. Suppose $P_1 = (V, \prec_1)$ and $P_2 = (V, \prec_2)$ and $\vec{G} = \tilde{P}_1 \cap \tilde{P}_2$. Then

(i) If (y, x) is not an arc of $\vec{G}$, then $x \prec_i y$ for $i = 1$ or $i = 2$.

(ii) If neither (x, y) nor (y, x) is an arc of $\vec{G}$, then either ($x \prec_1 y$ and $y \prec_2 x$) or ($y \prec_1 x$ and $x \prec_2 y$).

(iii) If both (x, y) and (y, x) are arcs of $\vec{G}$, then $x \parallel_i y$ for $i = 1$ and $i = 2$.

(iv) If (x, y) is an arc of $\vec{G}$, then $y \not\prec_i x$ for $i = 1, 2$.

The steps in the proof of the following proposition are justified using Remark 13.13. An example illustrating this proposition is given by the orders and digraphs in Figure 13.3.

Proposition 13.14. *(Bogart and Trenk, 2000) If $P_1 = (V, \prec_1)$ and $P_2 = (V, \prec_2)$ are ordered sets, then the graph underlying $\tilde{P}_1 \cap \tilde{P}_2$ is the incomparability graph of $P_1 \cap P_2^d$.*

Proof. Let $\vec{G} = \tilde{P}_1 \cap \tilde{P}_2$ and let G be its underlying simple graph. Then

$$xy \notin E(G) \iff (x \prec_1 y \text{ and } y \prec_2 x) \text{ or } (y \prec_1 x \text{ and } x \prec_2 y)$$
$$\iff x \text{ is comparable to } y \text{ in } P_1 \cap P_2^d.$$

Thus, $xy \in E(G)$ if and only if $x \parallel y$ in $P_1 \cap P_2^d$, and hence G is the incomparability graph of $P_1 \cap P_2^d$. □

Theorem 13.15. *(Bogart and Trenk, 2000) The following are equivalent.*

(i) $\vec{G}$ *is a bounded bitolerance digraph.*

(ii) *There exist interval orders P_1 and P_2 so that $\vec{G} = \tilde{P}_1 \cap \tilde{P}_2$.*

Proof. (i) $\implies$ (ii): Fix a bounded bitolerance representation of $\vec{G} = (V, A)$ in which $v \in V$ is assigned the interval $[L(v), R(v)]$ and the tolerant points $p(v), q(v)$. For each $v \in V$, let $I_1(v) = [L(v), q(v)]$ and $I_2(v) = [p(v), R(v)]$. Let $P_1 = (V, \prec_1)$ be the interval order represented by $\{I_1(v) \mid v \in V\}$ and let $P_2 = (V, \prec_2)$ be the *dual* of the interval order represented by $\{I_2(v) \mid v \in V\}$. Then $\vec{G} = \tilde{P}_1 \cap \tilde{P}_2$ since

$$(x, y) \notin A(\vec{G}) \iff L(x) > q(y) \text{ or } R(x) < p(y)$$
$$\iff I_1(y) \ll I_1(x) \text{ or } I_2(x) \ll I_2(y)$$
$$\iff y \prec_1 x \text{ or } y \prec_2 x$$
$$\iff (x, y) \notin A(\tilde{P}_1) \text{ or } (x, y) \notin A(\tilde{P}_2)$$
$$\iff (x, y) \notin A(\tilde{P}_1 \cap \tilde{P}_2).$$

(ii) $\implies$ (i): Conversely, suppose there exist interval orders $P_1 = (V, \prec_1)$ and $P_2 = (V, \prec_2)$ with $\vec{G} = \tilde{P}_1 \cap \tilde{P}_2$. Fix an interval representation of P_1 in which v is assigned interval $I_1(v) = [L(v), q(v)]$ and an interval representation of the *dual* of P_2 (which is also an interval order) in which v is assigned interval $I_2(v) = [p(v), R(v)]$. If necessary, add a fixed constant to the endpoints of each $I_2(v)$ so that $L(v) < p(v)$ and $q(v) < R(v)$ for all $v \in V$. Now the intervals $[L(v), R(v)]$ and points $p(v), q(v)$ give a bounded bitolerance representation of a digraph $\vec{H}$. We will show that $\vec{G} = \vec{H}$, thereby proving that $\vec{G}$ is a bounded bitolerance digraph, as desired.

The equivalences above (with the first line replaced by $(x, y) \notin A(\vec{H})$) show that $\vec{H} = \tilde{P}_1 \cap \tilde{P}_2$, which together with the initial hypothesis $\vec{G} = \tilde{P}_1 \cap \tilde{P}_2$ gives $\vec{G} = \vec{H}$. $\qquad\square$

The following is a corollary to Theorem 13.15. We do not know of a direct proof of this result using bounded bitolerance representations without replicating the steps of the proof of Theorem 13.15.

Corollary 13.16. *(Bogart and Trenk, 2000) If the digraph $\vec{G}$ is a bounded bitolerance digraph, then so is its reversal $\vec{G}^r$.*

Proof. If $\vec{G}$ is a bounded bitolerance digraph then by Theorem 13.15 there exist interval orders P_1 and P_2 so that $\vec{G} = \tilde{P}_1 \cap \tilde{P}_2$. Then their duals P_1^d and P_2^d are also interval orders, and $\vec{G}^r = \tilde{P_1}^d \cap \tilde{P_2}^d$, so $\vec{G}^r$ is also a bounded bitolerance digraph. $\qquad\square$

The next theorem characterizes symmetric digraphs that are bounded bitolerance digraphs.

Theorem 13.17. *(Bogart and Trenk, 2000) A symmetric digraph is a bounded bitolerance digraph if and only if it has a loop at each vertex and its underlying graph is an interval graph.*

Proof. ($\Longrightarrow$): Let G be a symmetric bounded bitolerance digraph and let G be its underlying simple graph. As noted earlier, bounded bitolerance digraphs have a loop at each vertex. By Theorem 13.15, there exist interval orders P_1, P_2 so that $\vec{G} = \widetilde{P}_1 \cap \widetilde{P}_2$. If $x \rightleftharpoons y$ in $\vec{G}$, then by Remark 13.13 we have $x \parallel y$ in P_1 and P_2. If x and y are not adjacent in $\vec{G}$, then $x \prec_1 y$ if and only if $y \prec_2 x$. Thus, $P_2 = P_1^d$ and $\vec{G} = \widetilde{P}_1 \cap \widetilde{P_1^d}$. By Proposition 13.14, G is the incomparability graph of $P_1 \cap P_1 = P_1$, so G is an interval graph.

($\Longleftarrow$): Conversely, suppose $\vec{G}$ is a symmetric digraph with a loop at each vertex and that its underlying simple graph G is an interval graph. Fix an interval representation of G and let P be the interval order that comes from this representation. Then $\vec{G} = \widetilde{P} \cap \widetilde{P^d}$ because

$$(x, y) \in A(\widetilde{P} \cap \widetilde{P^d}) \iff x \parallel y \text{ in } P$$
$$\iff xy \in E(G)$$
$$\iff (x, y) \in A(\vec{G}).$$

By Theorem 13.15, $\vec{G}$ is a bounded bitolerance digraph.   □

Theorem 13.15 and Proposition 13.14 are also useful in determining whether a nonsymmetric digraph $\vec{G}$ is a bounded bitolerance digraph in cases where the complement $\overline{G}$ of the underlying graph of $\vec{G}$ has only a few transitive orientations. By Theorem 13.15, $\vec{G}$ is a bounded bitolerance digraph if and only if there exist interval orders P_1 and P_2 so that $\vec{G} = \widetilde{P}_1 \cap \widetilde{P}_2$. If so, by Proposition 13.14, G is the incomparability graph of $P_1 \cap P_2^d$. Thus, $P_1 \cap P_2^d$ gives a transitive orientation of $\overline{G}$. For each transitive orientation P of $\overline{G}$ we can find all pairs of interval orders P_1, P_2 for which $P_1 \cap P_2^d = P$ and then check if $\widetilde{P}_1 \cap \widetilde{P}_2 = \vec{G}$. If no such pair satisfy $\widetilde{P}_1 \cap \widetilde{P}_2 = \vec{G}$, then $\vec{G}$ is not a bounded bitolerance digraph. We use this line of reasoning in Section 13.7.

13.5 The digraph hierarchy

In Section 5.2.1, we introduced subclasses of bounded bitolerance orders based on three kinds of restrictions. The same restrictions can be applied to representations of a bounded bitolerance digraph $\vec{G} = (V, A)$, and we repeat them below.

Table 13.1. *Three categories of restrictions on bounded bitolerance representations.*

interval length	p and q	t_l and t_r
1. unit	a. point-core	i. tolerance
2. proper	b. totally bounded	ii. bitolerance
3. arbitrary	c. arbitrary	

Restrictions on intervals I_v

Definition 13.18. (Unit) $\vec{G}$ is a *unit bitolerance digraph* if it has a bounded bitolerance representation $\langle \mathcal{I}, p, q \rangle$ in which $|I_x| = |I_y|$ for all $x, y \in V$.

Definition 13.19. (Proper) $\vec{G}$ is a *proper bitolerance digraph* if it has a bounded bitolerance representation $\langle \mathcal{I}, p, q \rangle$ in which I_x is not properly contained in I_y for all $x, y \in V$.

Restrictions on tolerant points $p(v)$, $q(v)$

Definition 13.20. (Point-core) $\vec{G}$ is a *point-core bitolerance digraph* if it has a bounded bitolerance representation $\langle \mathcal{I}, p, q \rangle$ in which $p(v) = q(v)$ for all $v \in V$. In this case, we let $f(v) = p(v) = q(v)$, call this point the *splitting* point of I_v, and denote the representation by $\langle \mathcal{I}, f \rangle$.

Definition 13.21. (Totally bounded) $\vec{G}$ is a *totally bounded bitolerance digraph* if it has a bounded bitolerance representation $\langle \mathcal{I}, p, q \rangle$ in which $p(v) \le q(v)$ for all $v \in V$.

Restrictions on left and right tolerance

Definition 13.22. (Tolerance) $\vec{G}$ is a *bounded tolerance digraph* if it has a bounded bitolerance representation $\langle \mathcal{I}, p, q \rangle$ in which $p(v) - L(v) = R(v) - q(v)$ for all $v \in V$. In this case we write $t_v = t_l(v) = p(v) - L(v) = R(v) - q(v) = t_r(v)$.

The restrictions are listed by category in Table 13.1. As we saw in Chapter 10, these categories of restrictions are independent. The restrictions can be combined by taking one from each column to give 18 classes of bounded bitolerance digraphs, although some of the classes turn out to be equal. As in Chapter 10, we often refer to classes using their abbreviation. For example, in Chapter 10, (3ai) refers to the class of point-core tolerance orders, and here it refers to the class of point-core tolerance digraphs.

The analogue of Remark 7.9 for digraphs is the following.

Remark 13.23. *(Beads on a wire).* Given a representation $\langle \mathcal{I}, p, q \rangle$ of a bounded bitolerance digraph $\vec{G}$, the arc set $A(\vec{G})$ is completely determined by the ordering of the endpoints and tolerant points. Thus, we may convert one bounded bitolerance representation to another by perturbing endpoints and tolerant points, as long as we do not change their ordering.

Figure 13.4 shows the 18 classes of bounded bitolerance digraphs listed by their abbreviations. The examples along the edges refer to classes of digraphs obtained from their order theoretic counterparts, which we will construct in

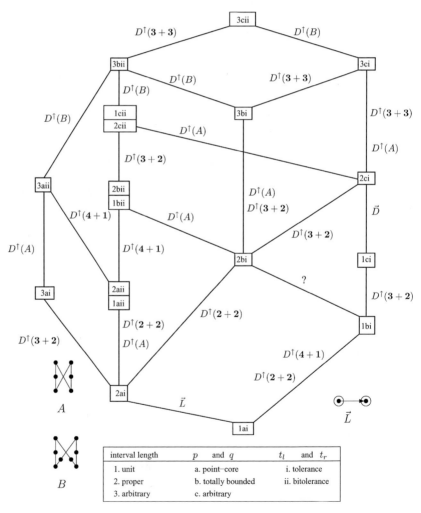

Figure 13.4. The hierarchy of classes of bounded bitolerance digraphs together with separating examples.

Lemma 13.27. As we have seen before, there are three types of results implied by Figure 13.4. (1) Classes that appear together in a box are equal. (2) A downward path from class T to class S means that class S is a subset of class T. (3) A digraph on an edge between two classes is a member of the larger class but not of the smaller class.

The rest of this section is devoted to justifying these results, which will be summarized in Theorem 13.30.

First we note that Figure 13.4 looks different from Figure 10.1. We list only abbreviations, rather than full names, in an attempt to keep the figure legible. The classes of unit bitolerance digraphs (1cii) and point-core bitolerance digraphs (3aii), whose analogues are equal in Figure 10.1, are not equal in Figure 13.4. Similarly, the classes of unit and point-core tolerance digraphs (1ci) and (3ai) are unequal. Lemmas 13.24 and 13.25 show the relation between these classes of digraphs.

Lemma 13.24. *Digraph $\vec{G}$ is a unit bitolerance digraph (1cii) if and only if its reversal $\vec{G}^r$ is a point-core bitolerance digraph (3aii).*

Proof. ($\Longrightarrow$): Fix a unit bitolerance representation $\langle \mathcal{I}, p, q \rangle$ of $\vec{G}$ in which $I_v = [L(v), R(v)]$ and $R(v) - L(v) = c$ for all $v \in V(\vec{G})$. We use the transformation from the proof of Theorem 5.26 which is illustrated in Figure 5.3.

Algebraically, this corresponds to creating a point-core bitolerance representation $\langle \mathcal{I}', f' \rangle$ of a digraph $\vec{H}$ with

$L'(v) = p(v),$
$R'(v) = q(v) + c,$
$f'(v) = R(v) = L(v) + c.$

Note that

$$(x, y) \in A(\vec{G}) \iff R(x) \geq p(y) \text{ and } L(x) \leq q(y)$$
$$\iff f'(x) \geq L'(y) \text{ and } f'(x) - c \leq R'(y) - c$$
$$\iff (y, x) \in A(\vec{H}).$$

Thus, $\vec{H} = \vec{G}^r$ and $\vec{G}^r$ is a point-core bitolerance digraph.

($\Longleftarrow$): For the converse, fix a point-core bitolerance representation $\langle \mathcal{I}', f' \rangle$ of $\vec{H}$ where $I'_v = [L'(v), R'(v)]$ and let c be the length of the longest interval in this representation. One can check that the following transformation produces a unit bitolerance representation of the reversal of $\vec{H}$.

$p(v) = L'(v),$
$q(v) = R'(v) - c,$
$R(v) = f'(v),$
$L(v) = R(v) - c = f'(v) - c.$ $\qquad\qquad\square$

Lemma 13.25. *Digraph $\vec{G}$ is a unit tolerance digraph (1ci) if and only if its reversal $\vec{G}^r$ is a point-core tolerance digraph (3ai).*

Proof. We apply the transformation used in the proof of Lemma 13.24. Note that

$$p(v) - L(v) = R(v) - q(v) \iff L'(v) - (f'(v) - c)$$
$$= f'(v) - (R'(v) - c)$$
$$\iff R'(v) - f'(v) = f'(v) - L'(v).$$

Hence, the transformation preserves the "tolerance" property. $\quad\square$

We have seen before (for orders in Theorem 10.3 and for graphs in Theorem 6.9) that corresponding "unit" and "proper" bitolerance classes are equal. The next theorem shows that the same is true in the case of digraphs.

Theorem 13.26. *The classes of unit and proper bitolerance digraphs are equivalent (1cii = 2cii), the classes of unit and proper totally bounded bitolerance digraphs are equivalent (1bii = 2bii), and the classes of unit and proper point-core bitolerance digraphs are equivalent (1aii = 2aii).*

Proof. This theorem follows from the observation in Remark 7.9 ("beads on a wire"). A formal proof can be obtained by using the proof of Theorem 10.3 and interpreting the representations as digraphs (rather than orders). $\quad\square$

The next lemma enables us to use separating examples from Figure 10.1 to generate separating examples for Figure 13.4.

Lemma 13.27. *Suppose there is an example P separating two classes of bounded bitolerance orders. Also suppose that all transitive orientations of the comparability graph $\overline{G}_P$ of P are isomorphic to P. Then there is an example $\vec{D}_P$ separating the corresponding classes of bounded bitolerance digraphs.*

Proof. Any bounded bitolerance representation of the order P also can be interpreted as a representation of a digraph $\vec{D}_P$, where the incomparability graph G_P of P is the (undirected) graph underlying $\vec{D}_P$. Thus, if P is a member of a class $\mathcal{X}$ of bounded bitolerance orders, then $\vec{D}_P$ is a member of the corresponding class $\mathcal{X}'$ of bounded bitolerance digraphs.

Any bounded bitolerance representation of the digraph $\vec{D}_P$ also can be interpreted as a bounded bitolerance representation of an order Q whose incomparability graph is G_P. Since Q is a transitive orientation of $\overline{G}_P$, by assumption, Q is isomorphic to P. Thus, if $\vec{D}_P$ is a member of a class $\mathcal{Y}'$ of bounded bitolerance digraphs, then P is a member of the corresponding class $\mathcal{Y}$ of bounded bitolerance orders.

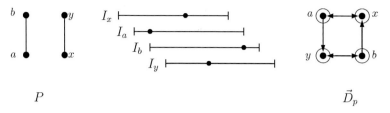

P $\vec{D}_P$

Figure 13.5. An order $P = 2 + 2$, a proper point-core bitolerance representation of it, and the resulting digraph $\vec{D}_P$.

Combining these, we conclude that if P separates $\mathcal{X}$ from $\mathcal{Y}$, then $\vec{D}_P$ separates $\mathcal{X}'$ from $\mathcal{Y}'$. □

Example 13.28. Figure 13.5 shows the order $P = 2 + 2$, a proper point-core bitolerance representation of it, and the resulting digraph $\vec{D}_P$. This digraph also appears as the middle digraph in Figure 13.9. One could not obtain the digraph on the left of Figure 13.9 in this way because it is not a proper point-core bitolerance digraph (see Exercise 13.6).

Remark 13.29. Note that the selection of digraph $\vec{D}_P$ in the proof of Lemma 13.27 depends upon a given $\mathcal{X}$-representation of P for the class $\mathcal{X}$. In Figure 13.4, when $D^{\uparrow}(P)$ appears on an edge below class $\mathcal{X}$ it indicates any of the $\vec{D}_P$ which belongs to $\mathcal{X}$ according to Lemma 13.27.

We now present the main result of this section.

Theorem 13.30. *The class hierarchy and the separating examples illustrated in Figure 13.4 are correct.*

Proof. The inclusions between classes follow directly from the definitions. The equivalences between classes in the same box are proven in Theorem 13.26. It remains to show that the examples shown along edges are separating examples.

The example $\vec{L}$ separating classes (1ai) and (2ai) is due to Lee (2002) and its justification is left as an exercise.

Next we discuss the digraph $\vec{D}$. Recall from Figure 2.8 that the Dartmouth graph D (shown in Figure 2.9) is a proper tolerance graph but not a unit tolerance graph. Thus, there is a proper tolerance representation of D which can be interpreted as a proper tolerance representation of a digraph $\vec{D}$ (whose underlying graph is D). This digraph $\vec{D}$ is not a unit tolerance digraph because its underlying graph is not a unit tolerance graph.

For each of the orders A, B, $2 + 2$, $3 + 2$, $4 + 1$, $3 + 3$, one can easily check that all transitive orientations of its comparability graph are isomorphic (see Exercise 13.8). Applying Lemma 13.27, since these orders appear as separating examples between classes of bounded bitolerance orders in Figure 10.1, their digraph counterparts provide separating examples between corresponding classes of bounded bitolerance digraphs in Figure 13.4.   □

13.6 Cycles

In Chapter 2, we saw that the chordless cycles C_n are not tolerance graphs for $n \geq 5$ (Lemma 2.14) but C_3 and C_4 are tolerance graphs, indeed they are unit tolerance graphs. Some orientations of the edges of C_3 and C_4 give unit bitolerance digraphs and some do not. In this section we follow Bogart and Trenk (1995) in showing that the classes of unit, proper and bounded tolerance and bitolerance digraphs are equal when the underlying simple graph is a chordless cycle. We give several characterizations of this class.

A digraph is called *weakly transitive* if it has no induced copy of the digraph in Figure 13.6.

Proposition 13.31. *Bounded bitolerance digraphs are weakly transitive.*

Proof. For a contradiction, suppose the digraph $\vec{C}$ shown in Figure 13.6 is a bounded bitolerance digraph. By Theorem 13.15, there exist interval orders $P_1 = (V, \prec_1)$ and $P_2 = (V, \prec_2)$ with $\vec{C} = \widetilde{P}_1 \cap \widetilde{P}_2$. Using Remark 13.13(i), we know that for some $i, j, k \in \{1, 2\}$ we have each of $x \prec_i y$, $y \prec_j z$ and $z \prec_k x$. Thus, at least two of i, j, k are equal and, without loss of generality, we may assume $x \prec_1 y$ and $y \prec_1 z$. By transitivity, $x \prec_1 z$ which contradicts $(z, x) \in A(\widetilde{P}_1 \cap \widetilde{P}_2)$.   □

Lemma 13.32. *The directed 3-path $\vec{G}$ with a loop at each vertex and the interval orders P_1 and P_2 shown in Figure 13.7 satisfy $\vec{G} = \widetilde{P}_1 \cap \widetilde{P}_2$. Furthermore, this representation is unique (up to reversing the roles of P_1 and P_2).*

Proof. It is easy to check that $\vec{G} = \widetilde{P}_1 \cap \widetilde{P}_2$ for $\vec{G}$, P_1, P_2 shown in Figure 13.7, hence we need only show uniqueness. Suppose $\vec{G} = \widetilde{Q}_1 \cap \widetilde{Q}_2$ where $Q_1 = (V, \prec_1)$ and $Q_2 = (V, \prec_2)$ are interval orders. By Remark 13.13(ii), we must have either $(x \prec_1 z$ and $z \prec_2 x)$ or $(z \prec_1 x$ and $x \prec_2 z)$. By reversing the roles

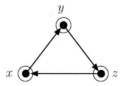

Figure 13.6. A digraph which is not a bounded bitolerance digraph.

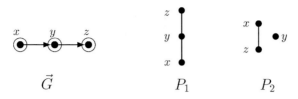

Figure 13.7. A digraph $\vec{G}$ and interval orders P_1, P_2 with $\vec{G} = \tilde{P}_1 \cap \tilde{P}_2$.

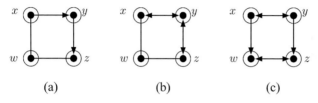

Figure 13.8. Forbidden subgraphs for a bounded bitolerance digraph with C_4 as underlying graph. An undirected edge indicates the presence of at least one arc between its endpoints.

of Q_1 and Q_2 if necessary, we may assume the former. Now consider the placement of y in Q_1 and Q_2. In Q_2, $y \prec_2 x$ contradicts $(x, y) \in A(\vec{G})$ and $z \prec_2 y$ contradicts $(y, z) \in A(\vec{G})$, thus Q_2 must be the order P_2 shown in Figure 13.7. Since $(y, x) \notin A(\vec{G})$ we must have $x \prec_1 y$ or $x \prec_2 y$, by Remark 13.13(i), and the latter is false, so $x \prec_1 y$. Similarly, since $(z, y) \notin A(\vec{G})$ and $y \not\prec_2 z$, we must have $y \prec_1 z$. Thus, Q_1 must be the order P_1 shown in Figure 13.7. □

Lemma 13.33. *Let $\vec{C}$ be a digraph with a loop at each vertex whose underlying graph is C_4. Then the following are equivalent.*

(i) *$\vec{C}$ is a bounded bitolerance digraph.*
(ii) *$\vec{C}$ does not contain an induced copy of any of the digraphs in Figure 13.8.*
(iii) *$\vec{C}$ is isomorphic to one of the digraphs in Figure 13.9.*

Proof. (i) $\Longrightarrow$ (ii): Let $\vec{C}$ be a bounded bitolerance digraph whose underlying simple graph $C = (V, E)$ has $V = \{x, y, z, w\}$ and $E = \{xy, yz, zw, wx\}$. By Theorem 13.15, there exist interval orders $P_1 = (V, \prec_1)$ and $P_2 = (V, \prec_2)$ so that $\vec{C} = \tilde{P}_1 \cap \tilde{P}_2$.

For a contradiction, first suppose that $\vec{C}$ has the form of a digraph shown in Figure 13.8(a). By Lemma 13.32, we may assume that P_1 and P_2 restricted to $\{x, y, z\}$ are as in Figure 13.7. By Remark 13.13(ii) we know (1) $w \prec_1 y$ and $y \prec_2 w$, or (2) $y \prec_1 w$ and $w \prec_2 y$. In case (1), $w \prec_1 z$ by transitivity and since $wz \in E(C)$ we cannot have $z \prec_2 w$. This forces P_2 to be the order $2 + 2$ which is not an interval order. In case (2) we get a similar contradiction.

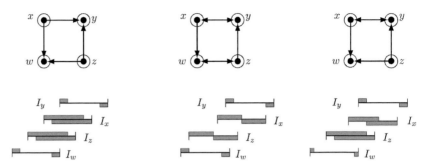

Figure 13.9. Unit tolerance digraphs whose underlying graph is C_4.

Next suppose that $\vec{C}$ has the form of a digraph shown in Figure 13.8(b). By Remark 13.13(iii), we know $x \parallel_i y$ and $y \parallel_i z$ for $i = 1, 2$. Without loss of generality we may assume $x \prec_1 z$ and $z \prec_2 x$ by Remark 13.13(ii), and we also know (1) $y \prec_1 w$ and $w \prec_2 y$ or (2) $w \prec_1 y$ and $y \prec_2 w$. In case (1), to make P_1 and P_2 interval orders, we need $x \prec_1 w$ and $w \prec_2 x$ which contradicts $wx \in E$. Case (2) leads to a similar contradiction.

Finally, suppose that $\vec{C}$ has the form of a digraph shown in Figure 13.8(c). By Remark 13.13(ii), we know (1) $w \prec_1 y$ and $y \prec_2 w$, or (2) $y \prec_1 w$ and $w \prec_2 y$. We consider case (1). Case (2) is similar. We must have $x \prec_1 w$ or $x \prec_2 w$ by Remark 13.13(i), and the former leads to $x \prec_1 y$, which is not possible by Remark 13.13(iii). Thus $x \prec_2 w$. Similarly, we must have $y \prec_1 z$ or $y \prec_2 z$ and only the latter is possible. We cannot have $z \prec_2 x$ because that would imply $y \prec_2 x$, thus by Remark 13.13(ii), we must have $z \prec_1 x$ and $x \prec_2 z$. But now P_1 is the order $\mathbf{2} + \mathbf{2}$ because Remark 13.13(iv) implies $z \not\prec_1 y$ and $w \not\prec_1 x$. This is a contradiction.

(ii) $\Longrightarrow$ (iii): This follows by considering cases depending on the number of double arcs in $\vec{C}$ (Exercise 13.10).

(iii) $\Longrightarrow$ (i): Bounded (bi)tolerance representations of the digraphs in Figure 13.9 are also given in that figure. $\qquad\square$

We are now ready to prove the result that characterizes those cycles that are unit, proper and bounded bitolerance digraphs.

Theorem 13.34. *(Bogart and Trenk, 1995) Let $\vec{C}$ be a digraph with a loop at each vertex and whose underlying simple graph is the chordless cycle C_n. Then the following are equivalent.*

(i) $\vec{C}$ *is a unit tolerance digraph.*
(ii) $\vec{C}$ *is a unit bitolerance digraph.*

(iii) $\vec{C}$ *is a proper bitolerance digraph.*
(iv) $\vec{C}$ *is a bounded bitolerance digraph.*
 (v) *Either $n = 3$ and $\vec{C}$ is weakly transitive, or $n = 4$ and $\vec{C}$ is isomorphic to one of the digraphs shown in Figure 13.9.*

Proof. (i) $\Longrightarrow$ (ii), (ii) $\Longrightarrow$ (iii) and (iii) $\Longrightarrow$ (iv) follow immediately from definitions.

(iv) $\Longrightarrow$ (v): Suppose $\vec{C}$ is a bounded bitolerance digraph whose underlying simple graph is C_n. By Remark 13.7(1), C_n is a bounded bitolerance graph and hence, by Remark 5.6, it is a cocomparability graph. Since C_n is not a cocomparability graph for $n \geq 5$ (Corollary 2.11) we must have $n = 3$ or $n = 4$.

If $n = 3$, then $\vec{C}$ is weakly transitive by Proposition 13.31. For $n = 4$, the result follows directly from Lemma 13.33.

(iv) $\Longrightarrow$ (i): There are six nonisomorphic, weakly transitive digraphs $\vec{C}$ in the case of $n = 3$ and it is not hard to construct unit tolerance representations of each of them (Exercise 13.11). In the case of $n = 4$ we note that the representations in Figure 13.9 are unit tolerance representations. $\qquad\square$

13.7 Trees

In Theorems 3.2 and 3.7 we characterize those (undirected) trees that are bounded tolerance graphs and tolerance graphs. In this section we follow Bogart and Trenk (1995) in characterizing those (directed) trees that are bounded bitolerance digraphs, unit bitolerance digraphs, and proper bitolerance digraphs. We use the term *tree* to refer to both graphs and digraphs, and in the latter case we mean a digraph whose underlying simple graph is a tree. As before, we indicate a loop at a vertex by drawing a circle around that vertex.

13.7.1 Trees that are bounded bitolerance digraphs

We begin by considering the case of stars. Figure 13.10 gives a bounded bitolerance representation of a star that has nine leaves, three joined to the central vertex by a double arc (x_i), three joined by a single arc pointing away from the central vertex (y_i) and three joined by a single arc pointing towards the central vertex (z_i). This representation can easily be generalized to cases with arbitrarily many leaves of each type. We record this as a lemma.

Lemma 13.35. *Any digraph that has a loop at each vertex and whose underlying graph is a star is a bounded bitolerance digraph.*

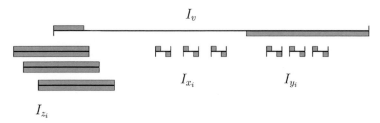

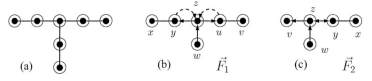

Figure 13.10. A bounded bitolerance representation of a directed star with nine leaves, three of each type.

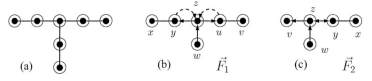

Figure 13.11. Forbidden digraphs for the class of bounded bitolerance digraphs.

In the main theorem of this section we give a forbidden subdigraph characterization of those trees that are bounded bitolerance digraphs. There are three types of forbidden subdigraphs and they are shown in Figure 13.11. As before, an undirected edge indicates the presence of at least one arc between its endpoints. The dashed edges in Figure 13.11(b) indicate that the arcs (y, z) and (u, z) may or may not be present. Given a digraph $\vec{G} = (V, A)$, recall that the notation $x \to y$ means that $(x, y) \in A$ and $(y, x) \notin A$, and $x \rightleftharpoons y$ means that both $(x, y) \in A$ and $(y, x) \in A$.

Theorem 13.36. *(Bogart and Trenk, 1995) Let $\vec{T}$ be a digraph with a loop at each vertex whose underlying graph T is a tree. Then the following are equivalent.*

(i) $\vec{T}$ *is a bounded bitolerance digraph.*
(ii) $\vec{T}$ *does not contain any of the digraphs in Figure 13.11 or their reversals as an induced subdigraph.*
(iii) T *is a caterpillar whose spine vertices can be labeled $v_1, v_2, \ldots, v_n$ where v_1 and v_n are leaves of T, and for $i \neq 1, n$, the leaf neighbors w of v_i are permitted as follows:*

  (1) $w \rightleftharpoons v_i$: *any number of such w can be present;*
  (2) $w \to v_i$: *any number of such w can be present if $v_{i-1} \to v_i$ or $v_{i+1} \to v_i$, and none otherwise;*
  (3) $v_i \to w$: *any number of such w can be present if $v_i \to v_{i-1}$ or $v_i \to v_{i+1}$, and none otherwise.*

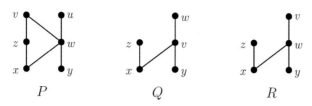

Figure 13.12. The orders P, Q, and R which give transitive orientations of $\overline{F}$ and F'.

Proof. (i) $\Longrightarrow$ (ii): Suppose $\vec{T}$ is a bounded bitolerance digraph whose underlying graph T is a tree. By Remark 13.7(1), T is a bounded bitolerance graph, and by Remark 5.6, T is a cocomparability graph. Now Theorem 3.2 implies that T is a caterpillar or equivalently has no induced T_2. This shows that any digraph whose underlying graph is T_2 (Figure 13.11(a)) is forbidden. It remains to show that the digraphs shown in Figure 13.11(b) and (c) and their reversals are not bounded bitolerance digraphs.

Let $\vec{F}$ be a digraph of the form shown in Figure 13.11(b) and let F be its underlying graph. For a contradiction, suppose $\vec{F}$ is a bounded bitolerance digraph. By Theorem 13.15, there exist interval orders $P_1 = (V, \prec_1)$ and $P_2 = (V, \prec_2)$ so that $\vec{F} = \tilde{P}_1 \cap \tilde{P}_2$, and by Proposition 13.14, F is the incomparability graph of $P_1 \cap P_2^d$. Thus, $P = P_1 \cap P_2^d$ gives a transitive orientation of the complement $\overline{F}$. One can check that $\overline{F}$ has only two transitive orientations – the order P shown in Figure 13.12 and its dual. We can switch the roles of P_1 and P_2 if necessary to ensure that P is the order shown in Figure 13.12.

Since $P = P_1 \cap P_2^d$, we know that P_1 contains all the comparabilities in P, plus additional comparabilities (since P is not an interval order). Similarly, P_2 contains all the comparabilities of the dual of P plus additional comparabilities. Now $(z, w) \notin A(\vec{F})$ so we must have $w \prec_1 z$ or $w \prec_2 z$ by Remark 13.13(i). If $w \prec_1 z$ then $y \prec_1 w \prec_1 z$ which contradicts $(z, y) \in A(\vec{F})$. Otherwise, if $w \prec_2 z$ then $u \prec_2 w \prec_2 z$ which contradicts $(z, u) \in A(\vec{F})$. Therefore, $\vec{F}$ is not a bounded bitolerance digraph. Finally, by Corollary 13.16, we can conclude that $\vec{F}r$ is not a bounded bitolerance digraph.

A similar contradiction is reached if one starts with a digraph $\vec{F}'$ of the form given in Figure 13.11(c). In this case, there are four possible transitive orientations, Q and R shown in Figure 13.12 and their duals (see Exercise 13.12).

(ii) $\Longrightarrow$ (iii): Now suppose $\vec{T}$ is a tree with a loop at each vertex and $\vec{T}$ does not contain any of the digraphs in Figure 13.11 or their reversals. Since digraphs of the form in Figure 13.11(a) are forbidden, the underlying graph T must be a caterpillar. Let $v_1, v_2, \ldots, v_n$ be the spine of caterpillar T where $v_i v_{i+1} \in E(T)$ for $i = 1, 2, \ldots, n - 1$, and without loss of generality we may assume v_1 and v_n are leaves of T. By Lemma 13.35, we need only consider the case $n \geq 4$.

For $i \neq 1, 2, n-1, n$, the leaf neighbors which are forbidden (by $\vec{F}_1$ in Figure 13.11(b)) are those w with $w \to v_i$ and $(v_i, v_{i-1}), (v_i, v_{i+1}) \in A(\vec{T})$. Thus $w \to v_i$ is permitted only if $v_{i-1} \to v_i$ or $v_{i+1} \to v_i$ as in condition (iii)(2). Similarly, the leaf neighbors forbidden by the reversal of $\vec{F}_1$ are those w with $v_i \to w$ and $(v_{i-1}, v_i), (v_{i+1}, v_i) \in A(\vec{T})$. Thus $v_i \to w$ is permitted only if $v_i \to v_{i-1}$ or $v_i \to v_{i+1}$ as in condition (iii)(3). Leaf neighbors of the form $w \rightleftharpoons v_i$ are always permitted.

Next we consider $i = 2, n-1$. Any of the leaves of T adjacent to v_2 could be labeled as the spine vertex v_1. To allow for the greatest flexibility in attaching leaves to v_i, we would like v_1 to be joined to v_2 by a single arc so that v_2 has both a single arc pointing away and an arc pointing towards it along the spine of T. Thus, we choose the vertex among v_2's leaf neighbors to be relabeled v_1 as follows. If all the candidates are joined to v_2 by a double arc, choose arbitrarily. If not, consider only single arc neighbors of v_2 to be labeled v_1. If $v_3 \rightleftharpoons v_2$ then choose arbitrarily among the remaining candidates. Otherwise, relabel v_1, if possible, so that $(v_1 \to v_2$ and $v_2 \to v_3)$ or $(v_3 \to v_2$ and $v_2 \to v_1)$. If this is not possible, choose arbitrarily among the remaining candidates. In a similar way, relabel v_n.

Now suppose $w \to v_2$ but condition (iii)(2) is not satisfied, so $(v_2, v_1) \in A$ and $(v_2, v_3) \in A$. Since w is a single arc neighbor of v_2, we know $(v_1, v_2) \notin A$, for then $v_1 \rightleftharpoons v_2$ and w would have been chosen instead of v_1 to label as v_1. Thus $v_2 \to v_1$. If $(v_3, v_2) \notin A$, then $v_2 \to v_3$ and again we would have chosen w to be labeled as v_1. Thus $v_2 \rightleftharpoons v_3$ and we have $\vec{F}_2$ of Figure 13.11(c) induced in $\vec{T}$, a contradiction. Similar contradictions are reached when condition (iii)(3) is not satisfied, and in the case $i = n-1$.

(iii) $\implies$ (i): Suppose $\vec{T}$ has the form given in (iii). Label the spine vertices $v_1, v_2, \ldots, v_n$ where v_1 and v_n are leaves of T chosen as above. The following intervals and tolerant points give a bounded bitolerance representation of the spine of $\vec{T}$:

$$I_{v_i} = [8i, 8i + 10],$$

$$p(v_i) = \begin{cases} 8i + 1 \text{ if } (v_{i-1}, v_i) \in A(\vec{T}) \\ 8i + 4 \text{ otherwise,} \end{cases}$$

$$q(v_i) = \begin{cases} 8i + 9 \text{ if } (v_{i+1}, v_i) \in A(\vec{T}) \\ 8i + 6 \text{ otherwise.} \end{cases}$$

It remains to attach leaves to the spine. Figure 13.13 shows a tree $\vec{T}$ and a portion of a bounded bitolerance representation of it which illustrates the construction in this proof.

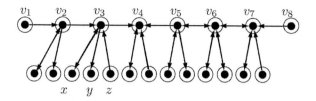

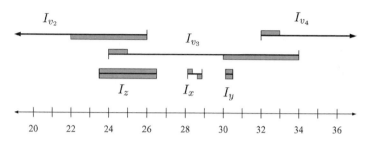

Figure 13.13. A bounded bitolerance digraph $\vec{T}$ and a representation of part of it.

(1) Leaf neighbors x with $v_i \rightleftharpoons x$

Such neighbors are always allowed for $i \neq 1, n$. Note that v_i is the only spine vertex whose interval intersects $[8i + 4, 8i + 6]$. And indeed no other intervals created in later parts of this representation will intersect that interval. Thus the intervals for any double arc leaf neighbors of v_i can be placed side by side in $[8i + 4, 8i + 6]$ with any tolerant points.

(2) Leaf neighbors z with $z \rightarrow v_i$

Suppose $\vec{T}$ has r such vertices, $z_1, z_2, \ldots, z_r$. If $v_{i-1} \rightarrow v_i$, let $L(z_j) = q(z_j) = 8i - \frac{j}{2r}$ and $R(z_j) = p(z_j) = 8i + 3 - \frac{j}{2r}$ for $j = 1, 2, \ldots, r$. Otherwise, by (iii)(2), if $v_{i+1} \rightarrow v_i$, let $L(z_j) = q(z_j) = 8i + 7 + \frac{j}{2r}$ and $R(z_j) = p(z_j) = 8i + 10 + \frac{j}{2r}$ for $j = 1, 2, \ldots, r$. One can check that this gives a bounded bitolerance representation of $\vec{T}$.

(3) Leaf neighbors y with $v_i \rightarrow y$

If $v_i \rightarrow v_{i-1}$ then no intervals constructed so far intersect $[8i + 3, 8i + 4]$ and no intervals constructed later in this representation will intersect that interval. Thus the intervals for these leaf neighbors of v_i can be placed side by side in $[8i + 3, 8i + 4]$ with any tolerant points. Otherwise, by (iii)(3), $v_i \rightarrow v_{i+1}$ and no intervals in the representation intersect $[8i + 6, 8i + 7]$ and the intervals for these leaf neighbors of v_i can be placed side by side in this interval, with any tolerant points. $\square$

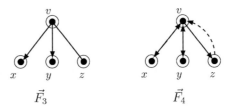

Figure 13.14. Digraphs which are not unit bitolerance digraphs.

13.7.2 Trees that are unit/proper bitolerance digraphs

Figure 13.14 depicts digraphs which are not unit bitolerance digraphs. As before, the undirected edge in $\vec{F}_3$ indicates that the arc between v and z can point in either (or both) directions. The dashed arc from z to v in $\vec{F}_4$ indicates that the arc (z, v) may or may not be present. Unlike the result in Theorem 13.36, the reversals of these digraphs *are* unit bitolerance digraphs, as we see in the next theorem.

Theorem 13.37. *(Bogart and Trenk, 1995) The following are equivalent statements about a digraph $\vec{T}$ with a loop at each vertex whose underlying graph T is a tree.*

(i) $\vec{T}$ *is a unit bitolerance digraph.*

(ii) $\vec{T}$ *is a proper bitolerance digraph.*

(iii) $\vec{T}$ *is a bounded bitolerance digraph and has none of the forbidden digraphs depicted in Figure 13.14.*

(iv) *T is a caterpillar whose spine vertices can be labeled $v_1, v_2, \cdots, v_n$ where v_1 and v_n are leaves of T, and for $i \neq 1, n$, vertex v_i may have only the following leaf neighbors y:*

    (1) $v_i \rightarrow y$*: none permitted;*

    (2) $y \rightleftharpoons v_i$*:*

        up to two such y if $v_{i-1} \rightarrow v_i$ and $v_{i+1} \rightarrow v_i$,

        at most one such y if exactly one of $v_{i-1} \rightarrow v_i$ and $v_{i+1} \rightarrow v_i$,

        no such y can be present if neither $v_{i-1} \rightarrow v_i$ nor $v_{i+1} \rightarrow v_i$;

    (3) $y \rightarrow v_i$*: any number of such y if $v_{i-1} \rightarrow v_i$ or $v_{i+1} \rightarrow v_i$, and none otherwise.*

Proof. (i) $\Longrightarrow$ (ii): This follows immediately from the definitions.

    (ii) $\Longrightarrow$ (iii): It is not hard to show that the digraphs depicted in Figure 13.14 are not proper bitolerance digraphs (Exercise 13.13).

    (iii) $\Longrightarrow$ (iv): By Theorem 13.36, T is a caterpillar and the digraphs $\vec{F}_1$, $\vec{F}_1^{\,r}$ and $\vec{F}_2$ from Figure 13.11 are not induced in $\vec{T}$. Let $v_1, v_2, \cdots, v_n$ be the spine of $\vec{T}$ with v_1 and v_n chosen as in the proof of (ii) $\Longrightarrow$ (iii) of Theorem 13.36.

First we establish (1). For $i \neq 1, 2, n - 1, n$, if $v_i \rightarrow v_{i-1}$ and $v_i \rightarrow v_{i+1}$, then leaves w with $v_i \rightarrow w$ are forbidden by $\vec{F}_3$ of Figure 13.14. Otherwise, $(v_{i-1}, v_i), (v_{i+1}, v_i) \in A$ and such w are forbidden by the reversal of the digraph $\vec{F}_1$ shown in Figure 13.11.

For $i = 2$, suppose there were a leaf y with $v_2 \rightarrow y$. By our rule for choosing v_1, we know the arc between v_1 and v_2 can not be a double arc, and since digraphs of the form $\vec{F}_3$ are forbidden, it must be the case that $v_1 \rightarrow v_2$. Now consider the arc between v_2 and v_3. If $v_2 \rightleftharpoons v_3$, then $n \geq 4$ by our rule for choosing which vertex to label as v_n, and we get an induced $\vec{F}_2$. If $v_2 \rightarrow v_3$, we get an induced $\vec{F}_3$, and if $v_3 \rightarrow v_2$, we contradict our rule for choosing which vertex to label v_1. The case $i = n - 1$ is similar.

Next we establish (2). Since $\vec{F}_3$ and $\vec{F}_4$ of Figure 13.14 are forbidden, there are at most two arcs that can point out from v_i. This proves (2).

Finally, we prove (3). If $i \neq 2, n - 1$, the constraint in (3) follows because $\vec{F}_1$ of Figure 13.11(b) is forbidden. For $i = 2$, suppose there were a leaf y with $y \rightarrow v_2$ but $(v_2, v_1) \in A$ and $(v_2, v_3) \in A$. If $v_2 \rightleftharpoons v_3$ then by our rule for choosing which vertex to label as v_n, we know $n \geq 4$. In this case, either $(v_1, v_2) \notin A$ and we get an induced $\vec{F}_2$ (Figure 13.11), or $(v_1, v_2) \in A$ and we contradict our rule for choosing v_1. The case $i = n - 1$ is similar.

(iv) $\Longrightarrow$ (i): We construct a unit bitolerance digraph which contains $\vec{T}$ as an induced subgraph. Use the same intervals and tolerances for the spine vertices as in the proof of Theorem 13.36. In the remainder of the construction, fix an ϵ with $0 < \epsilon < 1/2$.

If $v_{i-1} \rightarrow v_i$ we can represent a leaf z with $v_i \rightleftharpoons z$ by $L(z) = 8i - 1 - \epsilon$, $R(z) = p(z) = 8i + 9 - \epsilon$, and $q(z) = 8i + \epsilon$. And we can represent leaves $x_1, x_2, \cdots, x_{r_i}$ where $x_j \rightarrow v_i$ by $L(x_j) = q(x_j) = 8i - 2 + \frac{i}{2r_j}$ and $R(x_j) = p(x_j) = 8i + 8 + \frac{i}{2r_j}$.

Similarly, if $v_{i+1} \rightarrow v_i$ we can represent a leaf z with $v_i \rightleftharpoons z$ by $L(z) = q(z) = 8i + 1 + \epsilon$, $R(z) = 8i + 11 + \epsilon$, and $p(z) = 8i + 10 - \epsilon$. And we can represent leaves $x_1, x_2, \cdots, x_{r_i}$ where $x_j \rightarrow v_i$ by $L(x_j) = q(x_j) = 8i + 2 - \frac{i}{2r_j}$ and $R(x_j) = p(x_j) = 8i + 12 - \frac{i}{2r_j}$. $\square$

13.8 Unit vs. proper

The unit vs. proper question has appeared repeatedly throughout this book. It makes its last appearance in the next theorem which originally was proved in Shull and Trenk (1997b).

Theorem 13.38. *The following are equivalent.*

(i) $\vec{G}$ *is a unit bitolerance digraph.*
(ii) $\vec{G}$ *is a proper bitolerance digraph.*
(iii) $\vec{G}r$ *is a point-core bitolerance digraph.*

Proof. The equivalence of (i) and (ii) follows from Remark 13.23 and the proof of Theorem 10.3 since the latter transforms a proper bitolerance representation into a unit bitolerance representation without changing the order of any endpoints or tolerant points.

The equivalence of (i) and (iii) follows by using the same transformation illustrated in Figure 5.3 for orders. The intervals and tolerant points at the top of that figure give a unit bitolerance representation of the digraph $\vec{G} = (V, A)$ with $V = \{x, y, z\}$ and $A = \{(y, z), (x, x), (y, y), (z, z)\}$. The intervals and splitting points at the bottom of the figure give a point-core bitolerance representation of the digraph $\vec{G}' = (V', A')$ with $V' = \{x, y, z\}$ and $A' = \{(z, y), (x, x), (y, y), (z, z)\}$. In general, the transformation takes a unit bitolerance representation of $\vec{G}$ and returns a point-core bitolerance representation of $\vec{G}r$. A precise proof can be found by transforming the unit bitolerance representation $\langle \mathcal{I}, p, q \rangle$ of $\vec{G}$ with $I_v = [L(v), R(v)]$ and tolerant points $p(v), q(v)$ to the point-core bitolerance representation $\langle \mathcal{I}, f' \rangle$ of $\vec{G}r$ with $I'_v = [L'(v), R'(v)]$ and splitting points $f'(v)$. The transformation is achieved using the mapping: $L'(v) = p(v), R'(v) = q(v) + c, f'(v) = L(v) + c = R(v)$, where $c = R(v) - L(v)$ is the unit length of the intervals in the unit bitolerance representation.   □

In the remainder of this section, we present results from Prisner (1989) that give additional conditions equivalent to a digraph being a point-core bitolerance digraph. The new conditions have algorithmic consequences (see Corollary 13.41). We begin with some background.

An $x - y$ *chain* $\vec{C}$ in digraph $\vec{G}$ is a digraph with $V(\vec{C}) = \{x = a_0, a_1, a_2, \ldots, a_n = y\} \subseteq V(\vec{G})$ and $A(\vec{C}) \subseteq A(\vec{G})$ consisting of exactly one of $(a_i, a_{i+1}), (a_{i+1}, a_i)$ for each $i : 0 \leq i \leq n - 1$. An $x - y$ chain $\vec{C}$ is called *z-avoiding* if $z \notin V(\vec{C})$ and $(a_i, z) \notin A(\vec{G})$ for any i with $(a_i, a_{i+1}) \in A(\vec{C})$ or $(a_i, a_{i-1}) \in A(\vec{C})$. A set $\{x, y, z\} \subseteq V(\vec{G})$ is a *diasteroidal triple* if there is a z-avoiding $x - y$ chain, an x-avoiding $y - z$ chain, and a y-avoiding $x - z$ chain. The set $\{x, y, z\}$ in Figure 13.15 is a diasteroidal triple. Note that the $x - y$ chain with arcs $\{(x, v), (v, y)\}$ is *not* z-avoiding, but the $x - y$ chain with arcs $\{(x, v), (y, v)\}$ *is* z-avoiding.

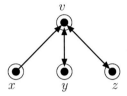

Figure 13.15. A digraph with diasteroidal triple $\{x, y, z\}$.

Theorem 13.40 shows that a digraph $\vec{G}$ contains a diasteroidal triple precisely when an associated (undirected) graph G^* contains an asteroidal triple. We next define this graph G^*.

Given a digraph $\vec{G} = (V, A)$, define G^* to be the undirected graph with $V(G^*) = V \cup \tilde{V}$ where $\tilde{V} = \{\tilde{x} \mid x \in V\}$ is disjoint from V, and

$$E(G^*) = \{xy \mid (x, y) \in A \text{ or } (y, x) \in A\}$$
$$\cup \{xy \mid \exists z \in V \text{ with } (x, z), (y, z) \in A\}$$
$$\cup \{x\tilde{y} \mid (x, y) \in A \text{ or } x = y\}.$$

We make two remarks about G^*.

Remark 13.39. (i) The vertices in $\tilde{V}$ form an independent set in G^*.
(ii) The neighborhood $\mathcal{N}(\tilde{y})$ is a complete subgraph of G^* for each $\tilde{y} \in \tilde{V}$.

The next result appears in Prisner (1989), although in that paper, point-core bitolerance digraphs have their loops removed and are called *interval catch digraphs*.

Theorem 13.40. *Let* $\vec{G} = (V, A)$ *be a digraph with a loop at each vertex and* G^* *be the associated undirected graph defined above. The following are equivalent.*

(i) $\vec{G}$ *is a point-core bitolerance digraph.*
(ii) G^* *is an interval graph.*
(iii) G^* *has no asteroidal triple.*
(iv) $\vec{G}$ *has no diasteroidal triple.*

Proof. (i) $\implies$ (ii): Note that Lemma 5.18 also holds in the directed case, thus we may fix a point-core bitolerance representation $\langle \mathcal{I}, f \rangle$ of $\vec{G}$ in which all endpoints and splitting points are distinct. Let $\mathcal{I} = \{I_x \mid x \in V\}$ and let J_x be the smallest interval that contains all the splitting points in I_x. Next, for all $x \in V$, let $J_{\tilde{x}}$ be the interval consisting of the single point $f(x)$. It is easy to check that whenever two intervals in the collection $S = \{J_x \mid x \in V\} \cup \{J_{\tilde{x}} \mid \tilde{x} \in \tilde{V}\}$

intersect, they contain a common splitting point. Now one can verify that the intervals in S give an interval representation of G^*.

(ii) $\Longrightarrow$ (i): Let $\vec{G} = (V, A)$ and suppose $\{I_x \mid x \in V\} \cup \{I_{\tilde{x}} \mid \tilde{x} \in \widetilde{V}\}$ is an interval representation of G^*. Let J_x be the union of the intervals $I_{\tilde{y}}$ that intersect I_x and let $I'_x = I_x \cup J_x$, that is, we extend I_x so that it includes all $I_{\tilde{y}}$ which it intersects. Fix a point $f(x) \in I_{\tilde{x}}$ for each $x \in V$. It remains to show that the intervals $\{I'_x \mid x \in V\}$ and points $\{f(x)\}$ give a point-core bitolerance representation of $\vec{G}$.

First note that for all $x \in V$, we have $x\tilde{x} \in E(G^*)$, so $I_x \cap I_{\tilde{x}} \neq \emptyset$ and thus I'_x includes all of $I_{\tilde{x}}$. Therefore, $f(x) \in I_{\tilde{x}} \subseteq I'_x$. If $(x, y) \in A$ then $x\tilde{y} \in E(G^*)$ so $I_x \cap I_{\tilde{y}} \neq \emptyset$ and $f(y) \in I_{\tilde{y}} \subseteq I'_x$ as needed. If $(x, y) \notin A$ we must show $f(y) \notin I'_x$. Suppose $f(y) \in I'_x$. Then $I_x \cap I_{\tilde{y}} \neq \emptyset$ so $x\tilde{y} \in E(G^*)$ and $(x, y) \in A$, a contradiction.

(ii) $\Longrightarrow$ (iii): This follows from Theorem 1.3.

(iii) $\Longrightarrow$ (ii): Let $\vec{G} = (V, A)$ and suppose G^* has no asteroidal triple. If G^* has no induced cycle C_n with $n \geq 4$ then G^* is an interval graph by Theorem 1.3. So, for a contradiction, suppose G^* has an induced cycle C with vertex set $\{v_1, v_2, \ldots, v_t\}$ and $t \geq 4$. By Remark 13.39, $\mathcal{N}(\tilde{x})$ is complete for all $\tilde{x} \in \widetilde{V}$. Since $\mathcal{N}(v_i)$ is not complete we know $v_i \in V$ for each i.

We now show that for each i there exists a $\tilde{z} \in \widetilde{V}$ whose neighbors on the cycle are precisely v_i and v_{i+1}. Each edge $v_i v_{i+1}$ of the cycle C arises either from an arc (v_i, v_{i+1}) or (v_{i+1}, v_i) in $\vec{G}$ or because v_i and v_{i+1} have a common out-neighbor z in $\vec{G}$. Moreover, in the latter case, this z can not be the out-neighbor of any other vertex in the cycle C for otherwise there would be a chord in G^*.

If $(v_i, v_{i+1}) \in A$ and $(v_{i+1}, v_i) \notin A$ then let $\tilde{z}_i = \tilde{v}_{i+1}$, if $(v_i, v_{i+1}) \notin A$ and $(v_{i+1}, v_i) \in A$, then let $\tilde{z}_i = \tilde{v}_i$. It is not possible for both (v_i, v_{i+1}) and (v_{i+1}, v_i) to be arcs of $\vec{G}$ without causing a chord somewhere in C. Otherwise, if neither (v_i, v_{i+1}) nor (v_{i+1}, v_i) is an arc of $\vec{G}$, then there exists a $z \in V$ with $(v_i, z), (v_{i+1}, z) \in A$ and we let $\tilde{z}_i = \tilde{z}$.

In each of these cases, $v_i \tilde{z}_i\, v_{i+1} \tilde{z}_i \in E(G^*)$. Furthermore, if $\tilde{z}_i$ were adjacent to any other vertex in C, one can verify that there would be a chord in C. Thus, $\mathcal{N}(\tilde{z}_i) = \{v_i, v_{i+1}\}$ and the $\tilde{z}_i$ are distinct. Now $\tilde{z}_1, \tilde{z}_2, \tilde{z}_3$ is an asteroidal triple in G^*, a contradiction.

(iii) $\Longrightarrow$ (iv): Let $\vec{G} = (V, A)$ and suppose G^* has no asteroidal triple. For a contradiction, assume $\vec{G}$ has a diasteroidal triple x, y, z. Let $\vec{C}$ be a z-avoiding $x - y$ chain in $\vec{G}$ with $V(\vec{C}) = \{x = v_0, v_1, v_2, \ldots, v_t = y\}$ and exactly one of (v_i, v_{i+1}) and (v_{i+1}, v_i) is an arc of $\vec{C}$ for each i. We will construct an $\tilde{x} - \tilde{y}$ path in G^* that does not intersect $\mathcal{N}(\tilde{z})$.

Let P be the path $[\tilde{x}, x = v_0, v_1, \ldots, v_t = y, \tilde{y}]$. We obtain P' by replacing v_i by $\tilde{v}_i$ whenever $(v_i, z) \in A$. (If x or y is replaced by $\tilde{x}$ or $\tilde{y}$, then remove the duplicate copy of $\tilde{x}$ or $\tilde{y}$.) Note that if $(v_i, z) \in A$ then since $\vec{C}$ is z-avoiding, we must have $(v_{i-1}, v_i) \in A$ and $(v_{i+1}, v_i) \in A$, so $v_{i-1}\tilde{v}_i$ and $v_{i+1}\tilde{v}_i$ are edges of G^*. Thus, P' is also a path in G^* from $\tilde{x}$ to $\tilde{y}$. Path P' does not intersect $\mathcal{N}(\tilde{z})$ since all v_i for which $v_i\tilde{z} \in E(G^*)$ have been replaced by $\tilde{v}_i$.

Similarly, there is a $\tilde{y} - \tilde{z}$ path that does not intersect $\mathcal{N}(\tilde{x})$ and a $\tilde{x} - \tilde{z}$ path that does not intersect $\mathcal{N}(\tilde{y})$. So $\tilde{x}, \tilde{y}, \tilde{z}$ is an asteroidal triple in G^*, a contradiction.

(iv) $\Longrightarrow$ (iii): Given that $\vec{G} = (V, A)$ has no diasteroidal triple, for a contradiction, assume that G^* has an asteroidal triple. Whenever x is part of an asteroidal triple in G^*, it can be replaced by $\tilde{x}$ since $x\tilde{x} \in E(G^*)$ and all neighbors of $\tilde{x}$ are neighbors of x. Thus, we may assume G^* has an asteroidal triple $\tilde{x}, \tilde{y}, \tilde{z}$.

Let $P = [\tilde{x} = w_0, w_1, \ldots, w_n = \tilde{y}]$ be an $\tilde{x} - \tilde{y}$ path in G^* that does not intersect $\mathcal{N}(\tilde{z})$. We wish to convert P into an xy-chain $\vec{C}$ in $\vec{G}$.

First consider the case $w_i, w_{i+1} \in V$. If $(w_i, w_{i+1}) \in A$, use this arc for $\vec{C}$, if $(w_i, w_{i+1}) \notin A$ but $(w_{i+1}, w_i) \in A$, use this arc for $\vec{C}$. Otherwise, there exists $u \in V$ with $(w_i, u), (w_{i+1}, u) \in A$ and we use these arcs in $\vec{C}$ and add vertex u.

For the endpoints, replace $\tilde{x}, \tilde{y}$ by x, y and use the arcs $(w_1, x), (w_{n-1}, y) \in A$ for $\vec{C}$. Since the vertices in $\tilde{V}$ form an independent set, the remaining case occurs when $w_i \in \tilde{V}$ and $w_{i-1}, w_{i+1} \in V$. Replace w_i by $u_i \in V$ where $\tilde{u}_i = w_i$. By definition of G^*, $(w_{i-1}, u), (w_{i+1}, u) \in A$ and we use these arcs for $\vec{C}$.

If an element of V is repeated after these replacements, remove the section from the first occurrence to the last. We show that the result $\vec{C}$ is a z-avoiding $x - y$ chain in $\vec{G}$. The vertices of $P \cap V$ are not neighbors of $\tilde{z}$ in G^*, thus there are no arcs from any of these vertices to z in $\vec{G}$. The vertices u of $\vec{C}$ that were not originally in P were constructed so that all arcs of $\vec{C}$ incident to u point towards u. Thus $\vec{C}$ is a z-avoiding $x - y$ chain in $\vec{G}$. Similarly, there exists an x-avoiding $y - z$ chain and a y-avoiding $x - z$ chain in $\vec{G}$. So x, y, z is a diasteroidal triple in $\vec{G}$, a contradiction. $\square$

Theorem 13.40 has algorithmic consequences. Given a digraph $\vec{G} = (V, A)$ with a loop at each vertex, the associated undirected graph G^* can be constructed in polynomial time. It is well-known that interval graph recognition can be accomplished in polynomial time (see, for example, Golumbic, 1980), thus condition (ii) of Theorem 13.40 can be checked in polynomial time. This, together with Theorem 13.38, leads to the following corollary.

Corollary 13.41. *The classes of unit bitolerance digraphs, proper bitolerance digraphs, and point-core bitolerance digraphs can be recognized in polynomial time.*

13.9 Exercises

Exercise 13.1. Show that if $\vec{G}$ is a symmetric Ferrers digraph, then its underlying simple graph G is a threshold graph.

Exercise 13.2. Calculate the functions f_1 and f_2, defined in Proposition 13.2, for the digraph given in Figure 13.2.

Exercise 13.3. Prove Lemma 13.9.

Exercise 13.4. In Example 13.12, we exhibit interval orders P_1 and P_2 so that $\vec{G} = \widetilde{P}_1 \cap \widetilde{P}_2$ for the third digraph in Figure 13.9. Do the same for the first and second digraphs in Figure 13.9.

Exercise 13.5. Prove Remark 13.13.

Exercise 13.6. Prove that the digraph on the left in Figure 13.9 is not a proper point-core bitolerance digraph.

Exercise 13.7. Show that the digraph shown on the edge between classes (1ai) and (2ai) in Figure 13.4 separates those classes.

Exercise 13.8. Show that for each of the orders A, B, D, $2 + 2$, $3 + 2$, $4 + 1$, $3 + 3$, all transitive orientations of its comparability graph are isomorphic.

Exercise 13.9. If possible, find unit, totally bounded tolerance representations for the digraphs in Figure 13.9. Otherwise, show it is impossible.

Exercise 13.10. Complete the proof of (ii) $\Longrightarrow$ (iii) of Lemma 13.33.

Exercise 13.11. Give a unit tolerance representation for each of the six weakly transitive digraphs whose underlying graph is C_3.

Exercise 13.12. Complete the proof of (i) $\Longrightarrow$ (ii) of Theorem 13.36 by showing that any digraph of the form $\overline{F'}$ in Figure 13.11 is not a bounded bitolerance digraph.

Exercise 13.13. Show that the digraphs depicted in Figure 13.14 are not proper bitolerance digraphs.

Exercise 13.14. Use Theorems 13.37 and 13.24 to find a digraph that separates the class of unit bitolerance digraphs from that of point-core bitolerance digraphs. (Hint: underlying graph $K_{1,3}$.)

Chapter 14

Open questions and further directions of research

In this book, we have presented the major results on tolerance graphs, tolerance orders and many related topics. A number of basic unanswered questions have been raised in the text, and we collect them here along with other open questions. There are also a number of topics that we have not been able to address, and we mention them in this chapter as well.

1. We have presented hierarchies for the following classes of graphs and orders:

Graph classes	Figure	Complete?
perfect graphs	2.8	yes
interval probe graphs	4.2	yes
bounded bitolerance orders	10.1	no
NeST graphs	11.11	no
ϕ-tolerance chain graphs	12.1	yes
bounded bitolerance digraphs	13.4	no

As indicated above, three of these hierarchies are complete, that is, all containment relations between classes are shown. Find separating examples and/or inclusions to make the remaining three hierarchies complete.

2. Find polynomial time recognition algorithms for any of the classes of tolerance graphs other than bounded bitolerance graphs, or prove that recognition of the class is NP-complete.

3. Chapter 3: The longstanding question from Golumbic and Monma (1982) as to whether there are any (unbounded) tolerance graphs which are not cocomparability graphs is still open. As we mentioned following Question 3.1, such a graph could not be a trapezoid graph.

249

4. Chapter 3: Characterize the class of bipartite tolerance graphs (see Section 3.2).
5. Chapter 4: Determine the complexity status of the sandwich problem for classes of perfect graphs, tolerance graphs, strongly chordal graphs and chordal bipartite graphs (see Figure 4.7). What is the complexity of the graph sandwich problem for threshold tolerance graphs?
6. Chapter 4: In Section 4.5, we referred to the polynomial time algorithms that recognize an interval probe graph with respect to a fixed given partition.

    What is the complexity of recognizing interval probe graphs in the general case where no partition is given in advance?
7. Chapters 5 and 10: Characterize by minimal forbidden ordered sets any of the 18 classes of bounded bitolerance orders discussed in Chapters 5 and 10. These classes are equal (Theorem 10.5) and the characterization is known in the case of bipartite orders, as discussed in Fishburn and Trotter (1999). Can we extend the forbidden poset characterization to apply to larger classes of orders?
8. Chapters 7 and 10: For which of the 18 classes of bounded bitolerance graphs in Chapter 10 is membership a comparability invariant? Are there any which are not?
9. Chapter 10: Show that the class of unit totally bounded tolerance orders is equivalent to that of proper totally bounded tolerance orders, or find a separating example between these classes (see Figure 10.1).
10. Chapter 11: A number of open questions on neighborhood subtrees are presented in Chapter 11.

    Characterize the neighborhood subtree intersection graphs and the neighborhood subtree containment graphs.

    Are the classes of NeST graphs and bounded NeST graphs equal, or is there a separating example?

    Is the class of tolerance graphs incomparable with the class of bounded NeST graphs or is the former contained in the latter?

    What is the relationship between the classes of chordal and strongly chordal graphs compared to the NeST and bounded NeST graphs?
11. Chapters 11 and 12: A recognition algorithm for threshold tolerance (TT) graphs is given in Monma, Reed, and Trotter (1988), and we have seen in Chapters 11 and 12 that the graphs P_4, C_4, and S_3 are threshold tolerance but the graphs $\overline{S_3}$, $2K_2$, T_2, $\overline{T_2}$, and $\overline{C_k}$ for $k \geq 4$ are not threshold tolerance.

    Threshold tolerance graphs are threshold graphs, but are they tolerance graphs?

We know that the class TT is contained in the class of bounded NeST graphs by Theorem 11.25. Are they bounded tolerance graphs?

Where in the hierarchy diagrams of Figures 11.11 and 12.1 is the class of threshold tolerance?

12. Chapter 11: In Section 11.5 we saw Hayward's tolerance free representations of NeST graphs. This can be applied to tolerance graphs as well. Can this approach be used to prove additional results about tolerance graphs?

13. Chapter 11: Prove or disprove Conjecture 11.7.

14. Chapter 12: What is the answer to the "unit vs. proper" question for max-tolerance graphs? For min-tolerance graphs the classes are different (Figure 2.8), and for sum-tolerance graphs the classes are the same (Theorem 12.32).

Other open problems involving max-tolerance and sum-tolerance graphs were given in Question 12.33.

15. Chapter 12: Garth Isaak has asked the question: Can one give general conditions describing a class of functions ϕ for which the ϕ-tolerance graphs are perfect?

16. Chapter 13: Prove Corollary 13.16 directly, that is, without using Theorem 13.15.

17. Chapter 13: Show that the class of unit totally bounded tolerance digraphs is equivalent to that of proper totally bounded tolerance digraphs, or find a separating example between these classes (see Figure 13.4).

18. Chapter 13: Polynomial time recognition algorithms are known for the classes of bounded bitolerance digraphs (Remark 13.11), unit, proper, and point-core bitolerance digraphs (Corollary 13.41). Can the other classes of bounded bitolerance digraphs shown in Figure 13.4 be recognized in polynomial time?

19. Chapter 13: (from Shull and Trenk, 1997b) Is there an efficient algorithm to determine the following? Given a bounded bitolerance graph G, which orientations of the edges of G yield bounded bitolerance digraphs? The same question can be asked for any of the other classes of bounded bitolerance graphs such as unit bitolerance graphs. (The question is answered in the cases of trees and cycles in Sections 13.6 and 13.7.)

20. In Chapter 13, we study bounded bitolerance digraphs and other directed versions of tolerance graphs. Interval digraphs provide a different directed graph analog of interval graphs which we have not discussed in this book. A digraph $\vec{G} = (V, A)$ is an *interval digraph* if each $v \in V$ can be assigned two real intervals S_v, T_v, so that $A = \{(x, y) | S_x \cap T_y \neq \emptyset\}$. Interval digraphs were introduced in Sen, Das, Roy, and West (1989) and studied

further in Sen, Sanyal, and West (1995) and West (1998). The relationship between interval digraphs and bounded bitolerance digraphs is explored in Shull and Trenk (2001).

21. Tolerance competition graphs are another variation on a theme of tolerance in graphs and are investigated in Anderson, Langley, Lundgren, McCenna, and Merz (1994), Brigham, McMorris, and Vitray (1995) and (1996). There are many interesting open problems in this area. Buck McMorris has asked whether there are graphs which are not min-tolerance competition graphs. The reader is also referred to McKee and McMorris (1999) for further topics.

References

Allen, J. F., 1983. Maintaining knowledge about temporal intervals. *Comm. ACM*, **26** 832–843.

Anderson, C. A., L. J. Langley, J. R. Lundgren, P. A. McCenna, and S. K. Merz, 1994. New classes of *p*-competition graphs. *Congress. Numer.*, **100** 97–107.

Andreae, T., U. Hennig, and A. Parra, 1993. On a problem concerning tolerance graphs. *Discrete Applied Math.*, **46** 73–78.

Benzer, S., 1959. On the topology of the genetic fine structure. *Proc. Nat. Acad. Sci.*, **45** 1607–1620.

Berge, C., 1973. *Graphs and Hypergraphs*. North-Holland, Amsterdam.

Berge, C. and V. Chvátal, editors, 1984. *Topics on Perfect Graphs*, Ann. Discrete Math., **21** North-Holland, Amsterdam.

Berry, A., J. R. S. Blair, and P. Heggernes, 2002. Maximum cardinality search for computing minimal triangulations. In *Lecture Notes in Computer Science*. Springer., **2573**, 1–12.

Bibelnieks, E. and P. M. Dearing, 1993. Neighborhood subtree tolerance graphs. *Discrete Applied Math.*, **43** 13–26.

Bogart, K., P. Fishburn, G. Isaak, and L. Langley, 1995. Proper and unit tolerance graphs. *Discrete Applied Math.*, **60** 37–51.

Bogart, K. P., 2000. *Introductory Combinatorics*. Academic Press.

Bogart, K. P. and G. Isaak, 1998. Proper and unit bitolerance orders and graphs. *Discrete Math.*, **181** 37–51.

Bogart, K., G. Isaak, J. Laison, and A. Trenk, 2001. Comparability invariance results for tolerance orders. *Order*, **18** 281–294.

Bogart, K., M. Jacobson, L. Langley, and F. McMorris, 2001. Tolerance orders and bipartite unit tolerance graphs. *Discrete Math.*, **226** 35–50.

Bogart, K. P. and A. N. Trenk, 1994. Bipartite tolerance orders. *Discrete Math.*, **132** 11–22. Corrigendum to Bipartite Tolerance Orders *Discrete Math.*, **145** (1995) 347.

1995. Trees and cycles that are bounded bitolerance digraphs. *Congress. Numer.*, **112** 17–32.

2000. Bounded bitolerance digraphs. *Discrete Math.*, **215** 13–20.

253

Booth, K. and G. S. Lueker, 1976. Testing for the consecutive ones property, interval graphs and graph planarity using pq-tree algorithms. *J. Comput. Syst. Sci.*, **13** 335–379.

Brandstädt, A., V. B. Le, and J. P. Spinrad, 1999. *Graph Classes: A Survey*. SIAM Monographs on Discrete Math. and Applications, Philadelphia.

Brandstädt, A., J. P. Spinrad, and L. Stewart, 1987. Bipartite permutation graphs are bipartite tolerance graphs. *Congress. Numer.*, **58** 165–174.

Brigham, R. C., F. R. McMorris, and R. P. Vitray, 1995. Tolerance competition graphs. *Linear Algebra and its Applications*, **217** 41–52.

1996. Two ϕ-tolerance competition graphs. *Discrete Applied Math.*, **66** 101–108.

Buneman, P., 1974. A characterization of rigid circuit graphs. *Discrete Math.*, **9** 205–212.

Chvátal, V., 1984. Perfectly ordered graphs. In Berge, C. and V. Chvátal, editors, Topics on Perfect Graphs, Ann. Discrete Math., **21** 63–65.

1985. Star-cutsets and perfect graphs. *J. Comb. Theory B.*, **39** 189–199.

Chvátal, V. and P. L. Hammer, 1977. Aggregation of inequalities in integer programming. *Ann. Discrete Math.*, **1** 145–162.

Cogis, O., 1979. A characterization of digraphs with Ferrers dimension 2. CNRS Research Report.

1982. Ferrers digraphs and threshold graphs. *Discrete Math.*, **38** 33–46.

Corneil, D. G. and P. A. Kamula, 1987. Extensions of permutation and interval graphs. *Congress. Numer.*, **58** 267–275.

Corneil, D. G., S. Olariu, and L. Stewart, 1997. Asteroidal triple-free graphs. *SIAM J. Discrete Math.*, **10** 399–430.

Dagan, I., M. C. Golumbic, and R. Y. Pinter, 1988. Trapezoid graphs and their coloring. *Discrete Applied Math.*, **21** 35–46.

Dahlhaus, E., P. L. Hammer, F. Maffray, and S. Olariu, 1994. On domination elimination orderings and domination graphs. *Lecture Notes in Comput. Sci.*, **903** 81–92.

Derigs, U., O. Goecke, and R. Schrader, 1984. Bisimplicial edges, gaussian elimination and matchings in bipartite graphs. In *Intern. Workshop on Graph-Theoretic Concepts in Comp. Sci.*, pp. 79–87, Berlin. Trauner Verlag.

Doignon, J. P., A. Duchamp, and J. C. Falmagne, 1984. On realizable biorders and the biorder dimension of a relation. *J. Math. Psych.*, **28** 73–109.

Duchet, P., 1984. Classical perfect graphs. In Berge, C. and V. Chvátal, editors, *Topics on Perfect Graphs*, Ann. Discrete Math., **21** 67–96. North-Holland, Amsterdam.

Dushnik, B. and E. W. Miller, 1941. Partially ordered sets. *Amer. J. Math.*, **63** 600–610.

Farber, M., 1983. Characterizations of strongly chordal graphs. *Discrete Math.*, **43** 173–189.

Felsner, S., 1998. Tolerance graphs and orders. *J. Graph Theory*, **28** 129–140.

Felsner, S., M. Habib, and R. Möhring, 1994. On the interplay between interval dimension and ordinary dimension. *SIAM J. Discrete Math.*, **7** 32–40.

Felsner, S., R. Müller, and L. Wernisch, 1997. Trapezoid graphs and generalizations, geometry and algorithms. *Discrete Applied Math.*, **74** 13–32.

Fishburn, P. C., 1970. Intransitive indifference with unequal indifference intervals. *J. Math. Psych.*, **7** 144–149.

1985. *Interval orders and interval graphs: a study of partially ordered sets*. John Wiley & Sons, New York.

Fishburn, P. C. and B. Monjardet, 1992. Norbert Weiner on the theory of measurement (1914, 1915, 1921). *J. Math. Psych.*, **36** 165–184.

Fishburn, P. C. and J. A. Reeds, 2001. Counting split semiorders. *Order*, **18** 119–128.

Fishburn, P. C. and W. T. Trotter, 1999. Split semiorders. *Discrete Math.*, **195** 111–126.

Földes, S. and P. L. Hammer, 1977. Split graphs. *Congress. Numer.*, **17** 311–315.

Gallai, T., 1967. Transitiv orientierbare Graphen. *Acta Math. Hungar.*, **18** 25–66.

Garey, M. R. and D. S. Johnson, 1979. *Computers and Intractability: A Guide to the Theory of NP-completeness.* W. H. Freeman, New York.

Gavril, F., 1974. The intersection graphs of subtrees in trees are exactly the chordal graphs. *J. Combin. Theory B*, **16** 47–56.

1978. A recognition algorithm for the intersection graphs of paths in trees. *Discrete Math.*, **23** 211–227.

Gilmore, P. C. and A. J. Hoffman, 1964. A characterization of comparability graphs and of interval graphs. *Canad. J. Math.*, **16** 539–548.

Golumbic, M. C., 1977. Comparability graphs and a new matroid. *J. Combin. Theory B*, **22** 68–90.

1980. *Algorithmic Graph Theory and Perfect Graphs.* Academic Press, New York. Second edition, forthcoming in 2003, Ann. Discrete Math., Elsevier, Amsterdam.

1984. Algorithmic aspects of perfect graphs. In Berge, C. and V. Chvátal, editors, *Topics on Perfect Graphs*, Ann. Discrete Math., **21** 301–323. North-Holland, Amsterdam.

1998. Reasoning about time. In Hoffman, F., editor, *Mathematical Aspects of Artificial Intelligence*, volume 55 of *Proc. Symposia in Applied Math*, pp. 19–53. American Mathematical Society.

Golumbic, M. C. and R. E. Jamison, 1985a. Edge and vertex intersection of paths in a tree. *Discrete Math.*, **55** 151–159.

1985b. The edge intersection graphs of paths in a tree. *J. Combin. Theory B*, **38** 8–22.

2003, January. *Rank tolerance graph classes.* Technical report, Caesarea Rothschild Institute, University of Haifa.

Golumbic, M. C., R. E. Jamison, and A. N. Trenk, 2002. Archimedean ϕ-tolerance graphs. *J. Graph Theory*, **41** 179–194.

Golumbic, M. C., H. Kaplan, and R. Shamir, 1994. On the complexity of DNA physical mapping. *Adv. Appl. Math.*, **15** 251–261.

1995. Graph sandwich problems. *J. Algorithms*, **19** 449–473.

Golumbic, M. C. and M. Lewenstein, 2000. New results on induced matchings. *Discrete Applied Math.*, **101** 139–154.

Golumbic, M. C. and M. Lipshteyn, 2001. On the hierarchy of tolerance, probe and interval graphs. *Congress. Numer.*, **153** 97–106.

Golumbic, M. C. and C. L. Monma, 1982. A generalization of interval graphs with tolerances. *Congress. Numer.*, **35** 321–331.

Golumbic, M. C., C. L. Monma, and W. T. Trotter, 1984. Tolerance graphs. *Discrete Applied Math.*, **9** 157–170.

Golumbic, M. C., D. Rotem, and J. Urrutia, 1983. Comparability graphs and intersection graphs. *Discrete Math.*, **43** 37–46.

Golumbic, M. C. and E. R. Scheinerman, 1989. Containment graphs, posets and related classes of graphs. In Bloom, G. S., R. L. Graham, and J. Malkevitch, editors, *Combinatorial Mathematics*, pp. 192–204, New York, N. Y. Academy of Sciences.

Golumbic, M. C. and R. Shamir, 1993. Complexity and algorithms for reasoning about time: a graph theoretic approach. *J. Assoc. Comput. Mach.*, **40** 1108–1133.

Golumbic, M. C. and A. Siani, 2002. Coloring algorithms and tolerance graphs: reasoning and scheduling with interval constraints. In *Lecture Notes in Computer Science*, pp. 196–207. Springer-Verlag.

Grötschel, M., L. Lovász, and A. Schrijver, 1981. The ellipsoid method and its consequences in combinatorial optimization. *Combinatorica*, **1** 169–197.

Habib, M., D. Kelly, and R. Möhring, 1991. Interval dimension is a comparability invariant. *Discrete Applied Math.*, **88** 211–229.

———— 1992. Comparability invariance of geometric notions of order dimension. Manuscript.

Hayward, R. B., 1985. Weakly triangulated graphs. *J. Comb. Theory B.*, **39** 200–209.

Hayward, R. B., C. T. Hoàng, and F. Maffray, 1990. Optimizing weakly triangulated graphs. *Graphs and Combinatorics*, **6** 33–35. Erratum to *ibid.*, **5** 339-349.

Hayward, R. B. and P. Kearney, 1993, December. Investigating NeST graphs. Technical Report TR-CS-04-93, University of Lethbridge, Canada. Available in postscript at `www.cs.ualberta.ca/~hayward/papers/trcs0493.ps`.

Hayward, R. B., P. E. Kearney, and A. Malton, 2002. NeST graphs. *Discrete Applied Math.*, **121** 139–153.

Hayward, R. B. and R. Shamir, 2002. A note on tolerance graph recognition. *Discrete Applied Math.* (to appear).

Hoàng, C. T., 1987. Alternating orientation and alternating colouration of perfect graphs. *J. Combin. Theory B*, **42** 264–273.

Hougardy, S., 1998. Inclusions between classes of perfect graphs. Humbolt University, Berlin. Available at `www.informatik.hu-berlin.de/~hougardy/paper/classes.html`.

Isaak, G., K. L. Nyman, and A. N. Trenk, 2001. A hierarchy of classes of bounded bitolerance orders. To appear in *ARS Combinatoria*.

Jacobson, M. S. and F. R. McMorris, 1991. Sum-tolerance proper interval graphs are precisely sum-tolerance unit interval graphs. *J. Combinatorics, Information and System Sciences*, **16** 25–28.

Jacobson, M. S., F. R. McMorris, and M. Mulder, 1991. An introduction to tolerance intersection graphs. In Alavi, Y., G. Chartrand, O. Oellermann, and A. Schwenk, editors, *Proc. Sixth Int. Conf. on Theory and Applications of Graphs*, Wiley Interscience, New York, pp. 705–724.

Jacobson, M. S., F. R. McMorris, and E. R. Scheinerman, 1991. General results on tolerance intersection graphs. *J. Graph Theory*, **15** 573–577.

Jamison, R. E. and H. M. Mulder, 2000a. Constant tolerance representations of graphs in trees. *Congress. Numer.*, **143** 175–192.

———— 2000b. Tolerance intersection graphs on binary trees with constant tolerance 3. *Discrete Math.*, **215** 115–131.

Johnson, J. L. and J. P. Spinrad, 2001. A polynomial time recognition algorithm for probe interval graphs. In *Proc. 12th ACM–SIAM Symp. on Discrete Algorithms*, pp. 477–486.

Kelly, D., 1986. Invariants of finite comparability graphs. *Order*, **3** 155–158.

Langley, L., 1993, June. *Interval tolerance orders and dimension*. Ph.D. thesis, Dartmouth College.

1995. Recognition of orders of interval dimension 2. *Discrete Applied Math.*, **60** 257–266.

Lee, A., 2002, June. Digraphs of Ferrers dimension 2. Undergraduate honors thesis, Computer Science Dept., Wellesley College.

Lekkerkerker, C. and D. Boland, 1962. Representation of finite graphs by a set of intervals on the real line. *Fund. Math.*, **51** 45–64.

Lipshteyn, M., 2001. Probe graphs. Master's thesis, Bar-Ilan University.

Lovász, L., 1972. A characterization of perfect graphs. *J. Combin. Theory B*, **13** 95–98.

Ma, T. and J. P. Spinrad, 1994. On the 2-chain subgraph cover and related problems. *J. Algorithms*, **17** 251–268.

Mackenzie, D., 2002. Graph theory uncovers the roots of perfection. *Science*, **297** 38.

Mahadev, N. V. R. and U. N. Peled, 1995. *Threshold Graphs and Related Topics*. Ann. Discrete Math., **56**, North-Holland, New York.

McConnell, R. M. and J. P. Spinrad, 1997. Linear-time transitive orientation. In *Proc. 8th ACM-SIAM Symposium on Discrete Algorithms*, pp. 19–25.

1999. Modular decomposition and transitive orientation. *Discrete Math.*, **201** 189–241.

2002. Construction of probe interval models. In *Proc 13th ACM-SIAM Symposium on Discrete Algorithms*, pp. 866–875.

McKee, T. A. and F. R. McMorris, 1999. *Topics in Intersection Graph Theory*. SIAM Monographs on Discrete Mathematics and Applications, Philadelphia.

McMorris, F. R. and E. R. Scheinerman, 1991. Convexity threshold for random chordal graphs. *Graphs and Combin.*, **7** 177–181.

McMorris, F. R. and D. R. Shier, 1983. Representing chordal graphs on $K_{1,n}$. *Comment. Math. Univ. Carolin.*, **24** 489–494.

McMorris, F. R., C. Wang, and P. Zhang, 1998. On probe interval graphs. *Discrete Applied Math.*, **88** 315–324.

Meyniel, H., 1987. A new property of critical imperfect graphs and some consequences. *Europ. J. Combinatorics*, **8** 313–316.

Monma, C., B. Reed, and W. T. Trotter, 1988. Threshold tolerance graphs. *J. Graph Theory*, **12** 343–362.

Narasimhan, G. and R. Manber, 1992. Stability number and chromatic number of tolerance graphs. *Discrete Applied Math.*, **36** 47–56.

Opatrný, J., 1979. Total ordering problems. *SIAM J. Computing*, **8** 111–114.

Peled, U.N., 2002, August. *Neighborhood subtree intersection graphs are strongly chordal*. Technical report CRI-02-01, Caesarea Rothschild Institute, University of Haifa.

Pevzner, P., 2000. *Computational Molecular Biology: An Algorithmic Approach*. MIT Press.

Pnueli, A., A. Lempel, and S. Even, 1971. Transitive orientation of graphs and identification of permutation graphs. *Canad. J. Math.*, **23** 160–175.

Prisner, E., 1989. A characterization of interval catch digraphs. *Discrete Math.*, **73** 285–289.

Roberts, F. S., 1969. Indifference graphs. In Harary, F., editor, *Proof Techniques in Graph Theory*, pp. 139–146. Academic Press, New York.

1976. *Discrete Mathematical Models with Applications to Social, Biological and Environmental Problems*. Prentice Hall.

Rusu, I. and J. Spinrad, 2001. Domination graphs: examples and counterexamples. *Discrete Applied Math.*, **110** 289–300.

Scott, D. and P. Suppes, 1958. Foundational aspects of theory of measurement. *J. Symbolic Logic*, **23** 113–128.

Sen, M., S. Das, A. B. Roy, and D. B. West, 1989. Interval digraphs: an analogue of interval graphs. *J. Graph Theory*, **13** 189–202.

Sen, M. K., B. K. Sanyal, and D. B. West, 1995. Representing digraphs using intervals or circular arcs. *Discrete Math.*, **147** 235–245.

Sheng, L., 1998, October. *Some graph theoretic approaches to problems of the social and biological sciences*. Ph.D. thesis, Rutgers University.

1999. Cycle free probe interval graphs. *Congress. Numer.*, **140** 33–42.

Shevrin, L. N. and N. D. Filippov, 1970. Partially ordered sets and their comparability graphs. *Siberian Math. J.*, **11** 497–509.

Shull, R. and A. N. Trenk, 1997a. Cliques that are tolerance digraphs. *Discrete Applied Math.*, **80** 119–134.

1997b. Unit and proper bitolerance digraphs. *J. Graph Theory*, **24** 193–199.

2001. Interval digraphs and bounded bitolerance digraphs. *Congress. Numer.*, **151** 111–127.

Spinrad, J. P., 1985. On comparability and permutation graphs. *SIAM J. Comput.*, **14** 658–670.

Spinrad, J. P., A. Brandstädt, and L. Stewart, 1987. Bipartite permutation graphs. *Discrete Applied Math.*, **18** 279–292.

Spinrad, J. P. and R. Sritharan, 1995. Algorithms for weakly chordal graphs. *Discrete Applied Math.*, **59** 181–191.

Sysło, M. M., 1985. Triangulated edge intersection graphs of paths in a tree. *Discrete Math.*, **55** 217–220.

Tamir, A., 1983. A class of balanced matrices arising from location problems. *SIAM J. on Algebraic and Discrete Methods*, **4** 363–370.

Tanenbaum, P. J., 1996. Simultaneous representation of interval and interval-containment orders. *Order*, **13** 339–350.

1999. Simultaneous intersection representation of pairs of graphs. *J. Graph Theory*, **32** 171–190.

Tarjan, R. E., 1985. Decomposition by clique separators. *Discrete Math.*, **55** 221–232.

Tarjan, R. E. and M. Yannakakis, 1984. Simple linear-time algorithms to test chordality of graphs, test acyclicity of hypergraphs and selectively reduce acyclic hypergraphs. *SIAM J. Computing*, **13** 566–579.

Trenk, A. N., 1998. On k-weak orders: recognition and a tolerance result. *Discrete Math.*, **181** 223–237.

Trotter, W. T., 1992. *Combinatorics and Partially Ordered Sets*. Johns Hopkins University Press, Baltimore.

Walter, J. R., 1972. *Representations of rigid cycle graphs*. Ph.D. thesis, Wayne State University.

Warren, H. E., 1968. Lower bounds for approximation by nonlinear manifolds. *Trans. Amer. Math. Soc.*, **133** 167–178.

Waterman, M. S., 1995. *Introduction to Computational Biology*. Chapman and Hall, London.

Weiner, N., 1914. A contribution to the theory of relative position. *Proc. Cambridge Philosophical Society*, **17** 441–449.

West, D. B., 1998. Short proofs for interval digraphs. *Discrete Math.*, **178** 287–292.

Yannakakis, M., 1982. The complexity of the partial order dimension problem. *SIAM J. Alg. Discrete Meth.*, **3** 351–381.

Zhang, P., 1994. Probe interval graphs and their application to physical mapping of DNA. Manuscript.

Zhang, P., E. A. Schon, S. G. Fischer, E. Cayanis, J. Weiss, S. Kistler, and P. E. Bourne, 1994. An algorithm based on graph theory for the assembly of contigs in physical mapping of DNA. *CABIOS*, **10** 309–317.

Zhang, P., X. Ye, L. Liao, J. Russo, and S. G. Fischer, 1999. Integrated mapping package – a physical mapping software tool kit. *Genomics*, **55** 78–87.

Index of symbols

$\prec_c$ (central extension), 103

$\|$ (incomparable), 13

$\alpha(G)$ (stability number), 137

$\mathcal{B}(\mathcal{I})$ (order of extreme corners), 128

$B(P)$, 126

$\overline{B}$ (Berlin graph), 39

$c(v)$ (center point), 86

C_k (chordless cycle on k vertices), 9

$\chi(G)$ (chromatic number), 40, 136

$d(p, q)$ (distance in tree), 169

d_v (degree of the vertex v), 10

D_v (left diagonal), 92

D'_v (right diagonal), 92

$D^{\uparrow}(P)$, 233

diam(T)(diameter), 169

dim($B(P)$), 124

dim(P) (dimension), 14

$E(G)$ (the edge set of a graph G), 1

E^* (edge set of enhanced graph), 74

E^+ (completion edges), 75

E_f (forbidden edges), 77

E_0 (optional edges), 77

F dim($\vec{G}$) (Ferrers dimension), 222

F dim ≤ 2 (Ferrers dimension ≤ 2), 222

$G(V, E)$ (a graph G with vertex set V and edge set E), 1

$G^* = (P \cup N, E^*)$ (enhanced graph), 74

$\overline{G}$ (complement of graph G), 8

$\vec{G}^r$ (reversal of $\vec{G}$), 219

G_X (subgraph of G induced by $X \subseteq V(G)$), 5

$gr(G, \prec)$ (Grundy number), 42

I^* (normalization of I), 129

$I_x \ll I_y$ (relation between intervals), 12

$I_C(A)$ (covering intervals), 120

$I_N(A)$ (non-covering intervals), 120

idim(P) (interval dimension), 15, 124

$Inc(A)$ (incomparable to A), 111

$\langle \mathcal{I}, f \rangle$ (point-core bitolerance representation), 87

$\langle \mathcal{I}, p, q \rangle$ (bounded bitolerance representation), 85

$\langle \mathcal{I}, t \rangle$ (tolerance representation), 5

$\kappa(G)$ (clique cover number), 137

ℓ_x (extreme lower corner), 127

$L(v)$ (left endpoint), 30

$\mathcal{L}(x)$ (predecessor set), 126

$\mathcal{N}(v)$ (open neighborhood), 7

$\mathcal{N}[v]$ (closed neighborhood), 10

$\mathcal{N}^+(u)$ (out-neighbors), 219

$\omega(G)$ (clique number), 40, 136

$p(v)$ (left tolerant point), 85

$P = (X, \prec)$ (ordered set), 13

$P^d = (X, \prec^d)$ (dual), 13

$\widetilde{P}$ (less than or incomparable to digraph), 225

$\widetilde{P_1} \cap \widetilde{P_2}$, 225

Φ, 201

$Pred(A)$ (predecessor set), 111

$q(v)$ (right tolerant point), 85

$R(v)$ (right endpoint), 30

$\mathbf{R}^+$ (positive real numbers), 5

$R_i \ll R_j$ (relation between ribbons, trapezoids, parallelograms), 18

S_k (k-sun), 22

$Succ(A)$ (successor set), 111

$\mathbf{T}$ (host tree), 164

$T(c, r)$ (neighborhood subtree), 170

T_2, 33

T_3, 56
TT (tolerance threshold), 187
$t_l(v)$ (left tolerance), 85
$t_r(v)$ (right tolerance), 85
t_v (tolerance), 5

u_x (extreme upper corner), 127
$\mathcal{U}(x)$ (predecessors of all successors), 126
$V(G)$ (the vertex set of a graph G), 1
$x \rightarrow y$ (single arc from x to y), 219
$x \rightleftharpoons y$ (double arc between x and y), 219

Index

ϕ-tolerance chain graph, **195**, 215
ϕ-tolerance graph, **193**

50% tolerance graph, **47**, 48, 100
50% tolerance order, 87, 98

admissible, **42**, 43
admissible ordering, 42
alternately orientable, **38**, 39, 48
antichain, **13**
Archimedean ϕ-tolerance graph, **201**
Archimedean graph, **201**
Archimedean tolerance functions, **201**
asteroidal triple, **10**, 54, 245
AT-free, **11**, 20, 54, 60
autonomous set, 111
avoiding, *see* z-avoiding, 244

beads on a wire, 115, 230, 232
Betweenness Problem, **79**
bipartite graph, 53, 60
bipartite order, **152**
bitolerance graphs and orders , 84
block, 205
bounded bitolerance digraph, **223**, 225, 226,
    235, 237, 238
bounded bitolerance graph, **86**
bounded bitolerance order, **85**, 94, 110, 146,
    147
bounded bitolerance representation, **85**, **223**
bounded NeST, 176, 177
bounded NeST graph, 177
bounded tolerance digraph, 229
bounded tolerance graph, **5**, 29, 34, 48, 53, 60,
    68, 94, 136, 195
bounded tolerance order, 85, 88, 92, 110, 113

bounded tolerance representation, 47, **85**
box embedding, 127
BT-indexing condition, **153**

c-point, 57
cactus, **206**
caterpillar, **54**, 238, 242
caterpillar with toes, **54**, 58
central extension, 103
chain, **13**
chain cover, 107
chain graph, 195
chain ($x - y$ chain), 244
characteristic point, **57**
chord of a cycle, 7
chordal graph, **7**, 8, 11, 20, 36, 76, 165, 195
chordless cycle, 9
chordless sun, **201**
chromatic number, 136
clique, **4**
clique cover number, 137
clique number, 136
clone, 63
closed neighborhood, **10**
co-perfectly orderable graph, 42, 48
cocomparability graph, 10, 11, 13, 18, 20, 34,
    48, 53, 54, 60, 86, 136
coloring EPT graphs, 167
coloring intervals, 11
coloring of a graph, **4**
coloring probe graphs, 74, 82
coloring tolerance graphs, 136
coloring tolerance representations, 137
comparability graph, **9**, 10, 13, 33, 37, 109
comparability invariant, **109**, 113
comparable, **13**

complement, **8**
complete, **48**
complete bipartite graph, 208
complete hierarchy, 48, 67
complete subgraph, 4
complexity
    of Interval Graph Sandwich Problem, 79
    of Probe Graph Sandwich Problem, 81
    of recognizing bounded tolerance graphs,
        133
    of recognizing comparability graphs, 10
    of recognizing interval graphs, 11
    of recognizing permutation graphs, 133
    of recognizing tolerance graphs, 136
    of recognizing transitive orientations, 10
    of recognizing trapezoid graphs and orders,
        124, 133
    of recognizing weakly chordal graphs, 21
    of sandwich problems, 78
constant core, 47, 94, 95, 98, 151
constant tolerance NeST, 176, 179
constant tolerances, 31, 151, 194
containment graph, **168**, 183
containment NeST, 176, 182
core, 151
cotree, 53
coTT graph, **187**, 195, 197, 198, 216
covering, 13
crown, 154
cutpoint, 205
cutset, 41

degree, 10
degree partition, 23
diameter, 169
diasteroidal triple, 244, 245
digraph, *see* directed graph, 219
dimension, 14, 110, 113, 125
directed graph, 219
distinct endpoints, 90
distinct tolerances, 90
distinct tolerant points, 90
double arc, 219
dual, 13

edge intersection graph, 165
elementary reversal, 111, 112
embedded star, **188**
enhanced edge, 74, 75
enhanced graph, **74**
EPT graph, 166

equidistant centers, **188**
even pair, **41**
extreme lower corner, 127
extreme upper corner, 127

Ferrers digraph, **221**
Ferrers dimension, 220, **222**, 225
function diagram, 17
function graph, **18**

Gallai, 112
geometric interpretations, 91
graph sandwich problem, **77**
Grundy number, **42**

Hasse diagrams, 13
Helly property, 172
hereditary, 29
hereditary property, **5**
hierarchies, 48
hierarchy
    of ϕ-tolerance chain graphs, 195
    of bounded bitolerance digraphs, 230
    of bounded bitolerance orders, 147
    of containment graphs, 184
    of NeST graphs, 186
    of perfect graphs, 48, 49
    of probe graphs, 68
    of ribbon graphs, 17
hovering witness, 138

IGSP, **78**
implication class, 9
incomparability graph, 13, 224
incomparable, **13**
independent set, **4**
induced subgraph, **5**
intersection graph, 4, 165
interval catch digraphs, 245
interval completion, 65
interval containment graph, 32
interval dimension, **15**, 92, 94, 110, 124
interval graph, 1, **4**, 11, 31, 47, 48, 54, 68, 84,
        100, 195, 196, 198, 245
Interval Graph Sandwich Problem, 78, **79**
interval order, **13**, 14, 84, 98
interval probe graph, **65**, 66
interval realizer, **15**, 124
interval two-point digraph, **220**, 222
interval two-point representation, **220**
isolated vertex, **23**, 30, 196

left diagonal, 92
left tolerance, **85**
left-leaning representation, 92
left-right clique partition, **204**
linear extension, **14**
linear order, **13**
linear realizer, **14**, 113, 125
loop, **219**

m-star, **184**
max-tolerance, 194
max-tolerance chain graph, 195, 196
max-tolerance graph, 203
maximum cardinality search, 7
maximum weight stable set in a tolerance
    graph, 140
min-tolerance chain graph, 195, 196
min-tolerance graphs, 193

neighborhood, **7**
neighborhood subtree, 169, 170
neighborhood subtree intersection graph, 179
neighborhood subtree tolerance graph, **173**
NeST, **173**
    bounded NeST, 176, 177
    constant tolerance NeST, 176, 179
    containment NeST, 176, 182
    proper NeST, 176, 177
    unit NeST, 176, 177
nonassertive vertices, 56
normal Fdim ≤ 2 representation, **222**
normal representation, 57
normalized, **196**
NP-complete, 79, 81

odd chord, **21**
order autonomous set, 111, 112
ordered graph, **42**
ordered set, **13**
orders, **13**
out-neighbors, **219**

P-normalization, 129
parallelogram graph, **18**, 34, 48, 136
parallelogram order, 91, 92, 147
partitioned probe graph, 73
path graphs, 166
perfect elimination ordering, **7**
perfect graph, **40**, 43, 45, 48
perfectly orderable, 43, 44
perfectly orderable graph, **42**

permutation diagram, 16
permutation graph, **15**, 31, 32, 48, 54, 60
PGSP, **81**
phi-tolerance, *see* ϕ-tolerance, 193
point-core bitolerance digraph, **229**, 231, 244,
    245
point-core bitolerance order, **86**, 95, 146
point-core order, 87
point-core tolerance digraph, 232
point-core tolerance order, 98
polynomial time, 247
posets, 13
predecessor set, **125**
probe, 64
probe graph, **65**, 68
Probe Graph Sandwich Problem, **81**
proper ϕ-tolerance graph, 215
proper bitolerance digraph, **229**, 237, 242, 244
proper bitolerance order, **86**, 104, 147
proper interval graph, 12, **45**, 48, 67, 68
proper interval order, 147, 151
proper NeST, 177
proper NeST graph, 177
proper point-core bitolerance order, 150
proper point-core tolerance order, 151
proper probe graph, 67, 68
proper sum-tolerance graph, 195, 216
proper tolerance graph, **45**, 68
proper tolerance order, 147
proper totally bounded bitolerance order, 150

realizer, 14, 113, 125
reasoning about time, 25
recognition problems, see complexity, 10
regular representation, **30**, 37
reversal, **219**, 227, 231
ribbon diagram, 17
ribbon graph, **18**
right diagonal, 92
right tolerance, **85**

SBS-indexing property, 60, 61
separating examples, 200, 233
sign pattern, 209
simple elimination ordering, **21**, 22
simple vertex, **21**
simplicial, **7**
split graph, **8**, 165, 210
split interval order, 87, 148
split semiorder, 87, 148
splitting point, **87**

stability number, 137
stable set, **4**
star, 165, 237
star-cutset, **41**
Star-Cutset Lemma, 41
strong elimination ordering, **22**
Strong Perfect Graph Theorem, 40
strong umbrella vertex, **203**, 204
strongly chordal graph, **21**, 22, 195, 199
subtrees of a tree, 164
successor set, 125
successors, **219**
sum-tolerance, 194
sum-tolerance chain graph, 195, 197, 199, 216
sum-tolerance graph, 203
sun, 22
sun-free, 22
symmetric digraph, 228

temporal reasoning, 25, 135
threshold graph, 23, 48, 67, 68, 190, 195, 196
threshold tolerance graph, **187**, 188
tolerance chain graph, 212, 213
tolerance graph, 2, **5**, 29, 48, 66, 68, 195
tolerance orientation, 38
tolerance-free, 175
tolerant points, 89
totally bounded bitolerance digraph, **229**
totally bounded bitolerance order, 87, 147
trampoline, 22
transitive orientation, **9**, 33, 109
transitively orientable, 9, 33
trapezoid graph, **18**, 40, 48, 60, 91, 94, 136
trapezoid graph recognition, 124
trapezoid order, 91, 92, 94, 147
trapezoid representation, 91
tree, 57
trees
    that are ϕ-tolerance graphs, 207
    that are bounded bitolerance digraphs, 237
    that are bounded tolerance graphs, 54
    that are interval probe graphs, 71
    that are NeST graphs, 191
    that are tolerance graphs, 53, 58
    that are unit max-tolerance graphs, 216
triangulated graph, 7
TRO, **9**
two-pair, 21, **41**

umbrella vertex, **204**
underlying graph, **219**
unit bitolerance digraph, **229**, 231, 236, 242, 244
unit bitolerance order, **86**, 95, 104, 110, 115, 147
unit interval graph, 12, 45, 48, 67, 68
unit interval order, 147, 151
unit max-tolerance graph, 216
unit NeST, 177
unit NeST graph, 177
unit point-core bitolerance order, 150
unit point-core tolerance order, 151
unit probe graph, 66–68
unit sum-tolerance graph, 195, 216
unit tolerance digraph, 232, 236
unit tolerance graph, **45**, 47, 48, 66, 68, 102
unit tolerance order, 98, 100, 107, 110, 115, 119, 147
unit totally bounded bitolerance order, 150
unit vs. proper, 45, 67, 103, 150, 177, 215, 243
universal vertex, **23**, 175, 196

vertex intersection graph, **165**
VPT graph, 166

Warren's Theorem, 209
Weak Perfect Graph Theorem, 40
weakly chordal graph, **20**, 36, 38, 40, 48, 66, 184, 195
weakly transitive, 234
weakly triangulated graph, 20
width, **107**

z-avoiding, **244**